How to access the supplemental web resource

We are pleased to provide access to a web resource that supplements your textbook, *Qualitative Diagnosis of Human Movement, Third Edition.* This resource provides video clips for use in QMD Demonstrations and QMD Explorations, as well as forms for use with the Theory-Into-Practice Situations.

Accessing the web resource is easy!
Follow these steps if you purchased a new book:

1. Visit **www.HumanKinetics.com/QualitativeDiagnosisOfHumanMovement**

2. Click the <u>third</u> edition link next to the corresponding third edition book cover.

3. Click the Sign In link on the left or top of the page. If you do not have an account with Human Kinetics, you will be prompted to create one.

4. If the online product you purchased does not appear in the Ancillary Items box on the left of the page, click the Enter Key Code option in that box. Enter the key code that is printed at the right, including all hyphens. Click the Submit button to unlock your online product.

5. After you have entered your key code the first time, you will never have to enter it again to access this product. Once unlocked, a link to your product will permanently appear in the menu on the left. For future visits, all you need to do is sign in to the textbook's website and follow the link that appears in the left menu!

→ Click the Need Help? button on the textbook's website if you need assistance along the way.

How to access the web resource if you purchased a used book:

You may purchase access to the web resource by visiting the text's website, **www.HumanKinetics.com/QualitativeDiagnosisOfHumanMovement**, or by calling the following:

800-747-4457 . . . . U.S. customers
800-465-7301 . . . . .Canadian customers
+44 (0) 113 255 5665 . . . . European customers
08 8372 0999 . . . . Australian customers
0800 222 062 . . . . .New Zealand customers
217-351-5076 . . . . .International customers

For technical support, send an e-mail to:
support@hkusa.com . . . . U.S. and international customers
info@hkcanada.com . . . . Canadian customers
academic@hkeurope.com . . . . European customers
keycodesupport@hkaustralia.com . . . . . Australian and New Zealand customers

HUMAN KINETICS
The Information Leader in Physical Activity & Health

01-2013

HUMAN KINETICS WEB RESOURCE

Product: Qualitative Diagnosis of Human Movement Third Edition, web resource

Key code: KNUDSON-GTZN43-OSG

This unique code allows you access to the web resource.

Access is provided if you have purchased a new book. Once submitted, the code may not be entered for any other user.

THIRD EDITION

Qualitative Diagnosis of Human Movement

Improving Performance in Sport and Exercise

Duane V. Knudson, PhD

Texas State University–San Marcos

Human Kinetics

Library of Congress Cataloging-in-Publication Data

Knudson, Duane V., 1961-
  Qualitative diagnosis of human movement : improving performance in sport and exercise / Duane V. Knudson. -- 3rd ed.
    p. ; cm.
  Rev. ed. of: Qualitative analysis of human movement. 2nd.
  Includes bibliographical references and index.
  I. Knudson, Duane V., 1961- Qualitative analysis of human movement. II. Title.
  [DNLM: 1. Movement--physiology. 2. Athletic Performance--physiology. 3. Biomechanics--physiology. 4. Kinesiology, Applied--methods. WE 103]
    612.7'6--dc23
                          2012019083

ISBN-10: 1-4504-2103-2 (print)
ISBN-13: 978-1-4504-2103-4 (print)

The web addresses cited in this text were current as of July 2012, unless otherwise noted.

Acquisitions Editor: Myles Schrag; **Developmental Editor:** Christine M. Drews; **Assistant Editor:** Susan Huls; **Copyeditor:** Joyce Sexton; **Indexer:** Susan Danzi Hernandez; **Permissions Manager:** Dalene Reeder; **Graphic Designer:** Fred Starbird; **Graphic Artist:** Denise Lowry; **Cover Designer:** Keith Blomberg; **Photographer (cover):** Neil Bernstein; **Photographers (interior):** Neil Bernstein, Doug Fink, and Gregg Henness; **Visual Production Assistant:** Joyce Brumfield; **Photo Production Manager:** Jason Allen; **Art Manager:** Kelly Hendren; **Associate Art Manager:** Alan L. Wilborn; **Art Style Development:** Joanne S. Brummett; **Illustrations:** © Human Kinetics; **Printer:** Sheridan Books

We thank the Champaign Park District in Champaign, Illinois, for assistance in providing the location for the photo and video shoot for this book.

Printed in the United States of America   10 9 8 7 6 5 4 3 2 1

The paper in this book is certified under a sustainable forestry program.

Human Kinetics
Website: www.HumanKinetics.com

United States: Human Kinetics, P.O. Box 5076, Champaign, IL 61825-5076
800-747-4457
e-mail: humank@hkusa.com

Canada: Human Kinetics, 475 Devonshire Road Unit 100, Windsor, ON N8Y 2L5
800-465-7301 (in Canada only)
e-mail: info@hkcanada.com

Europe: Human Kinetics, 107 Bradford Road, Stanningley, Leeds LS28 6AT, United Kingdom
+44 (0) 113 255 5665
e-mail: hk@hkeurope.com

Australia: Human Kinetics, 57A Price Avenue, Lower Mitcham, South Australia 5062
08 8372 0999
e-mail: info@hkaustralia.com

New Zealand: Human Kinetics, P.O. Box 80, Torrens Park, South Australia 5062
0800 222 062
e-mail: info@hknewzealand.com

E5593

Contents

Foreword

Sport skill instructors face no more challenging task than identifying and correcting performance problems in those whom they serve. Yet this process of sifting through fleeting events embedded in complex patterns of movement and, after the visual manifestations have dissolved, deciding on a course of corrections is an inescapable obligation of teachers, coaches, exercise leaders, and therapeutic exercise specialists. The difficulty of successfully performing this combination of tasks—termed qualitative movement diagnosis (QMD) by author Duane Knudson—can be fully appreciated only by those who have tried to do it and failed.

Accurately perceiving motion—and human motion especially—is difficult in its own right; learning to see through the confusing display of moving arms, legs, and torso to pinpoint movement errors is even more so. But as readers of this book will come to appreciate, this is merely the first in a chain of exacting pedagogical actions that confront the would-be analyst. Performance errors are merely symptoms of underlying causes, and only those who are able to discern these causes will be of much help to their charges. Because the body is a system of independent yet linked body parts that share a complicated synchronicity, erroneous movement of one body part can produce a detrimental effect on another. The result is that the root cause of an observed problem—a slight imbalance in stance, improper positioning of a limb on the opposite side of the body, or a slight timing error in joint action—may be far removed from its effects. Thus the causes of a performance defect are not always evident, even to the experienced teacher or coach.

Making things even more difficult is the fact that performance deficiencies often have their basis in sources other than technique. They may be due, for example, to faulty perception or decision making. A batter's failure to hit a pitched ball may have little to do with his swing mechanics and everything to do with faulty visual tracking or late decision making. In other cases, errors may stem from poor joint flexibility or muscle weakness. The experienced physical therapist whose patient tilts her torso over the stance leg when lifting the opposite leg in the gait cycle will recognize this as Trendelenburg's sign and rightly conclude its cause to be a weak gluteus medius muscle on the same (stance) side.

Beyond this is the real possibility that an inept performance may be caused by underlying psychological and social factors. Can the athlete perform the skill correctly, or does he simply not want to do so? Is the athlete's performance limited by the constraints of her innate abilities? Performance breakdowns may plague an athlete who suffered a torn anterior cruciate ligament the previous year, not because he has forgotten the key elements of technique, but because he is afraid of reinjuring himself. Likewise, the biggest hindrance to a 10-year-old child learning to perform a back dive may not be movement errors per se, but fear of projecting herself backward from the diving board.

The final link in this chain of decision making is not any easier than those that preceded it. Regardless of how effective a coach might have been in detecting errors and pinpointing the root causes of those errors, there remains the formidable task of using this information in a pedagogical intervention. What information gained from the analysis should be relayed to the performer? How specific should it be? When should it be relayed? These and other considerations of course depend on the accuracy of data collected in the prior stages of QMD.

In light of all this, it is easy to see that accurate decisions all along the QMD error identification–intervention continuum require not only a "trained eye" but a wide-ranging knowledge of kinesiology. Professor Knudson rightly describes QMD as an interdisciplinary and integrative process. Identifying the causes of performance errors and remedying them require the professional to access several bodies of kinesiological knowledge simultaneously—knowledge that, for better or worse, traditional kinesiology curriculums continue to separate into such isolated courses as exercise physiology, motor learning, sport psychology, and biomechanics. Integrating this vast array of knowledge into effective clinical skills is both an art and a science; it depends not only on extensive practice but on employing a practical observational framework as proposed in this textbook.

After reading the chapters that follow, you will probably be convinced, as I was, that all students who embark on careers in the physical activity professions should be required to pass a course in QMD; just as differential diagnosis is a critical element in medical education, so should QMD be a critical element in kinesiology education. Until the initial publication of this text in 1997, however, there were few resource materials around which professors could design such courses. Obviously, biomechanics textbooks and courses are of some help in developing QMD skills but only in very limited ways. Now, thanks to Professor Knudson, students, teachers, and researchers have the benefit of not only a practical diagnostic framework, tutorials to guide them through the QMD process, and advice on how best to capture relevant information from motor performances, but also descriptions of intervention strategies. The unique, updated web-based features and high-resolution videos included in this instructional packet are invaluable tools in their own right, capable of sharpening the skills of even experienced diagnosticians.

Over 35 years ago while on the faculty at the University of Pittsburgh, I, along with graduate students, launched a series of exploratory studies on what Knudson has since labeled QMD. Our writings and research, though somewhat crude by today's standards, were helpful in calling attention to this subject, not only as an often-neglected element in the kinesiology curriculum, but also as a potentially fertile ground for research. Thankfully, scholars such as Knudson have taken it to much higher levels by updating and synthesizing research and clarifying the fundamental processes of QMD, as well as by turning much-needed attention to the importance of teaching kinesiology students how to develop proficiency in this critical pedagogical operation.

Shirl James Hoffman, EdD
Professor Emeritus of Kinesiology
University of North Carolina at Greensboro

Preface

One of the most important skills for all kinesiology professionals is known as qualitative movement diagnosis (QMD). QMD is critical to real-time professional intervention with clients to help them improve performance or reduce the risk of injury. Throughout the history of physical education, kinesiology, and exercise science, unfortunately, this professional skill has not been directly taught in most higher education curriculums. Previous editions of this book have tried to fill this gap and integrate the fragmented scholarship on QMD.

The book has been well received by scholars throughout the world and has been translated into five languages. Despite its success, the book is still seen by some as an applied biomechanics text. While biomechanics is the primary subdiscipline that provides knowledge guiding the modification of movement technique, the book has always argued for the integration of all the relevant subdisciplinary knowledge in kinesiology when one is diagnosing and modifying movement technique. This edition continues this tradition by expanding the review of knowledge from many subdisciplines of kinesiology and the disciplines of allied health and engineering that contribute information relevant to the qualitative diagnosis and improvement of human movement.

The most obvious change is the new title that focuses on the higher-order critical thinking processes of movement diagnosis. I have found that the term used previously, "analysis," tended to get in the way of the larger, integrated vision of movement diagnosis proposed. The reductionist interpretation of "analysis" was also problematic. The third edition, therefore, emphasizes the diagnostic and prospective treatment aspects of this important professional skill. Coaches, physical education teachers, and most fitness professionals are not to prescribe medical treatment; however, in this book, diagnosis refers to uncovering the causes of errors in movement and intervening with effective feedback in order to improve performance or safety.

One sad change is the absence of my coauthor, Dr. Craig Morrison, for this edition of the book. Dr. Morrison died in 2004. The untimely passing of my dear friend and colleague makes the completion of this next stage in the book a bittersweet achievement for me and for the field of kinesiology.

AUDIENCE AND SCOPE

This book is designed for upper-level undergraduate or graduate courses on the diagnosis and improvement of movement. This should be a capstone course in which experienced majors get to integrate and apply their subdisciplinary knowledge of kinesiology to help people improve their movement performance or move with a lower risk of injury.

ORGANIZATION AND FEATURES

The first edition, *Qualitative Analysis of Human Movement*, broke new ground as the first text devoted to the qualitative diagnosis of human movement. The second edition updated and improved the presentation of this important information through a reorganization of topics, additional application features, and the introduction of a

CD-ROM. The major organizational aspects and features of the second edition have been retained in the third edition.

The third edition includes more examples of the application of QMD along with high-resolution video images and clips for guided tutorials. More video clip examples of movement diagnosis, a transition to web-based practice features, and the expansion of the video and computer technology chapter are other improvements in this new edition. Over 80 new sources of research relevant to QMD have been added. The book continues a tradition of research-based knowledge organized into a simple theoretical structure, with numerous examples of application to improving human movement in a wide variety of performers and environments and for a wide variety of movements.

Part I introduces the interdisciplinary model of QMD developed in the book and summarizes the development of this approach as well as the perceptual factors relevant to movement diagnosis. Throughout the text, real-world examples show how the many subdisciplines of kinesiology contribute to each of the four tasks of QMD. Part II comprises four chapters that fully explore each of the four tasks of QMD. The book concludes with part III, on the application of QMD of human movement. These chapters guide the reader through real-world examples of QMD and show how the integration of video and computer technology can be used to improve QMD.

I hope that readers will find themselves thinking critically about their future or present professional practice. I also hope that kinesiology professionals will take steps to improve their skills in QMD to improve their service to clients and possibly to contribute to the knowledge base on the qualitative diagnosis of human movement.

To the Instructor

The third edition of *Qualitative Diagnosis of Human Movement* is a text for upper-level undergraduate or graduate courses on qualitative movement diagnosis (QMD). This edition has several updates and new features that should enhance your and your students' experience with this book. First, the web resource and e-book are now based on new movements that were recorded on high-definition (HD) video clips. Second, some of the movements have changed to reflect activities and sports of interest to current students. Third, new features—QMD Technologies and QMD Explorations—add to movement professionals' understanding and real-world application of QMD. Following are ideas for using the features in the book with your classes.

QMD Choices outline professional options in planning QMD. You can use these as discussion points or active learning strategies such as think–pair–share. Because these are often discretionary issues, there are often multiple correct answers with several trade-offs to discuss.

Movement professionals now use Internet, video, software, and other technologies to enhance qualitative diagnosis of human movement. New cameras and programs are being developed all the time. QMD Technologies explore the potential use of electronic technologies in assisting QMD. You are encouraged to keep students focused on the potential value of these technologies in improving observation, diagnosis, or presentation of intervention to performers. Getting students to think about evidence of benefit, time, and financial costs rather than the novelty or superficial appeal of new technology could be your key task in using these features.

Throughout most of the text, QMD Demonstrations use video to illustrate course content. You can use these in lectures and as movement examples for QMD.

The most complete guided practice of the qualitative diagnosis of human movement is provided in the QMD Practice features in chapter 9, Tutorials in Qualitative Movement Diagnosis. Students can observe, pause, and replay the video and compare their QMD with the suggested answers in the text. You can also use the video in lectures and discussions of QMD. The answers in the text can stimulate discussion and debate on each case study.

Two more features provide more open-ended examples of QMD in a variety of examples of human movement. QMD Explorations, also in chapter 9, provide video and unguided examples for students to use. You can use these examples in lectures, discussions, labs, or active learning exercises. Chapter 10 contains Theory-Into-Practice Situations. These short problems are real-life examples of movement issues relevant to QMD that kinesiology professionals face every day.

IMAGE BANK

For instructors, this text also comes with an image bank, which includes most of the figures, content photos, tables, and forms from the text, sorted by chapter. Images can be used to develop a customized presentation based on specific course requirements. A blank PowerPoint template is provided so instructors can quickly insert images from the image bank to create their own presentations. Easy-to-follow instructions are included.

Instructor resources can be accessed at www.HumanKinetics.com/Qualitative DiagnosisOfHumanMovement.

ACCESS TO VIDEOS AND FORMS

The web resource, which is a student resource, contains video clips related to the QMD Demonstrations that appear in chapters 1 to 8 and video clips for QMD Practices (tutorials) and QMD Explorations in chapter 9. The web resource also includes downloadable forms for the QMD Explorations in chapter 9 and the Theory-Into-Practice Situations in chapter 10. Instructors also have access to the web resource with all of these features. Students may access the content related to this text in a number of ways:

- Students who purchase a print book will receive access to the web resource via a key code.
- Students who purchase a used print book or an e-book that does not contain video will want to purchase the web resource so that they have access to the video clips.
- Students in your class may also use the enhanced e-book, which includes all of the video clips found on the web resource. These students do not need to purchase the web resource, but they will not have access to the download-able forms. The questions on the forms are in the e-book but not in a fillable format. You can provide the forms to them via the image bank or the web resource.

The web resource can be accessed at www.HumanKinetics.com/Qualitative DiagnosisOfHumanMovement.

To the Student

One of your most important skills as a future kinesiology professional will be the qualitative movement diagnosis (QMD) for your athletes, students, or patients. This book teaches you a holistic model of QMD that integrates what you know from the many subdisciplines of kinesiology in order to help these clients. Several features of this book offer you real-life examples and practice in QMD in a variety of activities and careers.

Use this book to develop skill in QMD and also to adopt a truly professional philosophy of continuous acquisition of knowledge. It is important to remember that each client deserves caring, evidence-based evaluation, diagnosis, and intervention. Some clients and peers might want to get by with uniform, easy answers to human movement problems, but that practice often limits the client's potential or increases the risk of injury. Take advantage of the vast knowledge from kinesiology regarding each movement, client, and situation. Doing your job at the highest level helps not only your clients but also your career.

ACCESS TO VIDEOS AND FORMS

The web resource contains video clips related to the QMD Demonstrations that appear in chapters 1 to 8 and video clips for QMD Practices (tutorials) and QMD Explorations in chapter 9. The web resource also includes downloadable forms for the QMD Explorations in chapter 9 and the Theory-Into-Practice Situations in chapter 10.

If you purchased a print book, you received access to the web resource via a key code. You can access the web resource at www.HumanKinetics.com/Qualitative DiagnosisOfHumanMovement.

If you purchased a used print book or an e-book that does not contain video, you will want to purchase the web resource so that you have access to the video clips.

Acknowledgments

I would like to thank several people who made significant contributions to the new edition and electronic features of this book. First, I am indebted to all the professionals at Human Kinetics who saw the need for the revision and several ways to improve my ideas. Specifically, I would like to thank Chris Drews, Myles Schrag, Doug Fink, Neil Bernstein, and Joyce Brumfield, who worked to make the new HD video and images a reality. Second, I would like to thank the many scholars whose work inspired the creation of this book. I especially appreciate that an influential leader in kinesiology and qualitative movement diagnosis, Dr. Shirl Hoffman, agreed to write a foreword for this edition. Finally, I appreciate all the contributions that remain in this book from my deceased coauthor and friend, Dr. Craig Morrison.

Credits

All photos © Human Kinetics unless otherwise noted.

Figure 2.1 Reprinted with permission from *Journal of Physical Education, Recreation & Dance,* March 1975, p. 49-52. JOPERD is a publication of the American Alliance for Health, Physical Education, Recreation, and Dance, 1900 Association Dr., Reston, VA 20191.

Figure 2.2 Reprinted with permission from *Journal of Physical Education, Recreation & Dance,* January 1982, p. 21-25. JOPERD is a publication of the American Alliance for Health, Physical Education, Recreation, and Dance, 1900 Association Dr., Reston, VA 20191.

Figure 2.3 Reprinted, by permission, from J. Hay, 1983, A system for the qualitative analysis of a motor skill. In *Collected papers on sports biomechanics,* edited by G.A. Wood (Perth, Australia: University of Western Australia Press), 97-116.

Figure 2.4 Reprinted, by permission, from J. Hudson, 1985, *POSSUM: Purpose/observation system for studying and understanding movement.* Paper presented at the AAHPERD national convention, April, Atlanta, Ga. (Chico, CA: California State University).

Figure 2.5 Adapted with permission from *Journal of Physical Education, Recreation and Dance,* June 1995, pgs. 54-55. JOPERD is a publication of the American Alliance for Health, Physical Education, Recreation and Dance, 1900 Association Drive, Reston, VA 22091.

Figure 2.6 Reprinted with permission from M.N. McPherson, 1990, "A systematic approach to skill analysis," *Sports Science Periodical on Research and Technology in Sport* 11(1): 1-10.

Figure 2.7 Illustrations courtesy of Steven Barnes, Multimedia Laboratories, Florida State University, Tallahassee.

Figure 2.8 Reprinted, by permission, from K. M. Haywood and N. Getchell, 2009, *Life span motor development,* 5th ed. (Champaign, IL.: Human Kinetics), 159.

Figure 2.9 Reprinted, by permission, from S.K. Gangstead and S.K. Beveridge, 1984, "The implementation and evaluation of a methodical approach to qualitative sports skill analysis instruction," *Journal of Teaching in Physical Education* 3: 60-70.

Figure 3.3 Reprinted from *Cognitive Psychology,* Vol. 24(3), S.E. Palmer, "Common region: A new principle of perceptual grouping," pgs. 436-447, copyright 1992, with permission from Elsevier.

Figure 4.3 Reprinted, by permission, from M.A. Lafortune and P.R. Cavanagh, 1983, *Effectiveness and efficiency during bicycle riding. In Biomechanics VIII B,* edited by H. Matsui and K. Kobayshi (Champaign, IL: Human Kinetics), 933.

Figure 4.5 Reprinted with permission from *Journal of Physical Education, Recreation and Dance,* January 2000, pg. 22. JOPERD is a publication of the American Alliance for Health, Physical Education, Recreation and Dance, 1900 Association Drive, Reston, VA 22091.

Figure 5.1 Reprinted from D.R. Bradley and H.M. Petry, 1977, "Organizational determinants of subjective contour: The subjective Necker cube," *American Journal of Psychology* 90(2): 253-62. By permission of H.M. Petry.

Figure 5.2 Adapted, by permission, from S.K. Gangstead and S.K. Beveridge, 1984, "The implementation and evaluation of a methodical approach to qualitative sports skill analysis instruction," *Journal of Teaching in Physical Education* 3: 60-70.

Figure 5.4, 8.4, 8.5 Courtesy of SIMI, Unterschleissheim, Germany.

Figure 7.1 Reprinted from H. Hatze, 1976, Biomechanical aspects of a successful motion optimization. In *Biomechanics VB*, edited by P.V. Komi. (Baltimore: University Park Press), 10. By permission of P.V. Komi.

Figure 7.4 Reprinted with permission from *Journal of Physical Education, Recreation and Dance*, January 1993, pg. 56. JOPERD is a publication of the American Alliance for Health, Physical Education, Recreation and Dance, 1900 Association Drive, Reston, VA 22091.

Figure 8.6 Courtesy of Siliconcoach, Dunedin, New Zealand.

Figure 8.7 Courtesy of Duane Knudson.

Figure 8.8 Photo courtesy of Kinesio Capture LLC.

Table 9.4 Adapted with permission from the *Journal of Physical Education, Recreation and Dance*, August 1996, pgs. 31-36. JOPERD is a publication of the American Alliance for Health, Physical Education, Recreation and Dance, 1900 Association Drive, Reston, VA 22091.

Table 9.5 Adapted with permission from *Strategies: A Journal for Sport and Physical Educators* 7(8): 19-22. Copyright 1994 by the American Alliance for Health, Physical Education, Recreation and Dance, 1900 Association Drive, Reston, VA 22091.

part

I

An Integrated Approach to Qualitative Movement Diagnosis

Qualitative movement diagnosis (QMD) is the main professional skill in many kinesiology professions for the improvement of movement technique of clients. QMD requires an interdisciplinary approach that integrates knowledge from all the subdisciplines of kinesiology. The consensus of the literature on QMD can be summarized in a simple four-task model, introduced in chapter 1. The tasks of preparation, observation, evaluation and diagnosis, and intervention address all the important issues in helping people move better or with greater safety. Chapter 2 reviews how this broader vision of QMD developed out of more fragmented, single-subdisciplinary approaches to observation or skill analysis. This section concludes by summarizing how both sensory and cognitive processes are involved in making judgments about human movement.

Interdisciplinary Nature of Qualitative Movement Diagnosis

There is a humorous fable about three golf coaches who tried in vain to agree on what would improve a golfer's swing. The conflict arose because each coach had a different perspective on how to help the golfer improve his swing. The sport psychologist focused on the stress the golfer experienced trying to change his technique, the biomechanist on how straight the lead arm was during the swing, and the motor learning expert on the size and direction of the error of the shot relative to the hole. They were insensitive to the variety of factors from outside their backgrounds that could affect a golf swing. Essential to a performer's improvement is an instructor's ability to observe, diagnose the movement, and intervene based on an integration of several subdisciplines of kinesiology or exercise science.

The introductory fable presents an analogy to the low-level skill or technique analysis often used by teachers and coaches who only detect certain "errors" and provide "corrections." A kinesiology professional, however, uses an interdisciplinary approach to qualitative movement diagnosis that results in a substantial improvement over the all-too-common low-level technique analysis. An interdisciplinary approach to qualitative movement diagnosis integrates knowledge from all the subdisciplines of kinesiology and enables professionals to identify the best way to help movers improve or reduce risk of injury.

Chapter Objectives

❶ Define *qualitative movement diagnosis* (QMD).

❷ Explain the difference between qualitative and quantitative assessments in kinesiology.

❸ Explain why QMD should be an interdisciplinary process.

❹ Describe the four tasks of an integrated model of QMD.

Physical education teachers, coaches, and many kinesiology professionals (e.g., athletic trainers, physical therapists, strength and conditioning coaches) face the challenge of accurate qualitative movement diagnosis (QMD) of human movement every day. QMD may be the most important skill that kinesiology professionals can use to help clients improve performance or reduce their risk of injury. This chapter introduces this important professional skill. You will also learn that for QMD to be the most effective, the approach needs to be **interdisciplinary**—integrating the knowledge from many subdisciplines of kinesiology.

QMD IN KINESIOLOGY

Kinesiology is the term used to refer to the academic discipline focused on the study of human movement, replacing what was often called *physical education* (Arnold, 1993; Hoffman and Harris, 2000; Newell, 1990). The premier, honorary scholarly society for this discipline in the United States (National Academy of Kinesiology) recommends the use of the term kinesiology to refer to this academic discipline in higher education. Kinesiology professionals work in a wide variety of careers (for example, teaching, coaching, dance, athletic training, sports medicine, physical therapy, fitness, ergonomics) and all use QMD to improve human movement. Athletic coaches use QMD to make judgments on technique, strategy, and team selection. Physical education teachers often use QMD to evaluate student performance, modify instruction, and assign grades. Allied health professionals use QMD to diagnose musculoskeletal pathologies and assist in prescribing treatment for recovery from injury. QMD may be the most important professional skill to help clients because it provides the immediate, field- or clinic-based guidance for instruction or intervention to improve human movement.

Definition of QMD

Qualitative movement diagnosis must be defined for the purposes of this text. This is necessary because many terms used in the kinesiology literature to refer to QMD are neither accurate nor all-encompassing. *Movement analysis, clinical diagnosis, skill analysis, error detection, observation, eyeballing, observational assessment*, and *systematic observation*, among other terms, have all been used to refer to QMD. This text defines QMD as the systematic observation and introspective judgment of the quality of human movement for the purpose of providing the most appropriate intervention to improve performance (Knudson and Morrison, 1996: 17). Notice the emphasis on diagnosis in QMD, where the movement professional critically selects the most appropriate intervention, just as a physician strives to prescribe the best treatment for a patient. The difference is that in most kinesiology professions, this diagnosis is not done for the purpose of identifying or treating injury, disease, or illness; instead it is done for the express purposes of improving movement performance and preventing injury. Since *observation, intervention*, and *performance* are used in this definition with particular meanings, it is necessary to define these terms also.

Observation is defined as the process of gathering, organizing, and giving meaning to sensory information about human motor performances. This definition is very similar to Sage's (1984) definition of *perception*, and the task of observation in QMD is closely related to perception. Intervention in QMD is defined as the administration of **feedback**, corrections, or other change in the environment to improve performance and prevent or treat an injury. Both observation and intervention are key tasks within the larger process of QMD. This book uses the term **performance** in a broad sense to mean both the short-term and long-term safety and effectiveness of a person's movement. This larger view of performance differs from that in motor learning, which traditionally limits the idea of performance to short-term movement effectiveness, using the term *learning* to refer to long-term ability. In this book, performance also includes the allied health professions' focus on improving movement technique to treat or reduce the risk of injury. "Improving performance" in this text means both improving movement potential and reducing the risk of potential injury during movement.

Observation in QMD is not limited to the use of vision only. All the senses that teachers and coaches can employ to gather information should be used. For example, a gymnastics teacher may rely on kinesthetic information from her arms in spotting early trials of a new skill. Since she is too close to the movement to make reliable visual checks of some phases of the skill, the information from her hand placement and the muscular effort she extends to assist the student in completing the skill are critical to QMD. Auditory information about the rhythm of students' impacts with the mat can also be important in QMD for a gymnastics teacher, a dance teacher, or a therapist evaluating gait. Sometimes therapists can use their sense of smell to determine if a wound may be infected and might be contributing to pain or a loss of motion.

> **» KEY POINT 1.1**
>
> Many words have been used to denote the process of the diagnosis of human movement. In this book, QMD is defined as the systematic observation and introspective judgment of the quality of human movement for the purpose of providing the most appropriate intervention to improve performance.

> **» KEY POINT 1.2**
>
> QMD is not limited to visual inspection of human movement. Good observation involves the use of all the senses to gather information about performance.

Qualitative Versus Quantitative Movement Assessment

Most observations, assessments, and evaluations of human movement by coaches, teachers, or allied health professionals are qualitative. Although a qualitative assessment of movement in QMD is by nature a subjective judgment, this does not mean that it is unorganized, vague, or arbitrary. In fact, to be most effective, QMD requires extensive planning, information from many disciplines, and systematic steps.

Figure 1.1 illustrates a continuum of human movement assessment. Any assessment of human movement can be located somewhere along the continuum from qualitative to quantitative. The qualitative end of the continuum involves the nonnumerical assessment of movement information or a judgment on the quality of an aspect of movement. The quantitative end of the continuum involves some measurement of performance. For example, QMD of baseball pitching could include statements by television commentators or technique reminders given to the pitcher by the pitching coach. Moving along the continuum are more quantitative assessments that use numbers like statistical breakdowns of pitches and their locations, radar measurements of ball speed, or complex biomechanical research.

Qualitative					Quantitative
Developmental level	Rating scale	Stride length	Velocity	Acceleration	Force

Figure 1.1   Sample continuum of human movement assessment for assessing running.

Information from traditional qualitative observational assessments in kinesiology falls near the left side of the assessment continuum. Most teachers or coaches use QMD in everyday practice situations to diagnose and correct errors. Other movement assessments are midway along the continuum. Evaluation of skill or developmental level, such as using a rating scale or timing a 40-yard dash, is at the beginning of quantified performance.

If performance can be expressed in numbers, then the assessment can be based on quantified data. Quantification of data (in seconds, feet, meters, degrees per second) moves the assessment farther to the right on the continuum. But even research measurements (quantification of a highly controlled nature) cannot be purely objective. There is some subjectivity in deciding where to place the tape measure or where to take a skinfold measurement. Quantification does not automatically ensure validity and reliability, and the lack of or a lower level of quantification in a QMD does not automatically mean that the assessment is less valid or reliable. Research and writings discussed later in this book indicate that there is a far richer decision-making base for QMD than is often thought (Morrison, 2000; Shiffrar, 1994).

The farthest levels of quantitative assessments in sport sciences, such as biomechanics and exercise physiology, are primarily being performed in university research settings or at the Olympic Training Centers for elite athletes. Biomechanics research measures instantaneous values of velocity, acceleration, or force for various parts of the body. Exercise physiology measures the time-varying values of oxygen consumption, body fat, or amounts of lactic acid in the blood. Most of these quantitative assessments of human movement, however, have been and will continue to be too expensive and difficult for widespread use in teaching and coaching settings.

A typical coaching situation in which a QMD would be helpful is illustrated in figure 1.2. An athlete has missed a fly ball during outfield practice. Many teachers might attempt to correct the hand position, as his is clearly not the best hand position for fielding balls above the waist. But a quantitative biomechanical assessment of the hand positions would be overkill and would not address the many factors that affect this performance. A good QMD of this situation would examine all factors affecting performance and then focus on the most important to help the player improve. The real question is whether or not QMD integrating all sources of information would suggest correcting hand position as the most important intervention.

In the real world, several other factors might contribute to missing a fly ball. The athlete could have poor vision and require glasses. Environmental factors such as the sun or wind may affect his performance. The psychological pressure of a critical time in the game may impair his concentration

Figure 1.2 One attempt to catch a fly ball. A QMD of this situation would help the instructor provide appropriate feedback to this performer. What do you think is the most appropriate feedback?

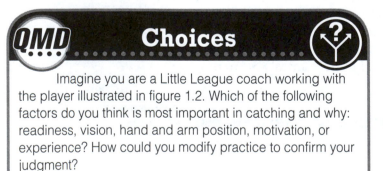

Choices

Imagine you are a Little League coach working with the player illustrated in figure 1.2. Which of the following factors do you think is most important in catching and why: readiness, vision, hand and arm position, motivation, or experience? How could you modify practice to confirm your judgment?

and perception. Maybe the player has an attention deficit hyperactivity disorder. If he is alert but does not attend to the right information (sound, ball trajectory, and spin), he could misjudge the ball. If he misses several fly balls in a row, the professional could begin to evaluate the size and direction of errors and decide whether lack of attention was the problem. This example illustrates the many factors that may have to be weighed in QMD and shows why QMD involves integration of the many subdisciplines of kinesiology.

INTERDISCIPLINARY NATURE OF QMD

Traditionally, kinesiology scholars have studied QMD from a single subdisciplinary perspective. For example, scholars interested in motor development have done extensive work in identifying and validating developmental sequences for fundamental movement patterns (Roberton and Halverson, 1984; Wickstrom, 1983). Psychology and motor learning specialists have studied how skills are learned—the practice conditions and feedback that are related to performance and learning (Ammons, 1956; Magill, 1994; Newell, 1976; Newell, Morris, and Scully, 1985; Schmidt and Wrisberg, 2008).

Experts in the field of biomechanics have formulated general principles of human movement for the purpose of QMD (Bunn, 1955; Groves and Camaione, 1983; Hudson, 1995; Knudson, 2007a; Kreighbaum and Barthels, 1985; Luttgens and Wells, 1982; Norman, 1975; Piscopo and Bailey, 1981). Biomechanists have also proposed specific methods of QMD, since this subdiscipline was commonly assumed to be a basis for QMD ability in students (Brown, 1982; Hay and Reid, 1982; Hudson, 1990a; Knudson, 2007b; Norman, 1975). Scholars interested in sport pedagogy have also recommended approaches to QMD (Barrett, 1979b, 1979c; Hoffman, 1974, 1977c; Pinheiro, 1994; Pinheiro and Simon, 1992).

The problem is that QMD in the real world requires the simultaneous integration of all these and other bodies of knowledge. *Integration* here refers to the simultaneous summing of all subdisciplines of kinesiology into a unified whole that can be bigger than the separate constituent parts. We will see in chapter 4 that kinesiology professionals strive to weigh information from experience and from the subdisciplines of kinesiology when planning for QMD. Qualitatively analyzing human movement from a single perspective will result in a fragmented and incomplete understanding of movement. Good QMD is an interdisciplinary process because all kinesiology subdisciplines contribute to the process. Simultaneously integrating many perspectives is the essence of an interdisciplinary process (see figure 1.3). In other words, QMD is not multidisciplinary, with unrelated subdisciplines contributing unique and discrete aspects to the whole (as in the story of the golf coaches in the chapter preview). The best QMD requires that the professional be knowledgeable about and simultaneously integrate information from all the subdisciplines of kinesiology as well as knowledge about the performer and movement context.

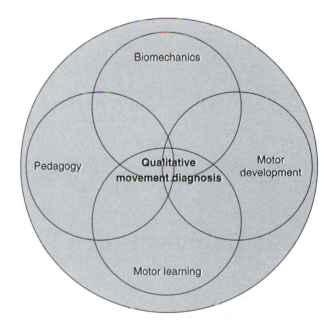

Figure 1.3 The interdisciplinary nature of QMD requires the integration of various subdisciplines within kinesiology. To prevent confusion, only four subdisciplines are illustrated.

> **»KEY POINT 1.3**
>
> Qualitative movement diagnosis requires an interdisciplinary approach, the integrated use of information from many subdisciplines of kinesiology. The golf coaches described in the chapter preview must combine information from their subdisciplines to help the golfer improve performance. A coach's knowledge of the performer's strength and flexibility (exercise physiology) must be integrated with knowledge of the various aspects of the technique (biomechanics), the practice required (motor learning), and the anxiety of the performer (sport psychology).

QMD that bases changes in technique solely on principles of biomechanics (the science of technique) is not ideal because it overlooks the relevant information that other disciplines contribute to these judgments. In the outfielder example, the athlete's hand position may not be typical of his performance and thus not the most important correction. Knowledge of strength, maturation, and stages of motor development may also apply to the situation being evaluated. Motor development literature would provide information about the typical stages people go through in learning to catch. Motor learning and pedagogy research would also help in shaping the communication of corrections and the appropriate practice schedule. Knowledge of psychology could help in the motivation necessary to sustain the practice and the short-term decrease in performance needed for long-term improvement. Professional QMD goes beyond a single subdisciplinary perspective to integrate all knowledge relevant to the movement, client, and situation.

The paradox of QMD in kinesiology is that it has emerged directly from the major subdisciplines of the field, yet it has not been widely recognized as a distinct area of inquiry in its own right. Some kinesiology scholars believe that the whole profession has suffered due to lack of recognition of the importance of QMD (Hoffman, 1974; Huelster, 1939; Lounsbery and Coker, 2008; Morrison, 2004; Norman, 1975).

It may seem odd that it has taken so long for QMD to emerge as an entity in kinesiology, but there are several plausible explanations as to why kinesiology has been so slow to address this important ability formally. First, inadequate and expensive technology (high-speed film, video equipment, computers) has made it difficult to create knowledge at a deep enough level to affect instructional practice. Because the human body is so complex, it is difficult to create biomechanical models with enough accuracy and with all the relevant factors to define optimal movement. There is also a dearth of agencies that fund research on basic questions about improving human movement.

Second, some kinesiology faculty have assumed that QMD ability arises from the undergraduate biomechanics course. While the National Association for Sport and Physical Education (NASPE) guidelines and standards for instruction in introductory biomechanics include an objective to develop QMD competence (Kinesiology Academy, 1980, 1992), there is criticism and evidence that this objective is not being met (see Knudson, Morrison, and Reeve, 1991). Biomechanics is the science within kinesiology most strongly related to movement technique, but it is not the only relevant subdiscipline. Since QMD is an interdisciplinary process, QMD training in kinesiology professional programs should not reside solely in the introductory biomechanics course. Morrison and Harrison (1997) have summarized several strategies for teaching QMD within the kinesiology curriculum.

Another problem in the last few decades is that most of the research in kinesiology has become highly specialized by subdiscipline. Many kinesiology faculty present and publish their research in specialized academic societies like the American Society of Biomechanics, American College of Sports

QMD Demonstration 1.1

Observe the video clips of the golf swing in QMD Demonstration 1.1 in the web resource at www.Human Kinetics.com/QualitativeDiagnosisOfHumanMovement and list the information about the skill and performer that you think is relevant to evaluating the performance. Classify the information by kinesiology subdiscipline or other source.

Medicine, or the North American Society for the Psychology of Sport and Physical Activity. Even faculty interested in the profession of general kinesiology who join the American Alliance for Health, Physical Education, Recreation and Dance (AAH-PERD) must select interest areas from multiple associations and committees. Since QMD is based on the integrated use of many of these subdisciplines of kinesiology, it has been difficult to work across these boundaries. An excellent example of how the development of QMD has been limited by the variety of different approaches, rather than reflecting an integrated approach, is the variety of terms used to refer to QMD (Barrett, 1979b; Hoffman, 1977c; Morrison, 2004; Radford, 1989).

It is likely that scholars from any three subdisciplines of kinesiology would use different terms for the diagnosis of human movement. Indeed, scholars have used many terms to refer to what this text calls *QMD*. Terms like *observation* (Barrett, 1979a), *error identification* or *detection* (Armstrong and Hoffman, 1979; Cloes, Premuzak, and Pieron, 1995; Vanderbeck, 1979), *qualitative assessment* (James and Dufek, 1993), and *clinical diagnosis* (Hoffman, 1983) have also been used to mean QMD. The terminology confusion is also apparent in the variety of words other than *qualitative* that are attached to *analysis*: for example, *skill analysis* (Gangstead and Beveridge, 1984; Hoffman, 1974; Wilkinson, 1992b), *movement analysis* (Biscan and Hoffman, 1976), *visual analysis* (Wilkinson, 1992a), *subjective analysis* (Arend and Higgins, 1976), and *technique analysis* (Lees, 2007, 2008; Lyons, 2003). Another terminology problem is the use of the term *systematic observation* in sport pedagogy to mean the observation and classification of teacher and student behaviors for the evaluation of teaching (Behets, 1996; Siedentop, 1991), and not the QMD of performance.

The consistent use of standardized QMD terminology across the subdisciplines of kinesiology would be useful in the development of QMD. The fragmented development of terminology makes the understanding of QMD's origins in kinesiology more important. The next section briefly previews the integrated and interdisciplinary model of QMD used in this book. Chapter 2 provides a closer look at how commonalities in previous models of QMD led to the current interdisciplinary model.

Choices

After the first game of the season, the coach of a junior high school basketball team had a discussion with parents of two of her players. The first parent thought there were a number of technique errors committed by the players on the team. He was concerned that the coach did not provide enough corrections about technique. The second parent expressed a concern about the psychological pressure the coach was putting on the players.

How would you respond to each of these parents? Does one game provide enough information to judge coaching style? Do the parents have an accurate idea of what is important in helping players learn the game? Would most parents understand the technical terminology used in teaching, coaching, and QMD? Could you explain a balanced approach to helping a player improve performance and help the team?

INTEGRATED MODEL OF QMD

This text uses a simple four-task model of QMD. Figure 1.4 illustrates this model and some of the important issues within each task. All four tasks of an integrated QMD should be viewed as equally important. A weakness in any one task diminishes the effectiveness of QMD as a whole.

Some important features of the four-task model of QMD should be apparent in figure 1.4. First, the model is circular, emphasizing the continuous learning and

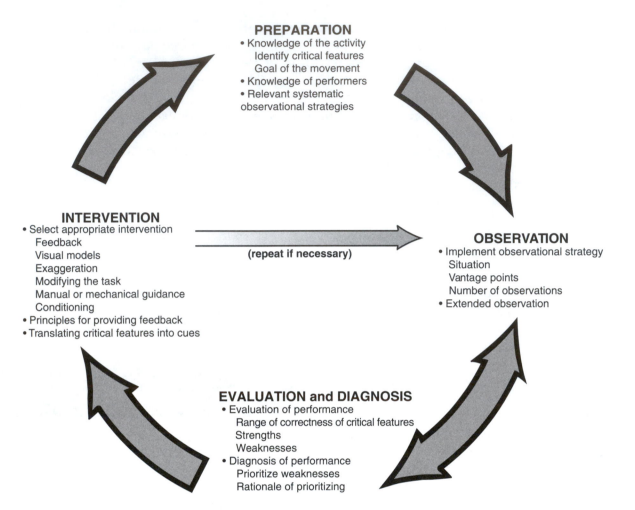

Figure 1.4 The comprehensive, integrated model of QMD. Some of the important issues for each task are listed.

improvement that are part of professional growth. Second, there is a way to move from intervention directly to observation. For example, a professional may provide feedback to a performer and immediately begin another observation to continue the QMD. Third, there can be movement in both directions between observation on the one hand and evaluation and diagnosis on the other. A skilled professional may adjust the observational strategy based on an evaluation of information in early observation of a performer.

The following chapters provide detail on the issues in each of the four tasks in an interdisciplinary QMD. In the first task of QMD, **preparation**, the professional needs to gather knowledge from research and professional opinion and to think critically about the key features of the movement, potential cues, and the common errors clients exhibit. The second task, **observation**, involves systematically gathering appropriate sensory information about the performance. The third task, **evaluation** and **diagnosis**, involves identifying strengths and weaknesses as well as prioritizing the possible ways to improve performance. The fourth task, **intervention**, often involves providing feedback or changes in practice conditions that will lead to improved performance.

>> **KEY POINT 1.4**

There are four important tasks in a comprehensive view of QMD: (1) preparation, (2) observation, (3) evaluation and diagnosis, and (4) intervention.

Any model of QMD can be viewed as part of the teaching process in most kinesiology professions. QMD is a key teaching skill that should be systematically addressed by the curriculum in teacher preparation and other kinesiology programs. Practicing professionals should continuously read and think critically about how they apply QMD in the classroom, field, or lab. They should also serve as mentors or teachers of QMD to students in the field. This cooperative training of future professionals, blending experience and scholarship, has been lacking in kinesiology (Siedentop and Locke, 1997).

QMD Demonstration 1.2

Observe the video clips of the tennis serve and forehand in QMD Demonstration 1.2 in the web resource at www.HumanKinetics.com/QualitativeDiagnosisOf HumanMovement and note the most obvious error. Observe the video clips again, this time with a larger vision of QMD, noting strengths and weaknesses. List all the kinesiology subdisciplines that provide knowledge relevant to noting all strengths and weaknesses.

The model presented here is broad enough to include all the major tasks of QMD yet simple enough not to be overwhelming. Like performers, professionals can be overcome with *paralysis by analysis* or information overload (figure 1.5), trying to apply a very complicated QMD approach or attend to unimportant information. A simple model based on critical movement features reduces the observational and assessment demand on the professional (although it may increase the demands of preparing for QMD). In the next chapter we will see that the use of a simple model is supported by the research on the validity and reliability of QMD.

Figure 1.5   Like a performer, a professional may be susceptible to paralysis by analysis. Both parties in QMD must focus their attention and be able to communicate.

PRACTICAL APPLICATIONS
Practical Application: A Tale of Three Coaches

Three Little League teams started practice for the summer season on the same day. The teams had the same number of players, the same level of skill and talent, and the same number of practices and games against the same opponents. The volunteer coaches appeared to be similar, but they had slightly different approaches to QMD.

Coach A had a treasury of time-honored baseball cues and a keen eye for finding problems in his players' performances. As soon as he observed an error, he provided the related correction. During batting practice, fielding practice, and competition, he bombarded players with a barrage of corrections and helpful pointers. Classic paralysis by analysis was occurring. The children on the team had difficulty improving despite Coach A's advice. Although they didn't win the championship or break any league records, coach and players had an enjoyable season. Parents were not sure why their children were not picking up all the gems of advice from Coach A. Why did it seem as though the children played below their potential while the other teams had most of the good luck?

Coach B had been a good ballplayer in his day. He was a natural athlete, and the game came easily to him. Coach B presented the fundamentals of the game as he remembered them from his playing days, but he didn't have Coach A's keen eye. He couldn't seem to find the problem in a player's throw or swing. Coach B often focused on errors he was familiar with. Unfortunately, he did not notice that these errors were symptomatic of other problems in technique or strength limitations. The parents were impressed with Coach B's demonstrations and believed the children were lucky to have such a skilled coach. Nevertheless, the team didn't make the playoffs, and the players never seemed to look as good as Coach B did when he was demonstrating a technique.

Coach C knew there was a lot the children needed to work on to become skilled players, and he knew they needed to work together to make a good team. He often wanted to help each child with personal coaching, but he knew this wasn't possible. Coach C thought he had better choose his words carefully to communicate effectively and to protect his players' self-esteem and confidence. Almost unconsciously, Coach C praised everyone's effort and often found only one thing each child should work on. The parents were concerned that Coach C didn't look as active, involved, or technical as Coach A. He certainly didn't look as good in action as Coach B. Nevertheless, the children learned quickly and trusted Coach C because he always seemed to find the one thing that helped them hit the ball or get the throw to the right spot. Coach C's players had a great season. They hit, ran, and fielded with confidence. Skill and luck came together, and the team won the postseason tournament. Somehow Coach C must have known that diagnostic skill was needed to evaluate technique and decide how best to help improve performance.

The astute reader may have noticed that Coach A was a skilled observer and diagnostician of movement errors, but he did not use knowledge from the disciplines of psychology and motor learning to provide the most appropriate feedback. Coach B was knowledgeable about baseball skills but lacked the knowledge and experience to develop his own observational and diagnostic skills. Coach C was a good example of an integrated approach to QMD. The story suggests that Coach C was skilled in all of the four tasks of an integrated QMD: preparation, observation, evaluation and diagnosis, and intervention. Neglect of any one of the four tasks limits a teacher's or coach's effectiveness in improving players' performances. The next section of this book details the four tasks of the integrated model of QMD.

SUMMARY

Many professions use QMD to improve human movement. Good QMD is interdisciplinary, requiring the integration of information from many subdisciplines of kinesiology. Unfortunately, most research and professional literature on QMD has not been interdisciplinary; it has been written from the perspective of one subdiscipline (as with the golf coaches in the chapter preview). This text is based on a circular model of QMD, integrating all the kinesiology subdisciplines into four tasks: preparation, observation, evaluation and diagnosis, and intervention. This larger vision of QMD uses all the senses to gather information about the strengths and weaknesses of the movement.

QMD Technology 1.1

Throughout this book, a short feature focuses on the potential use of electronic technologies to support or supplement QMD. Readers are encouraged to explore these issues and discuss them with other professionals. There are numerous benefits of these discussions, including new ideas and differences of opinion, if the potential benefits of technology outweigh the monetary and time costs. The time the athlete has to practice and the time of the professional are valuable. The same is true for learners and teachers and for clients and clinicians.

Discussion Questions

❶ Which kinesiology professions rely most heavily on QMD to improve human movement?

❷ How has terminology been a barrier to the development of QMD in kinesiology?

❸ Why is QMD an interdisciplinary process? Give examples.

❹ Why is *multidisciplinary* a poorer description of QMD than *interdisciplinary*?

❺ In what situations of human movement are qualitative assessments more appropriate than quantitative assessments? Why?

❻ What does an *integrated* approach to QMD mean?

❼ Which do you think is more important to preparing for QMD—professional experience or research?

❻ Why do you think interdisciplinary approaches to QMD have been rare in kinesiology?

Models in Qualitative Movement Diagnosis

As explained in chapter 1, good qualitative movement diagnosis should be interdisciplinary. We owe this vision of qualitative movement diagnosis to the work of previous scholars and teachers interested in this important professional skill. As Isaac Newton said, "If I have seen further, it is because I am standing on the shoulders of giants." This chapter reviews the research and writings that contributed to the development of the four-task model of qualitative movement diagnosis used in this book. These contributions have often occurred independently in the many subdisciplines within kinesiology and in various other academic fields related to human movement. In fact, even the scholarly and scientific support for the four-task model of qualitative movement diagnosis has not resulted in widespread agreement within kinesiology regarding the use of terminology or any model of qualitative movement diagnosis.

Chapter Objectives

❶ Identify the differences between comprehensive and observational models of QMD.

❷ Identify major contributors and their contributions to the development of QMD.

❸ Explain the contributions of kinesiology subdisciplines to the development of QMD.

❹ Summarize the change in the vision of QMD over the years.

❺ Discuss how the validity and reliability of QMD affect the model of QMD used.

T he four-task model of qualitative movement diagnosis (QMD) introduced in chapter 1 did not develop in a vacuum. Many teachers and scholars have contributed to the important professional skill of diagnosing human movement. This chapter classifies the various models of QMD, shows how the vision of QMD has changed, and traces the contributions of the kinesiology subdisciplines to the development of knowledge about QMD. The chapter concludes with a summary of the research on the validity and reliability of QMD and the ways in which this knowledge affects a professional's approach to QMD.

CLASSIFYING MODELS OF QMD

There are many approaches to or models for QMD in the kinesiology literature. This variety stems from the many subdisciplines of kinesiology that contribute to QMD. Researchers interested in motor development have documented approaches to QMD based on phases or levels of motor development. Biomechanics researchers have proposed models to apply the principles of mechanics to the diagnosis of human movement. Scholars in sport pedagogy and motor learning have also contributed models for QMD.

If there are many models of QMD, how do we know which one is best or which features of the models to combine? Before answering this question, it is important to classify the various models of QMD. In this book, the classification includes two categories: comprehensive and observational.

Comprehensive models deal with the big picture of QMD, laying the groundwork for the whole process. These models usually provide information on movement goals, preparation for observation, stages of motor development, observation, evaluation, diagnosis of errors, and appropriate feedback. Comprehensive models attempt to summarize all the important tasks relevant to QMD.

Observational models of QMD are focused on the task of observation within QMD. They therefore fit into comprehensive models. Observational QMD models tend to fit into what this text calls the *observation task* of a comprehensive model of QMD. We will see that some observational models include parts of other tasks (beyond observation) common in comprehensive models. One could even argue that a continuum probably exists from a strictly observational model to a comprehensive model of QMD.

Remember that the terminology, scope, and complexity of the QMD models vary. Scholars from a particular subdiscipline often emphasize the aspects of QMD to which their subdiscipline is strongly related. Table 2.1 categorizes selected QMD models in kinesiology as comprehensive or observational. The next section summarizes the development of these models of QMD.

> **»KEY POINT 2.1**
>
> Models that have been proposed for QMD within kinesiology have either focused on the process of observation or taken a more comprehensive view of the tasks within QMD.

OVERVIEW OF HISTORY AND MODELS OF QMD

Since the kinesiology literature has typically been fragmented into subdisciplines, this section presents the models and historical developments within a subdisciplinary framework. Subdisciplinary contributions are presented in alphabetical order. Although the section is organized into subdisciplines, remember that an interdisciplinary process of QMD should integrate all strengths of these models. Readers should take note

Table 2.1 Scope of Selected Models for QMD

Comprehensive	Observational
Allison, 1985b	Abendroth-Smith, Kras, and Strand, 1996
Arend and Higgins, 1976	Barrett, 1979a, 1983
Balan and Davis, 1993	Brown, 1982
Bartlett, 2007	Cooper and Glassow, 1963
Hay and Reid, 1982, 1988	Dunham, 1986, 1994
Hoffman, 1983	Gangstead and Beveridge, 1984
Huelster, 1939	Hudson, 1985, 1995
Knudson, 2007b	Pinheiro, 1994
Knudson and Morrison, 1996	Radford, 1989
Lees, 2008	Roberton and Halverson, 1984
McPherson, 1990	Rose, Heath, and Megale, 1990
Norman, 1975, 1977	Seefeldt and Haubenstricker, 1982
Pinheiro and Simon, 1992	

of the timing of these contributions, think about the development of QMD, and consider how the model presented in chapter 1 includes facets of the models reviewed.

Biomechanical Contributions

In the first half of the 20th century, the term *kinesiology* was used to refer to the course in the physical education curriculum composed of applied anatomy and possibly some mechanical analyses of human movement. Early leaders in this area of kinesiology or biomechanics were interested in the diagnosis of human movement and wrote textbooks for their courses emphasizing QMD (Broer, 1960; Cooper and Glassow, 1963; Scott, 1942). Early quantitative biomechanical analyses were labor-intensive, involving a great deal of hand tracing and calculation from 16-millimeter film tracings. Thus, qualitative analyses of movement were more common and more widely accepted than they are today.

The changing nature of the kinesiology course made for considerable confusion. Before the science of biomechanics developed, students preparing for careers in physical education took a course in kinesiology that was essentially functional anatomy. This course is now the introductory biomechanics course taken by all kinesiology majors. Biomechanists at the AAHPERD national conventions in the middle 1970s realized these problems and tried to address them by organizing the first National Conference on Teaching Undergraduate Kinesiology in 1977. The NASPE Kinesiology Academy (now the Biomechanics Committee) appointed a task force to survey kinesiology/ and biomechanics instructors and to draft guidelines for the teaching of undergraduate kinesiology. After several years of discussion and input, in 1980 the Kinesiology Academy approved the "Guidelines and Standards for Undergraduate Kinesiology."

This proposal of a standard content for undergraduate biomechanics coursework in physical education had two of three main objectives focusing on QMD. The standards that focus on QMD, which are still in effect today, recommend that the biomechanics course provide students with

1. the knowledge necessary to undertake a systematic approach to the analysis of motor skill activities and exercises, and
2. experience in applying that knowledge to the execution and evaluation of both the performer and the performance in the clinical and educational milieu (Kinesiology Academy, 1980: 19).

This essentially qualitative emphasis in undergraduate preparation supports the implementation of QMD in undergraduate biomechanics courses, and the emphasis is retained in the revised guidelines (Kinesiology Academy, 1992; NASPE, 2003).

As biomechanics began to be the dominant course content (after the 1970s), many instructors did not maintain the focus of the introductory biomechanics course on QMD but instead focused on quantitative biomechanical research (Knudson, 2003). Several biomechanists have continued an interest in QMD in professional preparation, calling for more emphasis on the principles and application of introductory biomechanics course content (Berg, 1975; Brown, 1982, 1984; Davis, 1984; Hay, 1983, 1984; Hudson, 1995; Knudson, 2001, 2003; Knudson, Morrison, and Reeve, 1991; Kreighbaum and Barthels, 1985; McGinnis, 2005; Norman, 1975, 1977). These advocates of QMD in biomechanics conceptualize the process in a variety of ways, so there has been very little systematic development of undergraduate preparation in QMD in kinesiology. Many kinesiology majors have primarily quantitative analysis experiences in introductory biomechanics (Garceau, Knudson, and Ebben, 2011).

One of the earliest calls from a biomechanist to focus introductory biomechanics instruction on application of QMD in the field was Norman (1975, 1977), who proposed that 10 biomechanical principles of motion be used to assess movement qualitatively (figure 2.1). These principles were generated from a decade of experi-

---

Summation of joint torques

Continuity of joint torques

Impulse

Reaction

Equilibrium

Summation and continuity of body-segment velocities

Generation of angular momentum

Conservation of angular momentum

Manipulation of moment of inertia

Manipulation of body-segment angular momentum

Reprinted by permission from Norman 1975.

Figure 2.1 Biomechanical principles for QMD.

ence striving to teach application in the introductory course. Norman's approach has become a part of the extensive Canadian coaching-effectiveness programs.

In effect, Norman's model for QMD is based on these underlying mechanical factors that create human movement. The first step is identifying the mechanical purpose or objective in the movement. The mechanical purpose should focus not just on the desired outcome (for example, the distance of a punt) but also on the mechanical cause of the outcome. In the javelin throw, the athlete wants to maximize release velocity and optimize release conditions (height and angle). QMD is based on identifying errors or violations of the biomechanical principles that limit performance. A diving instructor who observes an overrotated entry position in a dive must decide whether the diver's error was the generation of too much angular momentum or poorly timed manipulation of body segments and their moments of inertia. Many biomechanics textbooks base QMD on the evaluation of mechanical principles (Groves and Camaione, 1975; Kreighbaum and Bartels, 1985; Knudson, 2007a; Luttgens and Wells, 1982). A series of articles by Sanders and Wilson (1989, 1990a, 1990b) proposed 12 biomechanical concepts and their application in teaching and coaching motor skills. All these biomechanical concepts or principles are more specific than Norman's general principles.

Brown (1982) proposed 19 visual evaluation techniques that were developed with the Youth Sports Institute in preparing volunteer youth sport coaches. These techniques are essentially an observational model, organized into five areas: (1) vantage point, (2) movement simplification, (3) balance and stability, (4) movement relationships, and (5) range of movement (see figure 2.2). The observation and evaluation of movement are based on the application of these 19 techniques, first generally and then specifically. One observes multiple trials by first considering the vantage point, observing slower parts of the movement to simplify observation, and then focusing on faster, more complex parts of the movement. Brown's biomechanics paper concisely summarized the growing body of pedagogy literature of the time, which focused attention on the skill of observation.

Later that decade, at the second national symposium on teaching kinesiology and biomechanics in sports, several key developments occurred. Marett and colleagues (1984) reported on a national survey of kinesiology and biomechanics courses. Undergraduate kinesiology or biomechanics courses were placing a greater emphasis on teaching mechanical principles and applying biomechanics in QMD, at the expense of functional anatomy. Several other presentations focused on strategies for teaching QMD in the undergraduate biomechanics courses (Brown, 1984; Daniels, 1984; Hay, 1984; Hoffman, 1984; Hoshizaki, 1984; Kindig and Windell, 1984; Phillips and Clark, 1984; Stoner, 1984). The latest technological advance in interactive instruction, the interactive videodisk, was also demonstrated at this meeting (Zollman and Fuller, 1984). There seemed to be momentum for the goal of teaching the application of biomechanics with QMD. Unfortunately, this strong start has not grown into a clear consensus to emphasize QMD over quantitative problem solving in the introductory biomechanics course in the United States.

About this time, one of the most comprehensive models of QMD was proposed by Hay (1984). His model, which was the basis for a popular undergraduate biomechanics text (Hay and Reid, 1982, 1988), involves four steps.

1. Development of a deterministic biomechanical model of the skill
2. Observation of performance and identification of faults

Vantage Point

Select proper viewing distance

Observe from different angles

Observe the performance from a carefully selected angle

Select with few distractions

Select setting with vertical or horizontal references

Observe a skilled performer as a model

Movement Simplification

Observe slower-moving parts

Observe separate components of complicated skills

Observe the timing of performance components

Balance and Stability

Look at supporting parts of the body

Look at the height of the body and body parts

Movement Relationships

Look for unnecessary movements

Look for movement opposition

Observe motion and direction of swinging body parts

Look at the motion of the head

Note the location and direction of applied force

Range of Movement

Observe the range of motion of body parts

Look for stretching of the muscles

Look for a continuous flow of motion

Reprinted by permission from Brown 1982.

Figure 2.2 Visual evaluation points.

3. Ranking the priority of faults

4. Instructions to the performer

The Hay and Reid approach relies on a strong knowledge of biomechanics to break down motor skills and assess the movements of performers, and the authors provide several insightful examples of QMD of sport skills.

The first step in the Hay and Reid model for QMD is the development of a biomechanical **deterministic model** of the skill. First, the mechanical purpose or result is identified—for example, time in a 100-meter dash, horizontal distance for a javelin throw, or height in a vertical jump. Then the biomechanical factors that directly influence or determine the result are identified. Figure 2.3 illustrates a deterministic model of a long jump. Most biomechanics models of QMD emphasize the

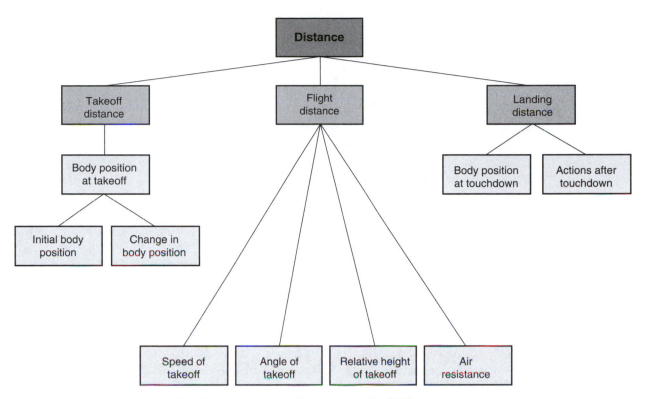

Figure 2.3 Deterministic model of a long jump used in preparing for QMD.

Reprinted by permission from Hay 1983.

importance of understanding the mechanical goal or purpose of movements to be assessed. A later model of QMD called *biomechanically based observation and analysis for teachers* (B-BOAT) expands the use of these deterministic models by adapting them to traditional organization (phases of the movement and skill level of the learner) of movement information (Abendroth-Smith, Kras, and Strand, 1996; Abendroth-Smith and Kras, 1999).

The second step in the model is the observation of performance and identification of faults. Hay and Reid outline many recommendations for using multiple senses and controlling the setting, vantage points, distances, and numbers of observations. They argue that the focus of observations generally should follow this pattern: Observe two or three trials to get a general impression (gestalt) and then focus on parts to be systematically observed in later trials.

Hay and Reid (1988) built on the work of Norman (1975, 1977) and suggested that there are two ways in which observers identify faults in human movement. The traditional approach is to break the movement down into phases and compare the performer's movement to a mental image of the appropriate movement. They call this the *sequential method*. They suggest that flaws in this method are the assumption of an ideal form, the lack of any valid rationale for determining ideal form, and the typical assumption that champion athletes use ideal form.

Hay and Reid propose that the better approach is to use the *mechanical method* to identify performance faults. The mechanical method is based on using deterministic, biomechanical models to systematically evaluate the model factors in order to see if performance can be improved. This evaluation begins by identifying the lowest factors on each path. These are the underlying determinants of the effectiveness of the performance. The factors that cannot be changed by the performer are eliminated from consideration, and the observer evaluates what factors can be improved. For example, if the coach feels that initial body position and speed of takeoff are the only critical changes relevant for a long jumper, he can focus subsequent practice on these factors.

Although Hay and Reid claim that the mechanical method does not rely on any visual model of ideal form, they do suggest that their method uses the form of skilled performers as a guide for what is effective and what correction may be appropriate. It appears that the mechanical method proposed is another structure or paradigm on which to base QMD. In the end it may still rely on the same mental image comparisons as traditional qualitative assessments of technique.

Hay and Reid also expand the vision of QMD to include the third step, ranking the priority of faults. They advocate that faults be prioritized using two rules: (1) faults are excluded if they are related to or result from other faults, and (2) faults should be corrected in the order that generates the most improvement in the time available. If an observer cannot prioritize some faults based on these rules, they should be ranked in the order in which they occur in the skill.

The final step in the Hay and Reid model for QMD is providing instruction to the performer. This is a critical step; the results of the QMD can be wasted if feedback is poorly delivered. Hay and Reid note that it is important to check the performer's understanding of the feedback. They also strongly suggest that feedback be limited to one fault at a time. They advocate direct corrections of faults or literal descriptions of what the performer should do. If this is unsuccessful, more figurative feedback that may indirectly lead to the correction should be used.

As already noted, Hay and Reid's text (1982, 1988) provides a strong comprehensive model of QMD as commonly used in coaching. Their approach relies on a strong knowledge of biomechanics to break down motor skills. Their book provides insightful examples of QMD of several sport skills using their method.

Another biomechanist, Hudson (1985), developed the POSSUM (purpose/observation system of studying and understanding movement) approach to qualitative assessment of technique. A strength of the Hudson model is that the purpose of a movement must be associated with some observable dimensions of the movement. These dimensions are the variables that the observer must evaluate visually. The Hudson model is based on selecting visual variables that are important. These must distinguish between skill levels, be qualitatively observable by the naked eye, and be subject to change by the performer.

Figure 2.4 illustrates how a purpose is linked with visually observable variables. Each visual variable has a continuum that assists in evaluating performance. The visual or spatial assessment of these dimensions is achieved with two kinds of visual foci. The observer may focus at the whole-body, or somatic, level or on the sectional (segmental) level.

Hudson continued to develop these teacher-friendly descriptions of biomechanical variables. In 1990 she proposed six generic dimensions of value and visual variables. Her research has also focused on examining how experienced and novice observers visualize variables in unfamiliar motor skills and on studying how long and in what phase of the movement

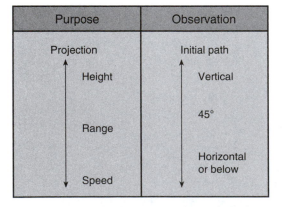

Figure 2.4 Examples of the principle of direction of force from the POSSUM model of QMD. The purposes of motor skills are linked with observable variables for the evaluation of performance.

Reprinted by permission from Hudson 1985.

visual variables are observed (Hudson, 1990a). Hudson (1995: 55) has expanded this idea and identified 10 core concepts of kinesiology (figure 2.5).

These concepts are the variables that a teacher or coach can evaluate and give feedback about in order to improve performance. The core concepts are, in essence, the technique knobs that a professional can turn in helping performers improve their movements. How this approach to QMD is applied in the real world of physical education is well illustrated in Hudson's work (2006).

An excellent comprehensive model of QMD based on previous biomechanical models was proposed by McPherson (1990). Her model of a systematic approach to skill diagnosis has four steps, preobservation, observation, diagnosis, and remediation, as shown in figure 2.6. Note how the four steps are similar to the four steps of the Hay and Reid (1982, 1988) model. McPherson's article provides excellent discussions of the importance of mechanical principles and their relationship to critical features. Her discussions on observational plans and diagnosis of errors are also very thorough.

At the third national symposium on teaching kinesiology and biomechanics in sports, fewer papers were presented on QMD in undergraduate biomechanics than at the second symposium. Research on the effect of biomechanics instruction and QMD (Dedeyn, 1991; Knudson, Morrison, and Reeve, 1991) and introductory experiences in QMD (Bird and Hudson, 1990) were presented.

Range of motion

Speed of motion

Number of segments

Nature of segments

Balance

Coordination

Compactness

Extension at release

Path of projection

Spin

Adapted by permission from Hudson 1995.

Figure 2.5   Core concepts of kinesiology.

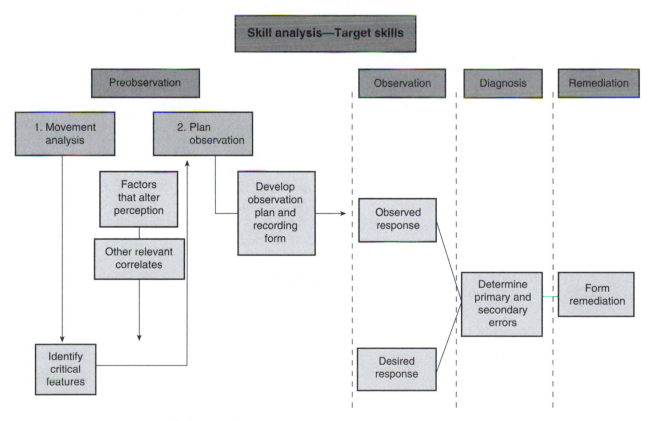

Figure 2.6   The McPherson QMD model.

Reprinted by permission from McPherson 1990.

QMD Choices ❓

Biomechanical models of QMD range from observational models to comprehensive models. When might an observational model approach to QMD be more appropriate than a more comprehensive model?

Other presentations described experiences of teaching biomechanical principles in a method aligned with QMD (Wilkerson, Kreighbaum, and Tant, 1991). After a discussion session at the symposium, a subcommittee was formed to revise the guidelines and standards. In 1992 the revised standards were adopted by the Kinesiology Academy (soon to be renamed the Biomechanics Academy) of NASPE, and they retained the QMD emphasis for undergraduate biomechanics instruction (Kinesiology Academy, 1992). Although the standards retained the emphasis on QMD, many new biomechanics faculty were more interested in teaching quantitative biomechanical problem solving.

At the fourth national conference on teaching biomechanics in sports (Wilkerson, Ludwig, and Butcher, 1997), there was again a noticeable decrease in papers on the diagnosis of human movement. A survey of instructors at the meeting showed a continued decrease (from 1977 and 1983) in the percentage of class time devoted to the application of biomechanics (Satern, 1999). It is likely that the percentage of courses teaching application to QMD is even lower. Most students in introductory biomechanics classes do not have experiences in the application of biomechanics in QMD. The professional skill of QMD does not appear to be a priority for the majority of biomechanics faculty. At the fifth national biomechanics teaching conference in 2010, no papers on QMD were presented.

» KEY POINT 2.2

Early leaders in biomechanics developed models of QMD and guidelines and standards for the introductory biomechanics course that emphasize this important skill. Not all biomechanics instructors agree with this philosophy of instruction in QMD in the introductory biomechanics course.

Biomechanical contributions to QMD include scholarly papers (Ae et al., 2007; Hay, 1984; Hudson, 1985, 1995; Knudson, 2001, 2007b; McPherson, 1990; Norman, 1977; Schleihauf, 1983), textbooks designed for implementing the qualitative nature of the guidelines and standards (Adrian and Cooper, 1995; Hall, 2007; Hay and Reid, 1988; Kreighbaum and Barthels, 1985; McGinnis, 1999), and several models of mechanical methods of QMD (Bartlett, 2007; Hay and Reid, 1988; McGinnis, 2005). It is clear that biomechanics has contributed to the development of QMD; however, it is not clear that the majority of biomechanics instructors in the United States follow the recommended emphasis on QMD in the NASPE guidelines. The most recent national survey reported that 17% of biomechanics faculty consider movement assessment an important topic in the introductory biomechanics class (Garceau, Knudson, and Ebben, 2011).

Motor Development Contributions

Scholars interested in motor development have documented the typical changes in movement patterns as children mature and become more skilled in many movements. Interest has recently expanded to life span changes in motor skills. Motor development research on the typical changes of individuals at all ages is invaluable for the kinesiology professional qualitatively assessing a movement. For example, a youth sport coach would benefit from knowing when growth spurts may influence changes in strength and coordination, or when most children attain mature levels of many fundamental movements.

The nature and description of these developmental changes in movements have been addressed in several ways. The developmental changes in movement have been called **developmental sequences** and tend to be described using either a whole-body approach or a movement-components (legs, trunk, arms, and so on) approach. These motor development sequences have naturally been applied as observational models of QMD.

There are several examples of the whole-body developmental sequence observational models of fundamental movement patterns (Branta, Haubenstricker, and Seefeldt, 1984; Haubenstricker, Branta, and Seefeldt, 1983; Seefeldt and Haubenstricker, 1982; Ulrich, Ulrich, and Branta, 1988; Wickstrom, 1983). A classic example of a whole-body developmental sequence was proposed for overarm throwing by Wild (1938). Wickstrom (1983) has reviewed much of this early work. The development of overarm throwing in children from two to seven years old generally shows four stages. Stage I involves a flexion and extension of the elbow, essentially in a sagittal plane and without a change in foot position. Stage II generally occurs from age three to five and has several characteristics. There still is no foot movement, but some transverse plane trunk rotation is added to an elbow flexion and extension that may be in a more oblique or horizontal plane. Stage III is usually seen in five- and six-year-olds. The movement begins with a step on the same-side leg (which limits the potential trunk rotation) and arm preparatory movements that are usually straight back. Stage IV is commonly found in boys and girls around six and a half years of age. This mature throwing movement involves a forward step with the opposite-side foot, a downward arm backswing, trunk rotation, horizontal adduction of the upper arm, and elbow extension.

Figure 2.7 illustrates stage II and stage IV in the whole-body development of the overarm throw. A physical education teacher can use this approach to know what actions to look for and the typical changes to expect in young throwers.

There are many motor development observational models of QMD using the movement-components approach (Halverson, 1983; Mosher and Schutz, 1983; Oslin, Stroot, and Siedentop, 1997; Painter, 1994; Roberton, 1983; Williams, 1980).

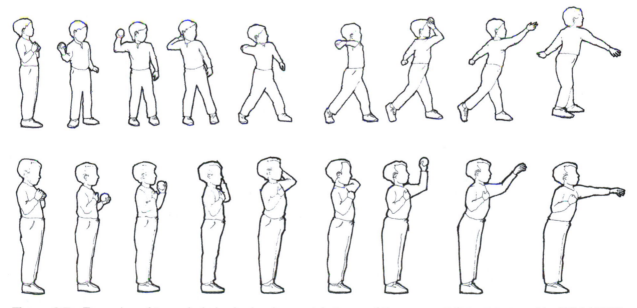

Figure 2.7 Examples of two whole-body developmental stages of the overarm throw proposed by Wild (1938). Which of the four developmental stages are illustrated and what characteristics help you know which stages these are?

Illustration courtesy of Steven Barnes.

The text by Roberton and Halverson (1984) summarizes many of the components developmental models of fundamental movement patterns studied at the University of Wisconsin. Roberton and Halverson studied the development of overarm throwing longitudinally and cross-sectionally and have validated developmental levels in six components of the overarm throw. A child's developmental level for various body parts of overarm throwing can be observed from the point of view of the trunk, backswing, humerus, forearm, stepping, and stride actions. Roberton and Halverson propose that the trunk action be evaluated first in throwing and striking QMD. They suggest that the key developmental changes in the other components of throwing and striking are timed to changes in the use of the trunk. Recent research focuses on the timing of the humerus and forearm components for various ages and skill levels (Southard, 2002, 2009), the confirmation of the arm component categories by 3-D biomechanical measurements (Stodden et al., 2004), and the reliability of test batteries of gross motor development (Barnett et al., 2009).

Since the overarm throw is important to many sports, the components model of motor development has also been applied to other sport skills. Rose, Heath, and Megale (1990) developed a components model for observation of the tennis serve, a skill related to the overarm throw. The tennis serve was broken down into developmental levels within six components of the movement. Messick (1991) also reported a prelongitudinal screening study of the motor development of the tennis serve based on six components. Observational models of components of sport skills are quite common. The problem is that these approaches do not point out to the professional all the factors and subdisciplines to consider in QMD.

Both whole-body and components observational models are effective in elementary physical education settings. Like youth sport coaches, physical educators need to know developmental milestones of many movements and the typical variability in children reaching these milestones. Maturation rates are typically plus or minus two years of chronological age. Physical educators must individualize teaching and feedback, because children of a given chronological age differ in developmental level and physiological age.

Research in motor development has been expanding the scope of the field. Studies in the 1990s focused on the application of developmental sequences to more specific tasks, such as sport skills (Haywood, Williams, and Van Sant, 1991; Messick, 1991; Rose and Heath, 1990; Rose, Heath, and Megale, 1990). Motor development theory and research has also focused on changes in development across the life span by documenting movement changes in older adults (Haywood and Williams, 1995; Williams, Haywood, and Van Sant, 1996). The text by Haywood and Getchell (2009) provides a logical decision tree for rating the developmental level of many movements and provides video clips for training in this kind of evaluation (figure 2.8). Motor development scholars have advocated for increasing the emphasis on motor skill performance throughout the life span, through professional QMD, to promote greater levels of physical activity to improve health (Stodden et al., 2008).

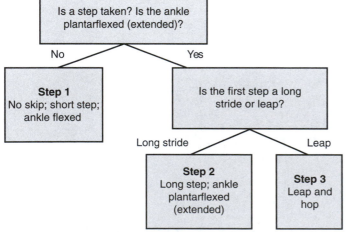

Figure 2.8 Logical decision tree.

Reprinted by permission from Haywood and Getchell 2009.

Motor development research has made significant contributions to QMD by documenting changes in motor skills and increasing the understanding of ways (whole-body vs. components) to observe and rate movement. Motor development has also strongly contributed to our understanding of the validity and reliability of visual observation. Future motor development research could be of even greater value to QMD if (1) comprehensive models of QMD were used for evaluating developmental level, and (2) it could be learned what corrections and practice are most effective in accelerating motor development. For example, research needs to determine if intervention should focus on the next developmental level or on the mature form to create the fastest skill development.

> **»KEY POINT 2.3**
> Motor development scholars have contributed to QMD by carefully documenting the developmental stages of fundamental movement patterns and observational models for identifying these stages or levels.

Motor Learning Contributions

Research in motor learning has focused on issues related to the speed and permanence of learning new motor skills. This research has often used simple motor tasks with performance goals that can be precisely quantified. More and more motor learning research is being conducted in practical and real-world conditions rather than on single-joint movements in lab conditions, expanding the opportunity to generalize the research. The major contribution of motor learning research to QMD in kinesiology is the wealth of information on the effect of practice schedules and many kinds of feedback on learning motor skills.

Feedback is the primary mode of intervention in QMD. Teachers of discrete movements (for example, a golf swing) may provide immediate feedback as part of QMD. An understanding of motor learning research on summary feedback might help a track coach provide good feedback in continuous events like sprinting or distance running. Several good review articles deal with movement feedback (Annett, 1993; Bilodeau, 1969; Kluger and DeNisi, 1996; Lee, Keh, and Magill, 1993; Magill, 1993, 1994; Newell, 1976; Newell, Morris, and Scully, 1985), and readers can consult the text on motor learning by Schmidt and Wrisberg (2008).

Recent research in motor learning has focused on contextual interference in learning (Schmidt and Wrisberg, 2008). This theory predicts that long-term learning of motor skills is enhanced with the use of practice tasks that vary rather than blocks of similar tasks. The prescription of practice is an important intervention strategy within QMD. The chapter on intervention discusses the use of modifications of practice to improve performance. It is likely that in the future, kinesiology professionals will have much more guidance on providing intervention, beyond feedback and to a variety of ways to modify practice conditions to improve performance.

Motor learning knowledge contributes strongly to the diagnosis and intervention tasks of QMD. Motor learning research, however, often relies heavily on quantitative measurements of human movement to test theories of learning movement. While motor learning scholars have not been leaders in the development of QMD models, the subdiscipline of motor learning provides critical knowledge to integrate into planning intervention (discussed in chapter 7) within QMD.

Pedagogy Contributions

Pedagogy scholars authored some of the first calls for greater emphasis on QMD in professional preparation in kinesiology. One of the earliest authors to comment on the need for QMD of movement was Huelster (1939). She suggested that courses like anatomy, body mechanics, and kinesiology were not enough to give physical education graduates the ability to do real-time QMD. Later research proved her right on this point, and many of Huelster's suggestions were well ahead of her time. Unfortunately, only a few of her suggestions have been incorporated into the current physical education curriculum.

The next major motivation within pedagogy for the development of QMD came from research suggesting that good performers do not necessarily make good visual assessors of movement (Kretchmar, Sherman, and Mooney, 1949). This view that the kinesthetic sense of skilled athletes may not transfer into QMD ability has been validated in the research of Girardin and Hanson (1967), Osborne and Gordon (1972), and Armstrong and Hoffman (1979).

A surge in pedagogy research on QMD in physical education occurred in the late 1960s and the 1970s (Armstrong, 1977a, 1977b; Biscan and Hoffman, 1976; Girardin and Hanson, 1967; Hoffman and Armstrong, 1975; Hoffman and Sembiante, 1975; Landers, 1969; Moody, 1967; Osborne and Gordon, 1972). In 1974 Shirl Hoffman wrote a watershed article about the inability of people who had taken the undergraduate biomechanics class to assess movement qualitatively. Other scholars have agreed with Hoffman's position that traditional biomechanics courses alone do not adequately prepare students for QMD and that specific instruction is necessary (Barrett, 1979a, 1979b, 1979c; Locke, 1972). This position has been supported by some research (Knudson, Morrison, and Reeve, 1991; Morrison and Reeve, 1988a), although there is evidence that applied and conceptual instruction in biomechanics can affect the physical education teacher's selection of feedback to students (Dedeyn, 1991; Knudson, Morrison, and Reeve, 1991; Nielsen and Beauchamp, 1992). Nielsen and Beauchamp (1992), for example, found that conceptual instruction (generic biomechanical principles like those of Norman or Hudson) improved the amount of corrective, accurate, and specific feedback provided by students who qualitatively assessed videotaped performances of volleyball and team handball skills.

Other important pedagogy developments in the 1970s were the scholarly reviews of observation literature by Barrett (1977, 1979a, 1979b, 1979c) and a comprehensive model of human movement diagnosis by Arend and Higgins (1976). The Arend and Higgins model presented a comprehensive picture of movement diagnosis, both qualitative and quantitative, and may have been the first formal call for an interdisciplinary approach to QMD. Their model was also detailed in a subsequent book (Higgins, 1977).

The Arend and Higgins (1976) paper was important because it provided a comprehensive summary of an interdisciplinary approach to the diagnosis of human movement. Issues from biomechanics, pedagogy, motor development, and other kinesiology subdisciplines are included in their plan for diagnosis. Their model also provided for several strategies based on whether the purpose of the diagnosis was skill or performance. They saw skill assessment as an evaluation of learning, or how human movement changed over time. Performance assessment was the evaluation of the execution of a task or subphase of a task. The Arend and Higgins model was so comprehensive that it was designed to accommodate any kind of assessment and diagnosis of human movement. For example, their model can be used for subjective, anatomical, or quantitative diagnoses of human movement.

The Arend and Higgins comprehensive model breaks QMD down into three phases: preobservation, observation, and postobservation. Preobservation has three levels of decomposition, or levels of the factors important in the movement. Each level provides more specific background information on the movement being diagnosed. The third level of decomposition identifies the precise biomechanical, tactical, and morphological factors related to the movement. This model and the paper were important because previous articles on QMD had focused on the observation task, devoting little discussion to the preparatory activities or the diagnostic activities after observation. Arend and Higgins also were early proponents of the concept of critical features, the key features of a movement that are necessary for optimal performance. We examine the concept of critical features more closely in chapter 4.

Siedentop (1991) identifies the 1970s as a time when the pedagogy literature changed from research on teaching methods to initial research documenting teaching and learning behaviors. Remember that the term *systematic observation* refers to the observation and classification of teacher and student behaviors in the process of evaluating learning, not the QMD of performance. Although the science of sport pedagogy was taking a major turn in focus, many pedagogy scholars continued to contribute to the professional skill of QMD. By the end of 1979, demand for the implementation of QMD in the undergraduate physical education curriculum had slowly grown. The initial call to address this need had been made, research had begun, a theoretical basis had been framed, and specific changes in undergraduate preparation had been proposed.

Pedagogical research on QMD continued to grow in the 1980s. Bayless (1980, 1981) reported two studies indicating that several styles of instruction could improve the error-detection abilities of undergraduate students. It appears that QMD ability can be developed using a variety of instructional approaches that include guided practice. Hoffman, Imwold, and Kohler (1983) applied the taxonomy developed by Fitts (1965) to assess children's movements. Also in 1983, two important QMD scholars summarized their research in key review papers (Barrett, 1983; Hoffman, 1983).

Barrett (1983) summarized her research on observational strategies and brought the professional skill of QMD to the attention of pedagogy scholars. Hoffman (1983) expanded his observational model of QMD into a more comprehensive vision called the *diagnostic prescriptive model*. The fundamental requirements were a good mental picture of what a performance should look like and a clear goal or purpose for the movement. The teacher focuses on the difference between the observed response and the mental image of the correct response. If a discrepancy exists between what is seen and what should be, the observer is charged with diagnosing (the extent of discrepancy and possible cause) and prescribing a remedy. Hoffman's model was further developed into a comprehensive model called the *hypothetico–deductive model*. Differences between observed and desired performance could be related to a lack of a critical ability, a deficiency in skill, or a psychosocial problem. The Hoffman model significantly extended the considerable work on observational models, hypothesizing on the evaluation and diagnosis of movement errors for a more comprehensive vision of QMD.

Pedagogy scholars with experience in traditional QMD models in biomechanics courses (Cooper and Glassow, 1963) developed important observational models of QMD. The Gangstead and Beveridge (1984) model was a true observational model, focusing on the observer's attention to the temporal and spatial aspects of the movement. The temporal foci are the preparation, action, and follow-through phases of the movement. The spatial foci are the performer's body weight, path of the hub (slowest-moving part), arms, legs, trunk action, head action, and impact or release parameters. Figure 2.9 presents this model as modified from Cooper and Glassow (1963). The Gangstead and Beveridge model is designed to concentrate the observer's

Body components	TEMPORAL PHASING		
	Preparation	Action	Follow-through
Path of hub			
Body weight			
Trunk action			
Head action			
Leg action			
Arm action			
Impact or release			

Figure 2.9 The Gangstead and Beveridge (1984) observational model of QMD.

Reprinted by permission from Gangstead and Beveridge 1984.

attention on the sequence and the critical features of the movement. This observational framework is useful for observers who have difficulty directing their attention to different parts of a movement.

Dunham (1986, 1994) proposed the use of task sheets for QMD and emphasized the importance of getting an overall feeling about the quality of the movement before observing specific components. This gestalt impression in the observational process is different from the first step in traditional observational models, which focus on temporal or spatial information first. Dunham instructed the observer to get an overall feel for the way the skill is performed. The basic ideas of a **gestalt** are that the whole is greater than the sum of its parts and that the best way to assess movement is to observe the whole and decide on the quality. If the quality is lacking, then evaluate the skill by temporal or spatial means, or use one of the other models proposed in this chapter to find the specific problem. Pinheiro and Cai (1999) found that preservice students were better at diagnosing errors in a badminton short serve when they were able to consult a criteria sheet containing skill information. Pinheiro and Cai went on to hypothesize that this approach to live movement assessment might be useful at all levels of expertise. Several of these approaches to organizing observation with task sheets (Dunham, 1994; Gangstead and Beveridge, 1984; Pinheiro, 2000) are summarized in chapter 5.

Research on QMD by pedagogy scholars after 1980 began to expand into visual perception, a natural outgrowth of the early emphasis on observation and improvements in technology. Researchers used eye-tracking recorders to document visual search patterns (Avila and Moreno, 2003; Bard et al., 1980; Hernandez et al., 2006; Petrakis, 1986, 1987, 1993) and psychological tests to determine the influence of perceptual style on QMD (Beveridge and Gangstead, 1984; Gangstead, Cashel, and Beveridge, 1987; Knudson and Morrison, 2000; Morrison and Frederick, 1998; Morrison and Reeve, 1989, 1992; Swinnen, 1984b). Research has shown that novices can improve visual perception of movement from participating in specific video training programs (Abernethy, Wood, and Parks, 1999).

We will see in the next chapter that visual observation is strongly affected by the observer's expectations and perception. Petrakis and Romjue (1990) examined the process of mental strategies of experienced observers to see if there were any similarities common to all observers. Procedural knowledge of QMD appeared to be the same for experienced observers, but their observation strategies appeared to be different. Some observers looked for specific parts of a skill to indicate proficiency; others scrutinized the overall performance. The former approach indicates an observer perceptual preference, while the latter points to a gestalt approach. Perceptual style, the way a person takes in and organizes sensory information for interpretation, does appear to make a difference in QMD. The effect of perceptual style has been a focus of later research on QMD (Knudson and Morrison, 2000; Morrison and Frederick, 1998; Morrison and Reeve, 1989, 1992). Chapter 5 shows several ways to organize an observational strategy for QMD.

Other QMD issues investigated by pedagogy researchers were the effects of professional experience (Allison, 1987a, 1987b; Barrett, Allison, and Bell, 1987; Imwold and Hoffman, 1983; Pinheiro, 1990), transfer to other skills (Morrison and Reeve, 1986; Nielsen and Beauchamp, 1992; Wilkinson, 1991), teaching QMD to classroom teachers (Morrison and Harrison, 1985), specific instruction in QMD (Gangstead, 1987; McPherson, 1988a; Morrison, Gangstead, and Reeve, 1990; Morrison and Reeve, 1988a, 1988b; Morrison, Reeve, and Harrison, 1992; Pinheiro, 1994; Wilkinson, 1990, 1992b), and retention of QMD ability (Wilkinson, 1992a; Morrison, 1994; Morrison and Harrison, 1985).

In a major theoretical leap, Kniffin (1985) took QMD to the next logical step. He showed that videotaped instruction in the QMD of movement could produce positive results in live QMD in the gymnasium. Although these results had been hypothesized by researchers in the field, this was the first real evidence that specific film and videotape training in QMD would lead to improvement in movement diagnosis in the classroom. Other research has shown insignificant differences in live QMD proficiency of kinesiology students after videotape training (Eckrich et al., 1994).

Interactive videodisks and CD-ROM technology have been used to store video for instruction in QMD (Chung, 1993; Harper, 1995; Kelly, Walkley, and Tarrant, 1988; Klesius and Bowers, 1990; Mathias, 1991; Walkley and Kelly, 1989; Williams and Tannehill, 1999). But the small number of studies and inconsistent results on the transfer of video training in QMD to live settings suggest that more research is needed. The growth of the Internet and the independence of web-based platforms suggest that future video presentation and replay for QMD are likely to be developed for the World Wide Web. Part III of this text and the web resource are good examples of this trend.

Radford (1988, 1989, 1991) added to the review literature on QMD by pedagogy scholars by defining movement observation and proposing ways of linking observation, feedback, and assessment. Several comprehensive models of QMD have emerged that have a pedagogical heritage (Balan and Davis, 1993; Pinheiro 1990, 1994; Pinheiro and Simon, 1992). Pinheiro has proposed a model that describes the overall processes of QMD (Pinheiro and Simon, 1992), as well as an observational model (Pinheiro, 1994). The comprehensive model of Pinheiro and Simon (1992) is based on an **information-processing** approach. The three levels in this model are cue acquisition, cue interpretation, and diagnostic decision. Cue acquisition is like the observation task in the integrated, comprehensive model of QMD presented in this text. Cue interpretation is analogous to our evaluation step. And the diagnostic decision is analogous to the diagnosis step within the evaluation and diagnosis task of QMD. All of these processes can also be viewed as part of information processing in QMD.

Balan and Davis (1993) presented an ecological task analysis approach to teaching physical education that was derived from the work of Davis and Burton (1991). Their approach includes QMD as an essential component of the teaching, learning, and evaluation process. Their model also differs from others in emphasizing performer responsibility and control of the observational environment. Responsibility for finding movement solutions is shifted to performers or students, which is appropriate for individualized styles of instruction. The professional is charged with controlling the physical (practice, equipment, and so on) and social environments to facilitate assessment of the movement. This model is almost like a style of teaching (the pedagogy connection again) and is reviewed more extensively in chapter 7.

Pedagogical contributions to the understanding of QMD have been significant. Scholars have focused attention on QMD instruction in the curriculum, the importance of competence in observation, and enlarging the scope of QMD inquiry beyond the observation of movement. Recently there has not been much progress in these areas from sport pedagogy. Recent papers have focused on observation (Schempp and Woorons-Johnson, 2006) or have explored computer-assisted instruction in simple visual error detection (McKethan et al., 2003). Lounsbery and Coker (2008) have called for greater emphasis in the kinesiology curriculum on QMD because more emphasis on motor skill development might contribute to higher levels of physical activity and lower rates of hypokinetic diseases.

> **» KEY POINT 2.4**
>
> Pedagogy scholars were early leaders in the call for greater emphasis on the skill of movement observation and teaching QMD in the kinesiology curriculum.

Other Disciplinary Contributions

Scholars in allied health and ergonomic disciplines are also interested in improving human movement and therefore have contributed to the development of QMD. Ergonomics or human factors focuses on creating safe work environments so that repetitive movements in work tasks maximize performance or productivity and do not lead to overuse or acute injury. Allied health professions like athletic training and physical and occupational therapy also use QMD to try to diagnose causes of musculoskeletal problems and effectively rehabilitate them.

Ergonomics professionals have for many years performed qualitative assessments of industrial and office work tasks, called time and motion studies. The movements examined, counted, or timed varied widely from relatively static postures or slow-lifting movements to very rapid meat-cutting and assembly line tasks. The focus of these studies is to document the body positions and motions used by the workers. Time and motion study data are then combined with other biomechanical and epidemiological data to determine the potential causes of occupational injuries and how they might be prevented. QMD in this discipline has not coalesced into a single approach but has rather remained focused on movement- or job-specific systems for assessing live or videotaped work tasks. Much of the validity and reliability of QMD literature reviewed in the next section come from ergonomics and human factors studies.

The literature in allied health and rehabilitation disciplines also provides knowledge relevant to QMD. The professions of athletic training, medicine, physical and occupational therapy, and strength and conditioning have a long history of systems for diagnosing likely musculoskeletal problems in patients and athletes. QMD of exercises and everyday movements of patients is used by clinicians to supplement other orthopedic tests to help guide rehabilitation. For example, athletic trainers and strength and conditioning professionals use QMD of exercise technique to try to diagnose strength imbalances or movement dysfunction (Hirth, 2007). There is a large body of research on the QMD of walking gait (Meyer, 2002; Kirtley, 2006; Toro, Nester, and Farren, 2003, 2007a; Whittle, 2007), given that the cost of quantitative gait assessment is so high. Readers interested in clinical biomechanics and quantitative assessment should explore the Gait and Clinical Movement Analysis Society (www.gcmas.org) and its journal *Gait & Posture*.

Clinical QMD systems more often focus on micro-level (anatomical structure and injury) movement dysfunction issues rather than coordination or technique issues of maximizing performance. These clinical diagnostic decisions about the biomechanical causes of movement problems are very difficult because of complexity of the human body and because there can be multiple causes of an observed movement deviation (Herbert et al., 1993).

This text primarily focuses on teaching and coaching sport movements, so only a few examples are presented throughout the book of the more clinically focused QMD. QMD in ergonomics and sports medicine do, however, have much in common, and attention to the knowledge in these areas of study is important to maximizing the effectiveness of QMD. Areas where the allied health professions have contributed much to QMD are in the tasks of diagnosis and documenting the validity and reliability of qualitative assessments of movement. Allied health sciences have been a leader in differential diagnosis procedures and systems for weighing evidence to treat clients (evidence-based practice) that should be emulated in QMD and professional practice (Amonette, English, and Ottenbacher, 2010; Knudson, 2005).

Ongoing Trends

A growing body of literature and interest in QMD originated from several subdisciplinary areas within kinesiology and other disciplines interested in human movement. Many subdisciplines of kinesiology have contributed to the development of QMD, but no consistent terminology or curricular implementation of QMD training has been agreed on. Unfortunately, much of the recent interest in QMD has focused on traditional error detection and correction or purely observational models of the process. The rest of this section discusses some trends in QMD within kinesiology and recommends opportunities for future research.

First, there appears to be a slowing of faculty interest in teaching and research in QMD. A number of doctoral dissertations have focused on QMD (Chung, 1993; Eckrich, 1991; Harper, 1995; Johnson, 1990; Kwak, 1994; Leis, 1994; Matanin, 1993; Pinheiro, 1990; Raudensky, 1999; Rush, 1991; Taylor, 1995; Williams, 1996) or on the use of subdisciplinary knowledge in improving performance (Lehmann, 2003). Professional journals like *Strategies* and the *Journal of Physical Education, Recreation, and Dance (JOPERD)* have a history publishing papers on how to qualitatively assess specific sport skills (for example, Jones-Morton, 1990a, 1990b, 1990c, 1990d, 1991a, 1991b; Hudson, 1990b; Knudson, 1991, 1993; Knudson, Luedtke, and Faribault, 1994; Knudson and Morrison, 1996; Tant, 1990). Unfortunately, these papers are few and far between and often are not based on a comprehensive and interdisciplinary model of QMD (Ciapponi, 1999; Coker, 1998; McKethan and Turner, 1999; Schempp and Woorons-Johnson, 2006; Wang and Griffin, 1998).

Second, there has been a growing emphasis on paying more attention to how kinesiology curriculum addresses professional preparation. The 1992 National Association for Sport and Physical Education/National Council for the Accreditation of Teacher Education (NASPE/NCATE) guidelines (No. 22) required that QMD be part of the undergraduate curriculum for physical education teacher preparation (NASPE, 1992). The current national physical education teacher education standards include two standards (Instructional Delivery and Management; Impact on Student Learning) that cover the major task of QMD (NASPE, 2008). Once accreditation agencies require training in QMD, kinesiology faculty will adjust their programs to meet these requirements. States in Australia (Victoria in 1996) and in the United States (South Carolina in 2000 and Utah in 1999) have picked up on these requirements and developed videotaped instructional units to assist their teachers in QMD. Despite these efforts, there is considerable local influence on physical education programs in the schools and physical education teacher programs in higher education. Research is needed on the effectiveness of systematic QMD training programs and the resulting improvements in student performance. Future research documenting the importance of QMD is needed to improve the standing of QMD in professional training and practice.

Third, there is a greater awareness of a need for interdisciplinary and cooperative approaches to teaching, learning, and research. Funding agencies and journals are encouraging interdisciplinary research. Unfortunately, this trend toward greater interdisciplinary efforts has been slow to develop (Harris, 1993). One problem is that true interdisciplinary and applied research is still not the norm and is often unfavorably reviewed by journals. Most journals focus on single disciplines; or, if the journal is large and multidisciplinary, articles that are interdisciplinary must still navigate peer reviews from a single discipline subsection of the journal. Journals devoted to applied and interdisciplinary papers (namely, *Journal of Interdisciplinary Research in Physical Education* and *Motor Skills: Theory into Practice*) have folded after only a few years in print. While many faculty may agree with an interdisciplinary vision of research and professional practice, they may also lack the skills to conduct this kind of research and continue with the specialized work they have been trained to do. Interdisciplinary cooperation on teaching and researching QMD, while the ideal approach, will likely take a few more years to develop acceptance. Discipline-specific journals and multidisciplinary journals *(Research Quarterly for Exercise and Sport, Medicine and Science in Sports and Exercise)* should develop review sections for interdisciplinary research so that interdisciplinary work on QMD will have high-influence outlets for research.

> **»KEY POINT 2.5**
>
> The time is ripe for the adoption of an interdisciplinary approach to research on and training in QMD. This larger vision of QMD is gaining acceptance very slowly because scholars tend to work in one subdisciplinary area and institutions tend to change slowly.

VALIDITY AND RELIABILITY OF QMD

Important questions in any assessment relate to the issues of validity and reliability. In QMD, **validity** typically refers to the professional's ability to correctly identify strengths and weaknesses of a performance. **Reliability** refers to the consistency of these subjective ratings. These issues are even more complex when put within the context of a comprehensive model of QMD. The validity of a QMD in the real world is affected by the diagnosis of the strengths and weaknesses of performance and by the intervention the professional selects. The more judgment and critical thinking used in a QMD, the greater the difficulty in showing its validity. The research has typically not dealt with this larger view of the validity of QMD, but rather with the ability to visually detect errors or rate movement technique factors. This section reviews the research on validity and reliability of QMD using studies from a variety of human movement disciplines.

Validity

Validity in QMD has two important levels. The first level is **logical validity**, which is established by logic and expert opinion on the movement being assessed. An example of logical or face validity would be seen if the QMD accurately identified the critical features of the movement being assessed in the opinions of experienced professionals. Most research on the validity of QMD has this low-level face validity in that expert opinion is used to establish the observation variables of interest and whether the assessments are accurate. This is a problem because the opinions of kinesiology professionals vary and few data confirm that experience alone improves the accuracy of qualitative assessments of movement. This text focuses on the other level of validity in QMD, criterion-referenced validity.

The highest level of validity evaluation is **criterion-referenced validity**. Criterion-referenced validity in QMD compares the qualitative evaluation of some critical feature of performance with a criterion measurement of that feature. For example, you could check the criterion-referenced validity for visual estimation of hip adduction in a squat (figure 2.10) by comparing ratings of coaches to high-speed video or goniometric measurements of hip angle. Some of the higher risk of anterior cruciate ligament injury in women compared to men has been attributed to the angles of the hip and knee in landing movements. To take advantage of this knowledge, a QMD of hip and knee angles in these movements during activity or in a screening (figure 2.10) must be valid to be of any value to the professional and the client. Another way to gauge criterion-referenced validity is to compare agreement rates (percentages) between visual ratings and measurements converted into ordinal categories (ranges of angles or distances). The following sections summarize the criterion-referenced validity research on the visual assessments of movement within QMD from several disciplines.

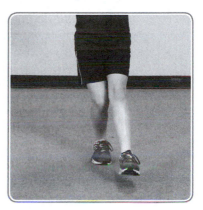

Figure 2.10 Logical validity in the QMD of hip adduction in a one-leg squat is based on the hypothesis that control of hip and knee motion is important in minimizing the risk of knee injury, while criterion-referenced validity is determined by the agreement between the amount of hip adduction rated by the observer and the actual hip adduction measured by biomechanical instruments.

Kinesiology Studies

Several kinesiology scholars have published studies documenting the criterion-referenced validity of visual assessments of movement relevant to QMD. Patla and Clouse (1988) may have been the first in kinesiology to document the criterion-referenced validity of visual assessments of videotaped gait, by five students. With 2 hours of training, these students were accurate in rating slow actions like foot placement but were less accurate in rating joint motions, especially during swing. Kilani and colleagues (1989) compared undergraduate students' perceptions of videotape replays of walking, jumping, and landing with biomechanical measurements. Students were asked to distinguish between real movements and false movements (generated by reversals of backward locomotion), and the results supported early perceptual research showing that general patterns of biological locomotion can be identified as normal or unusual, but this is different than evidence of the validity of visual perception of specific movement characteristics.

Knudson (1999b) reported two studies of the validity of visual assessments of experienced coaches, professors, and novice observers. One of six basketball coaches could make accurate ($r = 0.76$, errors 15-30%) visual estimates of trunk lean and knee angle in a videotape replay of the vertical jump. The second study used a visual analog scale for visual ratings of stick figures and video replay to compare experienced (professors) and inexperienced (majors) observers. Seventy percent of the majors and 20% of the professors gave relatively valid ($r = 0.76$ and 0.58, respectively) qualitative analog scale estimates of body angles of the vertical jump. Knudson and Morrison (2000) confirmed these results, reporting that 40% of untrained undergraduates could make accurate ($r = 0.75$) assessments of vertical jumps and that some of the difference between accurate observers and inaccurate observers might be related to a weak association with perceptual style measured by the Group Embedded Figures Test.

Stodden and colleagues (2004) reported that novice raters were able to visually rate most components of motor development of the overarm throw of 26 children, but there were differences up to one level for trunk and forearm action. Morrison and colleagues (2005) extended the research on validity of visual assessments of videotape replay of arm movements by examining the effect of reference or criterion movements. Eighty-six percent of kinesiology majors had visual estimates of elbow angular velocity significantly correlated (r = 0.84 to 0.94) with actual measurements and normally within 2% to 4%. Criterion movements did not improve visual estimates of the reversal angle for most students. However, the smaller percentage of students (17%) who could accurately rate reversal angle had small errors (4 degrees) relative to data from previous studies of estimating joint angles during replay of dynamic movement (Knudson, 1999b; Knudson and Morrison, 2000). More recently, Krosshaug and colleagues (2007) reported that trained, experienced observers underestimated (7-19 degrees) hip and knee angles in qualitative assessments of running and cutting.

> **QMD**   **Choices**   (?)
>
> What level of validity of QMD is most important to you? What level of validity of QMD should be the primary focus of future research in kinesiology?

Allied Health Studies

Allied health professions, particularly physical therapy, have reported validity research related to QMD. Orthopedic exams by health professionals often use visual assessments of range of motion rather than quantification from goniometers or inclinometers. Williams and Callaghan (1990) reported that physical therapists trained in these techniques could accurately (within 4 degrees) assess static shoulder range of motion in patients. Bruton, Ellis, and Goddard (1999) reported slightly larger 4- to 6-degree visual overestimations of true metacarpophalangeal joint angle in a sample of 40 physical therapists. Medial knee motion in squat and drop landings can be accurately assessed by a therapist (Stensrud et al., 2011). This is likely similar to the hip adduction in figure 2.10.

Health professionals also try to visually assess angles during dynamic movement, which tends to reduce criterion-referenced validity. One study found that observers could estimate step length in walking at slow to normal speeds to within 1.2 to 2.4 inches of accuracy (Stuberg, Straw, and Deuine, 1990). The best accuracy in visual assessment of step length occurred from close observational distances (0 to 3 meters), but errors were two to three times larger if faster walking speeds were evaluated. Saleh and Murdoch (1985) found that biomechanical measurement systems identified more gait abnormalities than did clinicians skilled in visual gait QMD. This could mean that fast-speed live movements are difficult to assess accurately (figure 2.11).

Clearly, the validity of QMD depends in part on the speed and complexity of the movement (Knudson and Morrison, 2000; Patla and Clouse, 1988). Knudson and Morrison reported that 2- to 9-degree errors in visual ratings of stick figures of vertical jumps grew to 20- to 30-degree errors when observers tried to rate videotape replays of vertical jumps. One possible reason that visual assessments of motion do not have consistently strong criterion-referenced validity is the perceptual limitations of observers. Many human movements are too fast, observer positions are too close, or backgrounds do not contrast sufficiently for accurate visual assessment of the motion. The perceptual limitations of observing the speed and complexity of motion are discussed in chapter 3. Consistent with the potential for rater bias discussed in chapter 3, validity complications in the allied health research include the tendency of assessments to be influenced by previous assessments (Eastlack et al., 1991; Miyazaki and Kubota, 1984) or expected pathology (Patla and Clouse, 1988).

Several allied health studies have attempted to develop valid QMD models for evaluating movement quality (Boyce et al., 1995). Physical therapy studies of QMD validity have found moderate to good validity for several kinds of visual assessments. Bernhardt, Bate, and Matyas (1998) found that 10 experienced physical therapists had moderately to highly accurate (correlations between 0.68 and 0.87) judgments of hand speed, jerkiness, and hand path using visual analog scales. There was less intertherapist agreement because each clinician had a different mapping of actual motion to ratings. McGinley and colleagues (2006) reported that physical therapists were able to visually rate the intensity of push-off in the gait of stroke patients.

Most research on the criterion-referenced validity of observational gait assessments has been mixed, even with formalized assessment protocols (Hugh Williamson Gait Laboratory Scale, Physician Rating Scale, Visual Gait Assessment Scale, Edinburgh Visual Gate Score, Salford Gait Tool). The review by

Figure 2.11 A side view is the best vantage point for observing the knee flexion angle in the volleyball player preparing to block. The fast speed of these movements however, typically means that the validity of accuracy of visual assessments of live motion will be moderate to poor.

Toro and colleagues (2003) and a validation study (Toro, Nester, and Farren, 2007a) summarize this work well. Physical therapists rating videos of patients with a variety of conditions range widely in the validity of their observational assessments of gait characteristics. Mean validity coefficients (Kappa) between 0.22 and 0.69 and mean agreement in category assignments of 58% have been reported (Toro, Nester, and Farren, 2003, 2007a).

Ergonomics and Human Factors Studies

Engineers in the field of ergonomics and human factors routinely develop observational models for evaluation of the physical demands of various work tasks. Ergonomics and human factors studies look at the optimum design of facilities, tools, and tasks for people at work. Ergonomics research has demonstrated that visual estimation of static body-segment angles in posture assessments are accurate to within 3 to 5 degrees (Douwes and Dul, 1991; Ericson et al., 1991). Note that this is more accurate than estimations in studies of posture in lifting (Chang et al., 2010) or jumping (Knudson 1999b; Knudson and Morrison 2000), where errors are 10 degrees or larger. In another study, untrained observers who visually estimated static shoulder angle from videotape replay had slightly greater errors (mean absolute errors of 9 degrees) (Genaidy et al., 1993). The subjects tended to overestimate the true angle in low shoulder angles and to underestimate the true angle in medium and high shoulder angles. This posture bias has also been reported in the visual ratings of arm postures

(Lowe, 2004). The overestimations of body angles were also consistent with results in ratings of the vertical jump (Knudson, 1999b; Knudson and Morrison, 2000). These studies confirm the visual perception research noted in chapter 3 that shows bias in normal perception of still objects or objects in motion.

A study by DeLooze and colleagues (1994) examined the agreement between two observers' visual estimations of body actions in a materials-handling task and measurements of the same variables. Coefficients of agreement were not acceptable for torso flexion or for the position of the arms and legs. Only subjective ratings of gross body posture had acceptable agreement ($K = 0.79$) with the criterion measurements. Paquet, Punnett, and Buchholz (2001) found that construction worker postures related to knee flexion and trunk bending and twisting could be accurately classified in three to nine categories. Similar results have been reported in poultry processing tasks (Juul-Kristensen et al., 2001). Observational assessments of trunk postures during work tasks are significantly ($r = 0.6$) but moderately correlated with simultaneous measurements, but there was considerable variation within subjects, suggesting that reliability was not strong (Burdorf et al., 1992). Village and associates (2009) reported that the validity of visual ratings using a low back observation instrument was acceptable for some technique variables but not for others with more specificity.

In summary, the ergonomics literature suggests that criterion-referenced validity of QMD exists for some variables but not for others. However, as in kinesiology, few ergonomics validity studies have included large numbers of raters who have been systematically trained in QMD. It is likely that specific training would be required to result in good criterion-referenced validity of visual ratings of body angles and movement.

Summary

Allied health, ergonomics, and kinesiology research on the criterion-referenced validity of QMD has tended to focus on novice or minimally trained observers; this research is in agreement that moderate validity exists for visual assessments, but only for some observers and some components of the movement. Moderate validity can be expected in visual ratings of static positions or slow movements, but faster or more complex movements are likely to have poor criterion-referenced validity.

Importantly, it is not known whether QMD of video compared to QMD of live action is less accurate, equally accurate, or more accurate. The 3-D perception of live action would logically be better, but video offers the possibility of freeze frame and replay, which could improve validity. There may be a point where the speed of motion being assessed is too great for accurate live action QMD, and QMD of videotape replay is therefore more accurate. It would be valuable to investigate the differences in perception of video compared with live action. The videos that accompany this book offer practice in QMD that should improve your diagnosis of live action.

More studies using a variety of human movements, conditions, and speeds of execution are needed to improve the evidence for the criterion-referenced validity of QMD. It is especially important that future research explore what movement variables can be accurately observed and rated, minimizing the threat of biases in perception. Studies are needed on the accuracy of the holistic ratings using visual analog scales, as well as more discrete, ordinal rating approaches to evaluation. Studies are also needed to document whether observers with specific training

> **» KEY POINT 2.6**
>
> Although QMD has been based on face or content validity, studies of criterion-referenced validity in several disciplines report moderate accuracy of visual ratings of static positions and slow movements, with lower accuracy for faster and more complex movements.

in a QMD system improve agreement with biomechanical quantifications of the same variables.

Reliability

More studies have been done on the reliability of QMD than on validity. Intrarater reliability is the consistency of the qualitative assessments performed by a single professional. Interrater reliability is the consistency or agreement of the qualitative assessments of several professionals assessing the same performer. There has been a great deal of reliability research on QMD in the allied health and kinesiology literature. Reliability issues are important to consider when planning QMD or interpreting qualitative assessments of others.

Kinesiology Studies

Motor development and adapted physical education studies of the reliability of QMD are noteworthy because many have used the *generalizability study* (G study) approach. This is important because a G study is designed to examine all variables that affect reliability of any measurement or assessment (observers, occasions, different performers). Roberton (1989) called for future motor development research to emphasize documentation of the validity and reliability of observational models based on developmental sequences.

QMD Demonstration 2.1

Observe the video clips of the vertical jump in QMD Demonstration 2.1 in the web resource at www.Human Kinetics.com/QualitativeDiagnosisOfHumanMovement and decide if the jumper used enough countermovement. Decide if the jumper bent the knees to at least 90 degrees. What level of QMD validity is being examined in this demonstration? Does the slow-motion version of the same jump make this visual assessment of knee angle easier?

QMD Technology 2.1

Observe one of the slow-motion video clips of the vertical jump from Demonstration 2.1 at www.Human Kinetics.com/QualitativeDiagnosisOfHumanMovement again and visually estimate the smallest angle of knee flexion at the reversal point. Use several replays and the pause feature to identify what you believe to be the smallest angle of knee flexion. Chapter 8 lists some software programs that allow observers to calculate (quantify) this joint angle to verify the qualitative assessment made. See chapter 8 for details on the limitations of these measurements and the sampling rate of video. It is possible that the true minimum knee angle in this jump was not captured by the still image you selected qualitatively from the video replay.

Ulrich (1984) reported acceptable reliability (Ks between 0.62 and 0.84) of a standardized assessment of 12 fundamental motor skills. Ulrich, Ulrich, and Branta (1988) reported a G study of the QMD using the Michigan State developmental stages of the hop, horizontal jump, and running. Subjects received minimal instruction (1 hour) in QMD of the whole-body developmental level of the three skills. One observer watching three trials could reliably assess the hop ($G = 0.88$). Reliably assessing the developmental level of the horizontal jump and running required multiple observers and trials. These two skills showed potential observer bias, with 20% to 30% of the variance in the ratings being related to interaction between subject and observer. The observers had been warned against this error, but there may have been a tendency to relate the age of the subject to the developmental level assigned. For example, some observers may have been inclined to rate a younger-looking subject lower or an older-looking subject higher.

Observer bias was also observed by Painter (1990, 1994) after training in QMD. This G study examined the effect of training, academic major, and kind of motor development model (whole-body or component) used to identify 20 female college students' developmental levels of hopping. A key finding was that focusing on hopping components resulted in greater reliability than a whole-body approach that required multiple observers. The components approach (arm or leg action of hopping) to QMD could reliably rate developmental level with single observers looking at one to five performances.

A study of a qualitative rating checklist for folk dance supports the need for training, even with use of a structured QMD checklist (Slettum et al., 2001). Good reliability for rating three of six skill components of folk dance was observed after physical educators received 4 hours of training. It appears that reliable qualitative assessments of complex movements like folk dance take considerable planning and practice.

Mosher and Schutz (1983) also found that slow actions of the overarm throw (foot placement and body rotation) could be reliably observed in one trial by one observer. Fast and complex movements are difficult to observe reliably. These studies found that the action of the arms during hopping (Painter, 1990) and throwing (Mosher and Schutz, 1983) was difficult to observe reliably. Painter concluded that one kinesiology student would need at least five trials to observe and rate the arm action in hopping. A single kinesiology student observer would have to observe 10 trials to reliably rate the whole-body developmental level of hopping.

Allied Health Studies

Allied health research in fields like physical therapy has supported the moderate reliability of QMD seen in the kinesiology literature. Physical therapy studies have examined the reliability of QMD of human gait and other clinical assessments. Observational gait assessments have had low to moderate interobserver reliability, with intraclass correlations between 0.6 and 0.7 (Eastlack et al., 1991; Goodkin and Diller, 1973; Krebs, Edelstein, and Fishman, 1985). The reliability of qualitative assessments of gross motor function in people with cerebral palsy, however, has been good (Gowland et al., 1995). A summary of these studies sheds some light on the issue of the reliability of QMD.

Attinger and coworkers (1987) studied the reliability of visual assessment of gait asymmetry with a jury of eight medical professionals. Reliable identification of the foot with the longest stance phase duration was 80% probable and rose to 87% probable when the asymmetry was greater than normal. The experts were unable to rate the amount of asymmetry, or reliably identify a limb that was loaded more than the other. Krebs and colleagues (1985) evaluated the inter- and intraobserver reliability of three expert observers with videotapes of 15 disabled children. Intrarater reliability was moderate. Mean agreement of repeated ratings was 69% (31% of assessments disagreeing with the same therapist's assessments made a month earlier). For clinicians to be 95% sure they had observed a change, there had to be a 10% difference in leg kinematics. Mean interrater agreement was 67.5%, with a mean intraclass correlation of 0.73. A 33% difference in walking kinematics would be required to detect differences in walking with different clinicians.

In a study of the interrater reliability of the QMD of gait reported by Eastlack and colleagues (1991), 53 physical therapists rated 10 gait variables (knee kinematics and temporal and spatial variables of the gait cycle) during the four phases of stance from videotapes of patients. Notably, this study was the first to allow the use of slow-motion replay by the observers. The interrater reliability of the gait variables was slight to moderate (<0.69). Joint kinematics had less agreement than temporal or spatial variables. These poor results may be related to the fact that 32 of the 53 physical therapists involved in the study were unfamiliar with any of the four major models for the QMD of gait in physical therapy. A recent study reported a new qualitative gait assessment tool that with minimal training showed good agreement (77%) between therapists rating video replays of children (Toro, Nester, and Farren, 2007b).

Other sources of information on the reliability of QMD in physical therapy include studies of visual estimation of joint range of motion and gross motor performance. Studies of the reliability of visual estimation of joint range of motion have had mixed results. Some studies have documented interobserver reliability of visual assessment with intraclass correlations above 0.82 (Watkins et al., 1991), whereas others have shown moderate ($0.34 < R < 0.83$) interobserver reliability (Youndas, Carey, and Garrett, 1991; Youndas, Bogard, and Suman, 1993). There is also evidence of poor agreement between visual estimation of ankle range of motion and goniometric measurement (Youndas, Bogard, and Suman, 1993). Gowland and colleagues (1995) reported good reliability for overall ratings ($R > 0.92$) and attribute ratings ($R > 0.84$) of cerebral palsy patients by 19 physical therapists. Poor to fair reliability was reported in visual assessments by physical therapists of a faster movement (violin playing; Ackermann and Adams, 2004), while good reliability was reported for visual ratings of computer use movements (James, Harburn, and Kramer, 1997).

Qualitative assessments of physical impairments and disabilities are common in physical therapy. Studies of these clinical assessments have shown results from poor ($K < 0.4$) to good ($0.7 < K < 0.9$) reliability depending on the training and nature of the movement being rated (Fife et al., 1991; Hendriks et al., 1997; Keenan and Bach, 1996; Terwee et al., 2005). The consensus of the physical therapy research on reliability suggests that there is potentially good intrarater reliability in many situations, while poor interrater reliability is likely due to differences in therapists' training or perception of motion.

PRACTICAL APPLICATIONS
Validity and Reliability of QMD

Kinesiology professionals can choose whatever model of QMD they want to use to help their clients. Naturally, they would want to know how accurate their diagnoses are and how consistently they are at this important skill. This text has summarized extensive research on these issues from a variety of disciplines interested in improving human movement. This research indicates that QMD can be accurate and consistently applied, but this is difficult and there are threats to both of these important properties of an assessment.

Imagine yourself as a kinesiology professional reevaluating your skill in QMD. How might you plan to document the validity and reliability of your QMD? What would be the ideal approach if you had few restrictions in time and money? What would be more economical and pragmatic?

Ergonomics and Human Factors Studies

Ergonomics research has also examined the reliability of qualitative assessments of technique in work tasks. Reliability of assessing the timing of various work techniques is generally good (Kazmierczak et al., 2006). Most, but not all, lifting variables can have good reliability (Kappas > 0.7) with trained raters (Village et al., 2009). Similar to what has been shown in work in other fields, reliability of qualitative ratings of posture improves to good agreement (0.84) when observers can freeze or use frame-

》KEY POINT 2.7

Research on QMD in a variety of disciplines has shown poor to moderate intrarater reliability and usually poor interrater reliability. Careful planning, standardization, and training may improve the reliability of QMD.

by-frame video replay to make their assessments (Kociolek and Keir, 2010). Bao and colleagues (2010) reported that observers could reliably identify potentially dangerous work technique, but the reliability of the observers in assessing the level of effort or exertion was poor.

Summary

The allied health, ergonomics, and kinesiology literature suggests that the reliability of QMD in actual practice (different professionals, subjects or clients, number of observations, vantage point, or observation conditions) may range widely from poor to acceptable depending on the conditions and model used. Most qualitative assessments will likely have poor to moderate reliability. There are, however, several strategies to increase the potential reliability of QMD. Specific training and practice using a QMD model is usually necessary. Increasing the number of observers or the number of trials observed (live trials or videotape replay) tends to increase reliability. Another approach is to increase the specificity of the system or model by observing discrete events and providing a simple rating for them (Kerner and Alexander, 1981). Identifying specific critical features and defining rules on how they will be evaluated can help build the potential for agreement in multiple observations of human movement. These strategies are incorporated into the discussion of the four tasks of QMD in chapters 4 through 7.

QMD Demonstration 2.2

Observe the video clips of the underarm throw in QMD Demonstration 2.2 in the web resource at www. HumanKinetics.com/QualitativeDiagnosisOfHuman Movement and decide if the throwing arm actions are correctly performed. How could you make this evaluation more reliable?

SUMMARY

The research and professional literature on QMD points to an interdisciplinary and integrated approach to QMD. This vision of QMD has grown out of contributions from scholars from most kinesiology subdisciplines and other fields related to human movement. This larger interdisciplinary vision of QMD has gained acceptance slowly because institutions change slowly. Several models of QMD can be used to accurately evaluate human movement, but this ability is not present in all observers and tends to be poorer for fast movements. Qualitative assessments have been shown to have poor to moderate interrater reliability with poor interrater reliability. Strategies to improve validity and reliability include training, increasing the specificity of the QMD model, and increasing the number of trials observed.

The models introduced in this chapter helped drive the development of the four-task model presented in chapter 1. This model has been used effectively and is quite adaptable to the technological advances that QMD is experiencing.

Discussion Questions

❶ What kinesiology subdisciplines have contributed most to the development of QMD theory? Research? Practice?

❷ How might kinesiology foster interdisciplinary cooperation in the development of QMD research?

❸ How should kinesiology programs teach QMD within the curriculum?

❹ What factors are most influential in increasing the potential validity of QMD?

❺ Can interrater reliability be as good as intrarater reliability in QMD?

❻ What model or approach to QMD would be most reliable?

❼ What model or approach to QMD would most likely lead to correct decisions about improving movement?

❽ What model or approach to QMD would most likely lead to correct decisions about preventing injury?

Role of the Senses and Perception in Qualitative Movement Diagnosis

A ballet instructor is working hard with her company to finish a piece she has choreographed for them. Movements in this piece must be precise so that dancers move safely and so that everyone is in time with the music and expressing the emotion of the dance. To make the piece come together, the instructor relies on many of her senses, her perception, and her decision-making ability. She watches for the dancers' placement as well as their body alignment and limb position. She listens carefully to make sure everyone is in time. Occasionally she participates in the class so that she can gather tactile and kinesthetic information about the progress of the piece. Without information from all of her senses, she would not have a complete picture of what happens in the new dance piece. From her interpretation of what has happened, she decides how to proceed with the piece.

Chapter Objectives

1. Describe the function of the sense organs associated with QMD.
2. Discuss some of the limitations of the senses and cognition in gathering information for QMD.
3. Describe the integration process of the senses used in QMD.
4. Define *perception*.
5. Describe four perceptual tasks as levels of decoding movement information.
6. List the components of perception and how they relate to QMD.
7. Summarize the perceptual research on QMD.
8. Discuss how a gestalt explanation of perception explains the interpretation of visual stimuli.

Threatical the ballet instructor observing her dancers in this chapter preview is busier than most of us realize. Qualitative movement diagnosis (QMD) in teaching requires a great deal of perceptual and cognitive activity—and most of it is not easy to study by scholars. This chapter examines only part of this total activity: gathering, integrating, organizing, and giving meaning to sensory information. This teacher must take stimuli from the environment (touch, kinesthetic, auditory, visual) using her sense organs, integrate that information, then organize it, and finally assign meaning to the stimuli. Once this information has been processed, it is ready for the evaluation and diagnosis task of QMD.

This chapter introduces some ideas on how scientists believe the complex processes of perception proceed during the observation of movement. The theories, research, and models presented in this chapter are our best estimate of how information is gathered and processed. Perhaps you can see how the current ideas have been integrated into QMD models or how your preferences for observation will guide your approach to systematic observation in QMD.

THEORETICAL BACKGROUND FOR SENSES AND PERCEPTION IN QMD

To understand how important the senses and perception are in QMD, let's review what is known about the senses and perception in a similarly complex perceptual activity, like flying jet fighters. Hartman and Secrist (1991) wrote a seminal article explaining that situational awareness in flying supersonic fighters involves more than exceptional vision and, in fact, requires considerable cognitive activity. The sense of vision is only a starting point for this complex perceptual activity, and many people are unaware of the cognitive element of perception. Pilots must take sensory information, organize it, interpret it, decide on the correct response, and then initiate the response.

Similarly, the gathering of information by the sense organs is only the beginning of observation within QMD. We must go through complex cognitive processes in QMD just as fighter pilots do. Fortunately, we have a bit more time to weigh our choices. The complexities of the senses and their contributions to QMD are staggering, even before we consider the senses' simultaneous interaction. The senses and cognition provide the information from which QMD decisions about human movement are made. The senses and perception are to QMD what biomechanical instruments are to the quantitative research on human movement technique; they gather and organize information about performance.

Perception of movement is the key to using sensory information. **Perception** is the organization and interpretation of stimuli from our environment, mediated by our senses. Perception involves organizing, or making sense out of, our sensory information (Sage, 1984). This cognitive component of perception is vitally important to what we truly see or hear. For example, if you knew virtually nothing about modern art, would you recognize a masterpiece mistakenly put on sale at a garage sale?

Throughout this chapter, this veiled cognitive component of perception is examined within the context of the QMD. Our integrated model has four major tasks that underpin the cognitive processes professionals use in QMD. Reviewing what is known about the senses and perception will help us understand the larger challenge of observation within QMD.

SENSES

Although this chapter examines the four major senses of QMD, which are vision, audition, touch (haptic), and kinesthetic proprioception, individually they work in an integrated fashion. This integration of information is part of the first step of perception: organizing sensory information.

This is analogous to the integrated way the senses should work together in the observation task of QMD (Hay and Reid, 1988; Hoffman, 1983; Radford, 1989). Different senses provide unique information that the observer puts together to improve the observation of performance. The kinesthetic proprioceptive sense an observer uses while spotting a gymnastics tumbling run may tell more about the forces being exerted by a performer than vision or audition. Audition may be the best sense for gathering temporal information on the timing of the tumbling, while vision is most sensitive to spatial changes in position of the body in flight. All this sensory input must be interpreted and evaluated to gather relevant information for the evaluation and diagnosis task within QMD. The most dominant sense is typically vision.

Vision

The primary sense used in QMD is vision, so most of the information in this chapter is on vision. Until the late 19th century, visual observation was the primary method for studying the biomechanics of movement. This section reviews important information about the capabilities and limitations of human visual perception of motion, as well as the 3-D expectations we cognitively impose on vision.

There are major limitations to our ability to see fast movements. In fact, the improvements in observational power created by photography and cinematography have dominated the science of biomechanics for the century since their development (Cappozzo, Marchetti, and Tosi, 1992). The initial furor created by the cinematographic photographs of moving animals made by Muybridge and Marey in the late 19th century is a testament to the limited perceptual power of the naked eye. Even today, sport fans feel cheated if they are not shown the slow-motion replay of most of the action. Some of the most popular shows on science-oriented cable programs (e.g., *MythBusters*) owe their popularity to the replay of high-speed video that brings invisible elements of motion into view for the first time for many people.

The sensory receptor concerned with vision is the eye. This receptor takes energy from the visual spectrum of electromagnetic radiation and converts it into nerve transmissions directed to the appropriate parts of the brain. The two main components of visual perception are focal and ambient vision. Focal vision uses a maximum acuity region of the retina (fovea centralis or fovea) for recognition of objects and fine detail. Ambient (peripheral) vision is used to orient yourself or your eye movements relative to motion (both yours and that of objects in the environment).

Functional Components of the Eye

Although the senses are clearly integrated with perception, we can still look at their many functional components separately. The eye has many parts that allow it to gather information from our surroundings. The major functions of the eye are accommodation, static visual

> **» KEY POINT 3.1**
>
> Good vision for QMD is a complex phenomenon. Focal vision and static visual acuity allow the recognition of objects and details. Ambient vision, dynamic visual acuity, perception of color, contrast, accurate eye movements, eye dominance, and peripheral vision all are important in following moving objects.

acuity, dynamic visual acuity, convergence and divergence, depth perception, eye dominance, tracking, peripheral vision, and fusion. The rest of this section reviews how the vision perception system works.

Visibility   The *visibility* of any object or event refers to its detectability by the human eye. This is a complex phenomenon because many levels of detectability are important in QMD. The lowest level could be just noting or recognizing an event. The latter part of this chapter examines the processes of perception, including detection, in greater depth. At higher levels, the ability to discern information about an event puts greater demands on visual perception. At the lowest level, the two most important factors in visibility are lighting conditions and contrast (McCormick and Sanders, 1982). These two factors are discussed in the following section on visual accuracy.

Because attention and cognitive processing are intimately involved with visual perception, is it possible to apparently be looking at an object or event and not even see it (inattentional blindness). Chapter 1 of *The Invisible Gorilla* (Chabris and Simons, 2010) summarizes extensive psychology research showing that people actually perceive far less of the visual field than they assume they do. When we focus our visual attention on something, we often don't see unusual activity or events in our line of sight. As we consider the physical limitations of visual perception, remember that there are also finite cognitive and attentional resources that affect our ability to see things.

Accuracy of Visual Perception   The ability to see details in an object is called visual acuity. This ability is strongly related to the adjustment of the eye's lens (accommodation) and the shape of the eye. The fine adjustments of the lens to focus the flow of light on the fovea of the retina are critical to visual acuity. Recall that cognitive attention influences perception, so visual acuity can be influenced by your attention to an object, which drives the unconscious control coordinating eye movements and lens accommodation. Even with cognitive attention to an object of interest, other objects in your apparent field of vision could be invisible due to attention, an eye blink, or your blind spot. Your blind spot, relative to any focus of visual attention (Sanny, 1999), comes from the one spot on your retina where blood and nerve supply does not allow light-sensing cells (rods and cones). All these factors can contribute to poor visual perception of events or motion.

There are many measures of visual acuity; the most common is minimum separable acuity. A good example of this is the common Snellen eye chart, in which the smallest features discernible (usually letters) are evaluated. This is the most common measure of **static visual acuity (SVA)**. Besides an individual's innate SVA, many factors affect visual discrimination, among them contrast, lighting, motion, time, color, and age. Although most QMD is concerned with **dynamic visual acuity (DVA)**, SVA is important because many movements have phases with minimal or no movement. Valenti and Costall (1997) indicate that there is a great deal of information to be gleaned from situations of limited movement. Their research suggests that the amount of force needed to overcome a resistance can be gleaned from static postures, facial expressions, and other body details.

Contrast is the percentage difference in illumination of the features of an object being viewed. The contrast between the black letters and the white paper of the Snellen eye chart (or this book) is very high. The contrast between a dark uniform and a poorly lit athletic field is considerably worse. If the contrast between an object and the background is low, the object must be larger to be as detectable as a smaller object with greater contrast. Professionals selecting colors for uniforms or clothing should think about the potential contrast among the colors selected, the environment, and the equipment.

The amount of light, or illuminance, is an important factor that interacts with contrast to determine the visibility of an object or event. There is extensive research on the effects of lighting on human performance in many tasks. In general, the more illumination, the better the performance (McCormick and Sanders, 1982). There can be too much illumination, however; glare or brightness reflected from the object and background make it difficult to see the object. It is also important to remember that the eye's sensitivity to light changes with illuminance. The transition of sensitivity from darkness to light is relatively quick (less than a minute), while adjustment from light to darkness can take 30 minutes or more (McCormick and Sanders, 1982). Much research in visual discrimination in varying lighting conditions is based on contrast sensitivity function (Kluka, 1991).

Another factor influencing visual discrimination is the perception of color. Recall that the retina has two types of light-sensitive receptors: rods and cones. A human eye has about 130 million rods that are primarily sensitive to light intensity and about 7 million cones that are primarily sensitive to the wavelength of light and are responsible for our perception of color (McCormick and Sanders, 1982). Some people have difficulty discriminating between red and green or between blue and yellow. True color blindness is rare, but color deficiency is usually found in 8% to 10% of males and less than 1% of females (Gavriysky, 1969). Although these are not large percentages of the population, it would be wise to plan sporting events with contrasting colors that are not combinations of red and green or blue and yellow.

The time available to focus on an object also strongly affects SVA. The greater the viewing time, the better the chance of making visual discriminations. If either the object or the observer is in motion, that reduces the time the eyes will be able to focus on the object. This brings us to DVA, which is the visual discrimination of an object when there is relative movement between the object and the observer. A person's DVA deteriorates rapidly as the eye's angular velocity exceeds 60 or 70 degrees per second (Bahill and LaRitz, 1984; Burg, 1966). Above these speeds, the eyes cannot smoothly rotate to keep the object on the fovea. Clearly, then, the faster the movement, the less time the object will be in our visual field, and the less able we will be to see and judge the motion of that object.

Our DVA increases from ages 6 to 20 and then tends to decrease (Burg, 1966; Ishigaki and Miyao, 1994; Morris, 1977). This is why it is developmentally appropriate that baseball and softball leagues for small children use a batting tee rather than machine- or human-pitched balls. As children's visual and hitting skills improve, coaches can incorporate pitched balls and more realistic batting tee drills. For example, older players practicing with a batting tee can focus their eyes forward, quickly saccade their eyes to the ball, and then hit the ball.

Unfortunately, good SVA does not guarantee that a person will have good DVA. Studies have found weak correlations ($R < 0.6$) between SVA and DVA at slow speeds

QMD Demonstration 3.1

Observe the two video clips of the ball bouncing in QMD Demonstration 3.1 in the web resource at www. HumanKinetics.com/QualitativeDiagnosisOfHuman Movement at normal speed. Can you see where the ball lands in relation to the line? One clip is regular-speed video (60 Hz) and the other is high-speed video (180 Hz). Does the ball strike the boundary line in each situation? Does the high-speed video make the assessment easier? Notice that in both clips, it is difficult to determine if the static ball is touching the boundary line. Notice also the 60 Hz variation in the gym lights in the 180 Hz vedeo.

(Burg, 1966; Kluka, 1994) and no correlation at faster speeds (Morris, 1977). The weak association between SVA and DVA is not surprising, since SVA is the ability to observe detail in ideal conditions (static, two dimensions, good lighting, and contrast), while DVA is observational ability in less than ideal conditions (motion, three dimensions, and poor contrast). It does appear that refractive (e.g., glasses or contacts) improvement of SVA can translate to improvements in DVA (Nakatsuka et al., 2006).

Research suggests that DVA can improve with training (Long and Rourke, 1989) and that there are large differences between individuals (Morris, 1977). Some people are velocity resistant and are not strongly affected by the relative motion of an object. Others are velocity susceptible: Their visual perception is easily disturbed by relative motion of the object (Morris, 1977). Clearly, relative motion and DVA influence the QMD ability of a teacher or coach. Research on ball catching has also shown that the temporal constraints of the environment, such as time available to view the trajectory and interaction of eye and head movements to track the object, all affect catching ability (Montagne, Laurent, and Ripoll, 1993).

There are many examples in sport where the relative motion of people or objects past the observer is so great that it cannot be observed reliably. Officiating in sports like basketball and American football has been controversial for many years. The National Football League and professional tennis tournaments use instant replays and 3-D ball tracking to make final judgments on disputed calls. In tennis, a controversy over calling balls in or out (Vincent, 1984) led to the development of photoelectric sensors to help call the service line in professional matches. Braden (1983) studied the accuracy of judging where tennis balls landed from various court positions. He found that the players were less reliable (11% error rate with a mean error of 5 inches or 13 centimeters) than the linespeople or umpires, who usually have a better angle to view shots near boundary lines. This research has been supported by recent work by Mather (2008).

Another problem related to time and DVA involves sporting events of very short duration. Examples are collisions or release events in high-speed sports. In most striking sports (for example, baseball and tennis), coaches use cues to have the players watch the ball until it hits the bat or racket. Because ball–bat collisions in baseball and softball last only 1 or 2 milliseconds, it is highly unlikely that any athlete can see the ball hit the bat. Seeing the ball hit the bat may be unimportant in light of our earlier discussion of ambient and focal vision. Focal vision would yield information about the bat–ball contact, whereas ambient vision would guide the bat to the ball. Guiding the bat to the ball for contact is more important than actually seeing the contact, so coaches should cue hitters to focus their attention on the early trajectory of the pitch, not watching the ball hit the bat. Watts and Bahill (1990) reviewed their studies of vision in baseball and concluded that batters cannot track the ball to the point of impact, even in slow-pitch softball! Ball speeds in most sports exceed the eyes' ability to track the trajectory smoothly (Ripoll and Fleurance, 1988). In the following section on eye movements, we will see how the eyes deal with tracking very fast objects.

Events occurring faster than about 1/4 of a second usually cannot be seen (Eastman Kodak Company, 1979). If humans had very fast vision, there would be no illusion of motion when we watch movies, which are really the flashing of 24 distinct pictures per second. There is a clear time limitation in our eyes' ability to perceive information from moving objects in our field of view. Knowledge of this limitation is critical to planning for QMD and what specific performance information is reliably observable. A complete description of eye motions to track moving objects is presented in the following section on important eye movements. After reading this section, you will know why officials sometimes appear to be looking right at a key play and still miss the call.

Important Eye Movements The eyes use many kinds of movements to view moving objects. These movements are coordinated to keep the two eyes working together. Kluka (1991) classifies eye movement into four types: saccadic, vestibulo-ocular, vergence, and smooth pursuit. Saccadic eye movements are for scanning rapidly and jumping to various points in the visual field. Vestibulo-ocular movements are coordinated with head motion to keep the eyes on an object. **Vergence** eye movements allow the eyes to focus on objects at different distances, while smooth pursuit eye movements are used to follow slow-moving objects.

The eye movements that make it possible to view objects up close and at a distance are convergence and divergence, respectively. Accommodation and convergence relate to the eyes' ability to focus quickly, smoothly, and accurately as objects approach. A basketball player defending an approaching opponent has his eyes converge to align both eyes with the opponent, and the lenses of the eyes use accommodation to keep the image in focus. This ability is especially important in sport because objects and individuals are always changing their relationships to us. It is equally important in QMD. These functions are achieved by changes in the tension of the muscles of the lens and the muscles that move the eyes.

To observe this aspect of vision, hold a pencil at arm's length and slowly move it toward your nose. As the pencil gets closer, your eyes move from an almost straight-ahead position in the sockets to a position where they seem to be touching the nose. You are now cross-eyed. Your eyes have converged. At the same time, the lens has changed shape to keep the pencil in focus. The same process takes place as we observe skills that involve movement close to us.

Depth perception is the ability to judge how far away objects are from you or the relative distances objects are away from you and each other. At a distance we generally judge depth by comparing object size, detail, texture gradient, closer objects, and linear perspective. Look out the window or into the distance and see how many of these factors you can detect. As objects get closer, this perception is mainly a function of eye position as sensed by kinesthetic proprioceptors in the eye muscles as the eyes move closer to the nose or farther outward toward the side of the head.

When something has our visual attention, we carefully focus both eyes on the object; this is **fixation**. We use a very narrow field (about 3 degrees) to stabilize our visual focus on the fovea of the retina. This focus to control the motion of the image over the fovea is quite complex and involves tiny rotations (microsaccades) to maintain a fixation (Kowler, 2011). To get a feel for how small this area of visual

QMD Technology 3.1

Perhaps the most influential technology advance supporting QMD is high-speed video. Many quality cameras can also capture high-definition and high-speed (greater than 60 pictures per second) video. Many limitations of visual perception of motion and events can be overcome by observation of video replay. High-speed video provides a window to the invisible world of motion that is faster than our visual perception. See chapter 8 for more information on high-speed video. Observe the high-speed video in QMD Demonstration 3.1 and notice the variation in lighting. This is the 60 Hz variation in light intensity related to electrical power that is not visible to real time observation. Another high-speed video clip in the web resource is QMD Technology 8.1. Other QMD explorations and demonstrations present slow-motion versions of regular HD video. Replaying high-speed video in real time looks like slow-motion replay because of the greater number of images of the movement.

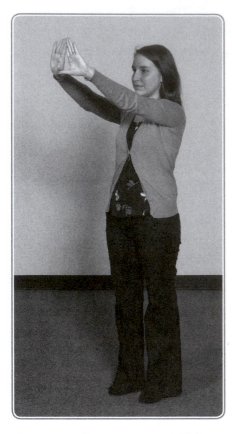

Figure 3.1 Simple test to establish eye dominance.

focus is, extend your arm forward. Hold it straight out with your thumb extended vertically. The width of your thumb in this position (the thumb rule) is a good approximation of the focus of your visual field (Groot, Ortega, and Beltran, 1994).

Because information from one eye reaches the brain faster and is processed more quickly, that eye becomes dominant (Kluka, 1991). The dominant eye guides the other eye in the direction of movement and fixations. The use of the dominant eye in sports that require accuracy has been studied for many years. The combination of eye and hand dominance has been a topic of studies of hockey, batting, and golf putting (Morrison, 1976; Steinberg, Frehlich, and Tennant, 1995; Tieg, 1983). It is easy to establish which eye is the dominant eye. Extend your arms forward, making a small (1 square inch [6 square centimeter]) hole between your hands (figure 3.1). Choose a distant object (such as a clock or spot on the wall) and center it in the hole formed by your hands. Without moving, close one eye at a time. The eye that still has the object lined up in the hole is your dominant eye.

The use of the eyes to track moving objects is a highly complex phenomenon. Eyes move to gather information for processing. This movement takes one of two forms: smooth pursuit or saccade. In the smooth pursuit, the eyes are able to rotate together to stay on the object they are tracking. In a saccade, the eyes jump from position to position to gather information from the object they are trying to track. The different forms of tracking affect the gathering of perceptual information. These tracking differences can limit the potential perceptual information that can be gathered.

- *Smooth pursuit.* When slow relative movement is occurring between an observer and an object, the eyes can smoothly move together in following the object until eye angular velocity reaches between 40 and 70 degrees per second (Bahill and LaRitz, 1984; Ripoll and Fleurance, 1988; Robinson, 1981). **Smooth pursuit** is the maintenance of foveal vision on slow-moving objects that is facilitated by unconscious learning through experience with moving objects (Kowler, 2011). People can become skilled at visually tracking objects and predicting their landing or bounce location. With training, angular velocities can reach 100 degrees per second (Buizza and Schmidt, 1986; Meyer, Lasker, and Robinson, 1985). A good way to illustrate the accuracy of visual tracking in estimating ball impact or intercept is to toss a tennis ball softly to a person 2 or 3 feet (0.6 to 0.9 meters) away. Have her catch the ball with one hand. After a couple of trials, ask her to close her eyes when you say "close" early or in the middle of the ball trajectory. Most people will be able to catch the ball with their eyes closed. Unfortunately, many sports or other movements require eye movements beyond our ability for smooth pursuit. In volleyball, for example, eye angular velocities of more than 500 degrees per second are needed to track the trajectory of a spiked ball (Kluka, 1991).

- *Saccade.* Sports like tennis, badminton, basketball, and baseball generate ball speeds that require another kind of eye movement to track the ball. The quick motion or repositioning of both eyes from one fixation to another is a **saccade**. While the eyes are rotating to the next fixation, they are essentially turned off to prevent a blur of light and images as they move. This downtime has been called saccadic suppression, omission, or visual masking (Campbell and Wurtz, 1978). Sometimes the term "change blindness" is used to refer to this phenomenon (Noe, Pesoa, and Thompson, 2000; Simons and Levin, 1997). Whatever the terminology, visual information during the saccade is cognitively ignored to maintain the perception of a stable world (e.g., prevent the perception of a rapid rotation of the environment as light moves rapidly across the retina). Coaches should understand how such terrible (or fortunate, depending on your team) calls are made in sporting events from an untimely saccade by an official.

Very little information is stored in short-term memory during saccades (Karn and Hayhoe, 2000; Noe, Pesoa, and Thompson, 2000; Simons and Levin, 1997, 1998). This means that visual information gathered between saccades is the only reference we have for mapping the succeeding saccade. It was formerly believed that a rich short-term memory operated during the saccade, acting as a visual map to the next item of interest in the visual field; but research has shown that this is not the case. In a saccade the eyes can reposition at angular velocities exceeding 700 degrees per second (Carpenter, 1988). Speeds of this magnitude probably negate the need for memory maps of the environment.

Recent research on saccadic eye movements in normal subjects has shown that there is no significant difference between men and women, but saccadic eye movement parameters decrease significantly with age (Wilson et al., 1993). Sport studies have shown anticipatory patterns of saccades relative to the kind of motion of the object that is being tracked (Bahill and LaRitz, 1984; Haywood, 1984; Hubbard and Seng, 1954; Ripoll and Fleurance, 1988). Interestingly, being a skilled athlete in many sports involves learning to suppress saccades to irrelevant visual stimuli and to focus on specific areas of the visual field (Lenoir et al., 2000; Vickers, 2011). To observe saccadic eye movement, simply watch someone's eyes in a park or mall. The person's eyes will dart around among the wide variety of visual targets.

- *Limitations of eye movements*. The limitations of eye motions in the tracking of moving objects, visual suppression during a saccade, and fixations all have implications for sports and QMD. The professional must realize that some high-speed events simply cannot be observed. If a key event occurs when the observer's eyes are in a saccade, it will not be seen. If an observational strategy is not followed, the observer's eyes may be drawn to and fixated on an extraneous action, causing the professional to miss an important error in performance. The selection of a viewing distance for observation in QMD has a major impact on the eye angular velocities required in tracking a performer. If it is important to gather visual information about a fast-moving skill or part of a skill, observers should increase their distance so that eye movement velocities are slower.

If vision is part of an intervention strategy, it is important not to ask performers to do things their eyes cannot do. The cues "Keep your eyes on the ball" and "Watch the ball hit your bat [racket]" clearly are impossible to adhere to (Watts and Bahill, 1990) and may be miscues (Kluka, 1991). *Sports Illustrated* and other sport magazines are filled with photos of baseball hitters or tennis players who usually have their eyes correctly focused forward of ball impact. Coaches of striking sports should select feedback about visual focus that take advantage of our knowledge of eye movements tracking fast moving objects.

The visual limitations of tracking fast-moving objects should also help professionals understand how decision errors occur in sports. For example, coaches can help players deal with terrible (only from their perspective, not the opponent's) calls that

QMD Choices

You are golfing with two friends. After watching your drive, they disagree on the quality of your swing. One of your friends is standing behind you, while the other is facing you square to your shoulders. One friend indicates that you lifted your head during the swing. The other believes you swung inside out. You want to accept feedback from one or both of these individuals. Using the information about vision, can you explain why they might disagree on what they have seen? If you perform the movement again, is there a chance they will see the same thing? Would slowing down your swing on the driving range be effective in reducing disagreement or creating the same technique?

QMD Demonstration 3.2

Observe the basketball shooting video clips in QMD Demonstration 3.2 in the web resource at www.Human Kinetics.com/QualitativeDiagnosisOfHumanMovement. Watch the flight of the ball and indicate what happened. While watching the ball, can you also gather information about the foot or hand position? How much information can you gather if you are tracking the flight of a projectile? Can you judge the qualities of the shooter's movement if you are focusing on object motion?

are part of competition. To try and minimize these errors in observation within QMD, the professional should increase the distance for observation to reduce the eye movements required to track fast objects.

Peripheral Vision   Peripheral vision is the ability to gather information from the environment other than the point of visual focus. This is a function primarily of the rods in the eyes, as they are situated in areas not central to where light is focused. This ability is particularly sensitive to slight movement, and its processing is faster than the perception of color by cones on the retina.

Peripheral vision directs our attention to movement in the environment around us so that we can process this information. It also sets the stage and orientation of events so that they can be mapped or matched to general backgrounds (Alfano and Michel, 1990). Restriction of peripheral vision, especially in QMD, can lead to fewer meaningful reference points and poorer observation. Peripheral vision contributes only to ambient vision and not to focal vision.

Fusion   Although the eye has often been compared to a camera, research on visual perception led Johannson (1975) to conclude that the eyes act more like motion-detection systems than they do still cameras. The eyes do not, in a sense, capture photos; rather, they constantly evaluate a changing flux of light focused on the retina to generate the perception of a 3-D image of the visual field. In normal vision, both eyes send information to the brain, where it is blended and interpreted as a 3-D phenomenon. This blending of each eye's essentially 2-D visual information into a 3-D whole is known as **fusion**.

Kluka (1987) presented a simple way to demonstrate this phenomenon. Tape two pieces of paper (8 1/2 inches by 11 inches), one white and one red, in a corner at eye level—the white paper in front of you and the red on the wall next to your left shoulder. Place a pocket mirror in front of your left eye, touching your nose, so that you can see the red piece of paper in the mirror. Look at the white piece of paper with your right eye. If you are the same distance away from both pieces of paper, you should see a single piece of pink paper instead of one white and one red. Another way to illustrate fusion is to hold your thumb vertically at arm's length and aligned with another object in the distance. By focusing your eyes on the thumb (seeing two distant objects) or the distant object (seeing two thumbs), you can see how information from two eyes is blended to create a 3-D representation. Remember that perception requires cognitive processing, so we are not just passive receivers of visual information.

Selective Nature of Visual Attention

Our senses have variable levels of sensitivity, and vision may be the most variable and selective of all the senses. Without a conscious effort to attend to one object, the eyes will dart about the visual field, moving to unusual or quickly moving objects. In short, normal vision pays attention to any number of things in a person's view. The discussion of perception in this chapter includes some examples of how the eyes search the visual field in order to make sense of visual stimuli.

The implications of this for QMD are interesting and varied. One perspective is to consider this selectivity a barrier to systematic observation and therefore plan a specific observational strategy to compensate. The other perspective is to use this sensitivity to locate and focus on unusual features of a movement. A problem with the second approach is that we are not always conscious of what our eyes are focused on. We may not be sure if we waste observational time looking at unimportant aspects of performance or focus on important aspects by ignoring the unimportant ones.

Focal and Ambient Visual Perception

To illustrate how closely tied the senses and perception are, we can use the example of two encompassing visual systems: ambient and focal. By examining these two systems, we can see that the senses are inseparable from perception. Although senses and perception are discussed as separate entities in this chapter, they are, in fact, intrinsically linked in their function.

There appear to be two visual perceptual systems working to provide information for decisions in movement and movement observation: focal and ambient (Schmidt and Wrisberg, 2008). Their function seems to explain many of the reasons why SVA and DVA are not strongly related, why batters cannot keep their eyes on the ball, and especially why biological motion—that is, the motion of animals—is seen differently than object motion (Shiffrar, 1994). Further work on these two systems may change the way we talk about vision, especially as it relates to biological motion.

Focal vision tells us *what* an object is and uses a small portion of the retina, whereas ambient vision tell us *where* an object is and about its motion, or about its motion relative to the observer. Ambient vision uses information from the whole retina (focal and peripheral). Focal vision uses information only from the most accurate point on the retina, the fovea. The fovea is a small point that is limited to about 3 degrees of the visual field (Kluka, 1991). As you read this book, you move the fovea across the words. Focus your eyes on a word in the text and notice that you cannot see words to either edge of the paper. Focal vision may have its greatest value when a particular detail of a movement must be seen, as opposed to determining relationships of body parts in a movement sequence.

Schmidt and Wrisberg (2008) have summarized the types of information provided by the optical flow (movement of objects across the retina) in ambient vision. The types of information provided by ambient vision that are useful for QMD are as follows:

- Velocities of the movement through the environment
- Direction of the movement relative to the position of fixed objects in the environment
- Movement of environmental objects relative to the observer
- Time until contact between the observer and an object in the environment (Schmidt and Wrisberg, 2008)

> **»KEY POINT 3.2**
>
> Ambient and focal vision are important to us in everyday life, but they are extremely important in the observation of movement. Realizing how ambient and focal vision work and what information they give us can enhance our visual data-gathering processes.

Biases in Visual Perception

There are several biases in visual perception that may be hardwired into the 3-D perceptual set of vision. Many of these phenomena are related to the geometry of the situation. The farther an object is from the viewer, the smaller it appears and the smaller the displacement past the observer, compared to a similar object moving at the same speed close to the observer. For example, on a late-night walk, nearby objects move past at the speed of your gait while distant objects appear to move past slowly. This is why people tend to overestimate the speed of objects close to them and underestimate the speed of objects at greater distances (Johansson, 1975).

The large horizontal perspective of our visual field also leads to a tendency to overestimate the lengths of vertical lines compared to horizontal lines (Prinzmetal and Gettleman, 1993). A novice coach, for example, might more readily perceive that the up-and-down motion of a runner is exaggerated before perceiving over-striding. Other biases are a tendency to underestimate object size with an inward shift of accommodation (Meehan and Day, 1995) and a tendency for dimmer objects to appear farther away (Kluka, 1991). Observers should be careful about evaluating the size of objects that approach or recede from their position. For example, an object of constant size that is brought closer to an observer (requiring either accommodation [refocusing of the lens] or convergence) will often be perceived as smaller than it actually is.

Many readers are familiar with some of the many images that can be interpreted as two different things (for example, a well-known illustration shows either two faces or a vase in the center). You can see only one interpretation of the image at a time. An example of the 3-D bias of vision is illus-

Choices

You are a cross-country skate-skiing instructor qualitatively diagnosing skate skiing in an activity class. Your students want to have fun, but you also have them focused on improving. What sensory information should you attend to the most? Are there specific senses that are most appropriate for the critical features of skiing?

Assume you are skiing behind a skier who needs a push-off correction. As you begin observing again, what senses are most relevant? A good approach would be to give feedback on the important sounds from the skis and then give verbal cues and prompts to signal the next weight shift or pole plant. Your auditory senses monitor the force and duration of the push-off, while your sense of vision provides information on balance during the next glide. You need to pay attention to these senses and integrate the sensory information to provide good QMD in this situation.

Figure 3.2 What objects do you see in this image? Do the lines intersect or are they parallel? Even simple drawings tend to be interpreted as 3-D objects.

trated in figure 3.2. What two things does it show? The image is literally two lines that touch at a point. The objects are typically interpreted as parallel lines (a road stretching to the horizon). Figure 5.1 in chapter 5 also illustrates our bias toward interpreting images three dimensionally.

Sport Vision

As the interest in vision in sports has grown, several articles have been written on visual skills in sports (Abernethy and Neal, 1999; Blundell, 1985; Fisk, 1993; Knudson and Kluka, 1997; McNaughton, 1986; Regan, 1997; Sherman, 1980; Vickers, 2007). Research has also been conducted on commercial programs for training DVA, such as "eyerobics" (Cohn and Chaplik, 1991; Long, 1994; MacLeod, 1991) and Dynavision (Klavora, Gaskovski, and Forsyth, 1994, 1995). The primary professional organization interested in vision in sport is the Sports Vision Section of the American Optometric Association. There is also a coaching interest group that publishes an online magazine, *Sportsvison Magazine* (www.sportsvisionmagazine. com/). Kluka (1991) summarized the research, noting that there are 14 important visual skills relevant to learning motor skills and other factors affecting visual perception. Another comprehensive summary of visual skills in sport was published by Vickers (2007).

Coaches need to observe how performers use their eyes in many sport skills. Knowledge of how the eyes work and their limitations is important for coaches so that they know what is observable and what feedback on the use of the eyes is helpful. For example, the miscue "Keep your eye on the ball" may not interfere with performance in some sports, but the cue "Watch the ball hit your bat" could adversely affect performance by encouraging head motion and less visual attention earlier in the trajectory of the ball. The trajectory of the ball is ascertained by ambient vision. Accurate placement of the bat concerns where the ball will be, not the state of contact between the bat and ball. Where the bat moves is a function of ambient vision, while the state of contact of the bat and ball is rarely identifiable with focal vision. Cues on ball tracking should emphasize focused attention, minimal or smooth head motion, and characteristics of the ball (seams, spin, and so on). A batting coach might say "Try to focus on the ball, looking for it out of the pitcher's hand." The adage to coach "from the eyes down" may be very important in some skills. It is also important that any intervention regarding vision be accurate and effective, not just a cliché.

Audition

The interaction of the ears and the brain allows us to perceive sound. From what we hear, we can make sense of the frequency of sound waves (interpreted as pitch) and the amplitude of sound waves (interpreted as loudness). The small muscles in the ear attached to the eardrum and ossicles can contract or relax to modify the sound. This is analogous to the way the muscles of the eye work with the lens to focus light in vision. This ability is demonstrated in the case of loud noises such as a jet aircraft departing. The eardrum muscles tighten to reduce vibration and sound transmission.

We can also best determine the tempo of activity from the input gathered by our ears. This information comes from alterations in frequency and pitch. Dance provides many examples of combining these elements of sound as factors for QMD. The desirable timing of actions in a dance

> **» KEY POINT 3.3**
>
> The information gathered by our ears (frequency and amplitude of sound) is vitally important to the QMD of rhythmic movements (dance, hurdles, and so on). Auditory information from discrete events is also important. The sound of impact in baseball and the sound of a gymnast landing are important sources of information about performance.

can be counted aloud (1, 2, 3, . . .). The loudness of each number can vary to show the emphasis placed on the action occurring at that time. This emphasis in sound connects to the cognitive aspect of the activity facilitating the whole movement.

Touch

>> **KEY POINT 3.4**

The tactile system (sense of touch) can provide a great deal of information about performance to the kinesiology professional and the athlete. This information is gathered primarily by pressure and touch receptors.

The haptic senses of touch (Meissner's and Ruffini's corpuscles, Merkel's disks) and pressure receptors (Pacinian corpuscles) send information to the brain when they are stimulated. These receptors are found in the skin and attached to hair follicles. Either the gentle pressure (touch) on the skin or the movement of the hair in the follicles triggers these receptors. When these stimuli breach the threshold of the nerve, signals are transmitted to the brain indicating direct contact with or close proximity to an object.

The pressure receptors (Pacinian corpuscles) have a higher threshold than touch receptors. They react when touch becomes pressure. Then information about the stimulus is sent directly to the brain. This pressure can help us determine whether or not a person can perform a movement independently or still needs spotting.

Kinesthetic Proprioception

The kinesthetic proprioceptors work to tell us about movement in our limbs and body. They do this by sending information to the brain from the stretch receptors in the muscles, the Golgi tendon organs (GTO) in the tendons, and the sense receptors in the joints. The stretch receptors tell the brain about speed of contraction and muscle length. The muscle spindle works by stimulating the nerve wrapped around it. It is highly sensitive to the slightest change in length. Information is sent to the brain by type Ia afferent nerve endings. The flower-spray endings associated with the spindle are less sensitive and harder to trigger. This information is sent to the brain by type II afferent fibers. The contrast between the signals of these two sensory receptors gives us information about the speed of movement.

>> **KEY POINT 3.5**

Kinesthetic feedback is another source of information about performance. It is gathered by GTO, muscle spindles, and other joint receptors. A coach might ask an athlete to identify how much knee flexion he used in the last trial and to focus on that sense in practice.

The GTO tells us primarily about the load on a muscle. It does this via the tendon's stimulation of the nerve endings. Joint receptors give general information to the brain that movement has taken place. The combination of this information allows us to understand the relationships between body parts and muscle tension.

USING SENSES TO UNDERSTAND MOVEMENT

We all know that eyes allow us to see and ears permit us to hear, and that we can sense movement and recognize touch. However, our senses provide more than general information via vision, audition, kinesthesia, and touch. Each sense can elicit qualities from the energy forms it interprets to provide highly specific information about what is occurring in our environment. The integration and interpretation of this information allow us to make decisions about how to proceed in QMD.

As you may recall, perception was defined earlier in the chapter as the organization and interpretation of stimuli from the environment, mediated by our senses.

Each of our senses provides us with a great deal of specific information from our environment, and this information becomes the basis for decisions. The following section gives examples of how the parts of the sensory system work together, that is, how vision interacts with touch and sound and maybe even with kinesthesia.

Part of the work of these four major sensory receptor groups is done by either electrical, chemical, or mechanical energy. The other part of their work involves perception. For example, think of tracking the flight of a kicked ball. Electrical energy is used in nerve transmission, chemical energy for color vision, and mechanical energy for transmitting sound waves through the eardrum to the bones in the middle ear. The perceptual component is easily demonstrated by the adjustments of the eye and head movement to the direction of the energy received by the senses.

For an example of how all the senses might be used to understand movement and provide feedback to an athlete, consider a coach spotting an athlete vaulting in gymnastics. The athlete runs the approach, hurdles onto a board, places her hands on the horse, and flies to a landing. In this skill the coach can use all of his senses to gain information about the performance to provide feedback.

Sound contributes to feedback decisions. The tempo, loudness, and pitch of the run-up approach, hurdle, hand placement on the horse, and foot landing on the mat all provide useful data. The tempo of the run-up, the different pitch from hitting the vaulting board, horse, or mat correctly or incorrectly, and the loudness of any of these tell the coach something about the quality of the skill.

While the athlete is moving, the coach's visual system can track her and provide information about body position, body placement in relation to the apparatus, and the relationship of body parts to one another. Interestingly, the linear and angular velocities may be so great when the athlete is on the board and horse that the coach cannot get any usable visual information. The situation forces him to make decisions about performance based on data from the other sensory systems.

The coach then has to rely on the systems of sound, touch, and kinesthesia to judge performance quality. During flight or landing, he will likely touch the athlete and may even push to help her in rotation or flight. At this point, touch and pressure information will be added to the input being processed by the coach. Proprioceptive kinesthetic information will flow to the central nervous system as the coach moves his body parts to support the gymnast. Clearly there is an incredible flow of data bombarding the brain. All of this information must be organized, given meaning, and then combined into a total picture to describe what has occurred.

To say the least, the perceptual process is an incredibly large and efficient system. It is even more amazing to think that we use all of these systems to gain information about a particular movement. Even if we use only two or three of these systems, we can imagine what the other systems would feel, look, or sound like. If the coach in our gymnastics example moved away from the horse, he could still imagine what a good vault would feel like, kinesthetically and haptically, as opposed to a bad vault.

INTEGRATION OF SENSES

As you can see, the perceptual system is continually bombarded by sensory input. How do we pay attention to what is important? How do we use all the information available from the senses to make decisions? The complex process of intersensory integration is the perception of an event, as measured in terms of one sensory modality, as changed in some way by the concurrent stimulation of one or more other

> ## »KEY POINT 3.6
>
> With a great deal of practice, the brain can automatically integrate sensory information and prioritize it based on its importance to the QMD being performed. A skilled coach might hear an unusual rhythm in an athlete's performance and then visually observe the movement. A diving coach concentrating on observing the dive might use the sound of the diver hitting the water to aid her visual interpretation of the athlete's entry.

sensory modalities (Welch and Warren, 1980). In other words, our perception of an event through one sense is affected by our perception of the same event through other senses.

Martino and Marks (1999) point out that senses are better categorized as interrelated than separate. These authors indicate that senses seldom work alone and that perception is suited to working with combined input. They hypothesize that there are two levels of processing when we are integrating information from the senses: prelinguistic and linguistic. At the prelinguistic level, perceptual information commingles before it is coded semantically. This occurs for visual and auditory stimuli. In the other level of processing, linguistic, input is combined from different senses semantically before it is processed for information. As you might guess, prelinguistic processing is faster than linguistic processing (Martino and Marks, 1999). This is another example of how cognitive processes are intertwined in perception.

It is even likely, as when spotting in gymnastics, that we use our haptic and kinesthetic proprioceptive senses as well as vision and audition. The sound of the block on the horse could be compared to the visual or proprioceptive information the coach has just observed. The approach and block sounded vigorous, but there still was not enough rotation. What sense might a gymnastics coach put the most confidence in when sensory information conflicts?

How does the brain deal with the input from competing senses, and how do the different elements in this flow of information from the different senses interact? The theories of cognition and information processing that attempt to explain sensory integration are beyond the scope of this book. It is important to know that we have the ability to integrate all our sensory information. We will learn later that this integration is similar to the psychological concept of a gestalt impression that some people like to use as a strategy for observing movement.

Sensory Detection of an Event

Relatively strong auditory and tactile stimuli take about 110 to 120 milliseconds to detect. Visual stimuli take slightly longer, about 150 milliseconds (Riggs, 1971). It appears that with intense training, however, people (like fighter pilots) can speed up their visual detection and integration of stimuli to 33 milliseconds (Secrist and Hartman, 1993). Note that reaction time is inversely related to stimulus intensity. That is, the more prominent the stimulus, the faster the response. It might take concentrated effort for a coach to use auditory observation in a competitive environment with crowd noise.

Stimuli in the observer's environment rarely act in isolation. The effect of other accessory stimuli (stimuli that we are not attending to selectively) on a primary stimulus (the sense we are using selectively) is one of either inhibition or facilitation. If the accessory stimulus is low to moderate in intensity, it generally facilitates perception of the primary stimulus. If the accessory stimulus has a high intensity, it may have an inhibiting effect (Shigehisa, Shigehisa, and Symons, 1973; Shigehisa and Symons, 1973). For this effect to be optimized, these stimuli must occur close together.

These effects may be due to physiological factors or selective attention (Welch and Warren, 1980). The physiological effects that seem to enhance detection relate to muscle tonus of sensory organs (the muscles of the ossicles in the ear or the muscles of the pupil in the eye). Selective attention may cause someone to pay more attention to a particular movement because of a secondary stimulus. Studies that have examined the effects of touch and hearing on vision support the idea of synesthesia for these senses (Marks, 1987; Martino and Marks, 1999). This synesthesia appears to broaden and enhance selective attention. Rather than just attend to one stimulus, the observer is able to attend to and use information from both senses.

QMD Demonstration 3.3

Observe the video clips of the tennis player and basketball player in QMD Demonstration 3.3 in the web resource at www.HumanKinetics.com/QualitativeDiagnosis OfHumanMovement. What did you notice? Why? How did your mind try to reconcile (logically fix or explain) the situation? How does your experience support the integrated use of cognition and your senses?

Spatial Stimuli

For spatial stimuli, the order of dominance changes. The visual sense predominates, followed by audition, proprioception, and then touch (Welch and Warren, 1980). Another major difference between vision and audition is that audition is accurate only in the horizontal plane, while vision is accurate in all planes. Sound received from the horizontal (left, right) is perceived more accurately than sound from above or below. It is interesting to note how closely auditory and visual perception interact in processing spatial information. Visual perception appears to provide a framework for auditory information (Platt and Warren, 1972; Warren, 1970). It seems that our visual memory provides a basic map to which auditory information can be applied.

Kinesthetic proprioception of hand movements can detect spatial differences

QMD Choices

You have to prepare for QMD of a beginning swimmer doing the breaststroke. Supporting this learner in the water will provide you tactile, kinesthetic, and visual stimuli about the quality of the movement. Which stimuli will be available for processing first? How can you enhance your sensory detection of technique events in this QMD? Which sense would you rely on most?

of about plus or minus 1.25 centimeters (Magill and Parks, 1983), which suggests that we can gain accurate information concerning movement from our kinesthetic receptors. We can sense the slightest movement of a body part of someone we are spotting. That is, we can sense an arm or leg movement of less than 1 inch (2.5 centimeters).

Temporal Information

When short-duration stimuli are processed, it seems that audition is most accurate, followed by touch and then vision (Welch and Warren, 1980). Auditory stimuli also appear to last longer than other stimuli (Behar and Bevan, 1961), so we can attend to auditory information for a longer time than other information. The temporal accuracy of kinesthetic proprioception relative to other senses is not clear. So in QMD of a long jump, the professional might want to have a visual focus (for technique), but would also try to attend to the sound of the approach (for the rhythm of the last few steps and takeoff) for supplementary information. This would be a good use of attentional energy, sensory abilities, and the integration of vision and audition for QMD of this skill.

KNOWLEDGE AND PERCEPTION

Along with the senses and skill knowledge, perception forms the basis of QMD. Knowledge of the components of a movement and their sequence is useless unless the professional can resolve meaning from the performance being observed. Essentially, this cognitive processing organizes and gives meaning to information from the sensory receptors. Remember that this is essentially the definition of perception (Sage, 1984).

> **» KEY POINT 3.7**
>
> Perception has two major components: organizing sensory information and assigning meaning to that information.

The process of perception appears to have two major parts. The first is the organization of sensory information, which involves sending information from the sensory receptors to the appropriate areas of the brain. The second deals with interpreting this information and assigning meaning to it. Once this meaning has been established, the QMD tasks of evaluation and diagnosis can follow.

Perceptual Tasks

The simple perception of natural forms is a delight.

Emerson

Emerson's statement is true in that perceiving natural phenomena (for example, animals, plants, rivers, and human motion) can be delightful. It may be wrong, however, to assume that perception is simple. In fact, perception is a highly complex process that we are only beginning to understand. Perception has been studied primarily from the point of view of the types of tasks used in perception. These tasks or experimental conditions allow us to understand what is required of the cognitive processes that attempt to decode environmental stimuli.

According to Proctor and Dutta (1995), these tasks (in order of increasing cognitive requirements) are detection, discrimination, recognition, and identification. People involved with QMD readily associate with the different tasks described here. Although the term *detection* was discussed earlier in relation to vision, it is used here to illustrate a level of cognition based on the amount of information available for a decision.

In detection tasks, a person need only indicate when a stimulus or event has occurred. A typical subject might be asked to indicate if a light flashed or if there was a sound. Usually studies dealing with detection attempt to ascertain the lowest threshold at which a stimulus can be detected. These studies are often conducted to see if the person's detection ability can be adapted to stimuli that are below the initial thresholds. With training, this effect is often achieved.

In sport, an example of this type of task might be this question: Did the offensive lineman (in an American football game) move his hand? To a game official, detection of this event would lead to a penalty if it occurred before the snap of the ball. Similarly, a teacher's or coach's perception might work at the level of detection in skill assessment. For example, a teacher might decide whether or not a child's elbow preceded the forearm in an overhand throw. Either it did or it did not. Could the elbow–forearm relationship be detected?

Discrimination tasks generally require a subject to attend to various stimuli and to distinguish among them. Are the stimuli the same or different? Or do they have more or less of some quality? A batter in baseball has to attend to the flight of the pitch and the spin on the ball, as well as discriminate among possible types of pitches in order to adjust his swing. Generally speaking, good batters have this perceptual ability developed to a higher level than poor batters. They can discriminate among pitches and adapt. This ability is important for teachers assessing movement technique. For example, did the hands land on the front, middle, or back part of the horse in a vault? The professional knows that the hands made contact with the horse but, in order to provide good feedback, must discriminate where contact occurred.

Recognition tasks require more perceptual processing than either detection or discrimination tasks. This type of task usually involves stimuli that have been presented previously, as opposed to those that have not. People can distinguish stimuli learned previously from stimuli that they are not conversant with. In a gymnastics floor routine, spectators would be able to recognize and name all of the tumbling moves they were familiar with. They would not be able to recognize moves they had not seen before.

Recognition implies that a good deal of knowledge is needed to help with the qualitative assessment. A spectator or coach who did not know the whole routine of a gymnast would not know if a part was missing or changed. In QMD, recognition is important because it is the level of perception that allows us to start perfecting skill performance. It is the level at which we look for complete matching of a performance and its critical features with a prototypic skill.

Identification tasks stress the perceptual processes even more than recognition does. These types of tasks require that people respond to a stimulus in a specific way, that they make a judgment in response to a stimulus. These responses can be the same for different or similar stimuli. A guard in basketball might encounter the stimulus of a certain defense set up by an opponent. She might respond by running this play or that play in order to beat that defense. Once the player has identified the stimulus (defense), she can decide on the appropriate response(s).

Similarly, a teacher or coach might identify a certain "common error" in a movement, such as stepping on the throwing-arm-side foot during a throw. His correction might differ based on the knowledge he has about the skill or performer. Skill-related feedback might be appropriate in one situation, while motivational feedback might be more effective in another. The process of identification is similar to the third task in QMD, that of evaluation and diagnosis.

> **» KEY POINT 3.8**
>
> Perceptual tasks have four levels: detection, discrimination, recognition, and identification. All four levels of perception are used in QMD.

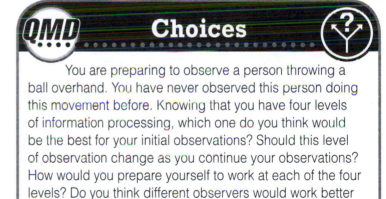

QMD Choices

You are preparing to observe a person throwing a ball overhand. You have never observed this person doing this movement before. Knowing that you have four levels of information processing, which one do you think would be the best for your initial observations? Should this level of observation change as you continue your observations? How would you prepare yourself to work at each of the four levels? Do you think different observers would work better at different levels? Is there a common, or agreed on, level comfortable to most observers?

As you can see, perceptual processes are complex and require a great deal of cognitive processing when applied in QMD. The complexities of QMD are far more intricate than most of us imagine. In truth, we have just begun to understand the cognitive–perceptual process in visual observations of movement.

Research on Perception in QMD

Evidence for the importance of perception in the diagnosis of movement has slowly accumulated over the years. The two major sources are (1) investigations that use perception or facets of perception directly in various qualitative assessments of movement and (2) those that examine perception in general. The information from these latter investigations can be applied indirectly to QMD. Investigations of the perception of motion are undoubtedly the basis for investigations of perception in QMD.

We can hypothesize that considerable and complex cognitive activity takes place in the observation of movement within QMD. The research on perception in QMD has dealt primarily with imagery, spatial ability, types of examples (good and bad), expert–novice comparisons, and perception of biological motion.

Imagery and QMD

It appears that vivid mental images of expected movement performance can be used to match information from observation. This process is similar to what has been called template matching (Anderson, 1990), feature integration (Triesman and Gelade, 1980), and recognition (Pinheiro and Simon, 1992). This matching process is also part of the evaluation of performance that determines its strengths and weaknesses. Because verbal information and visual information are stored differently, it is communication between the verbal and the visual abstract representations in the mind that allows for performance evaluation. What visual images of technique are called (verbal information) is cognitively integrated with visual imagery when we evaluate our perceptions of observed movements. This information is used in diagnosing performance and formulating potential intervention.

We are not sure how images are stored in the mind. Farah and colleagues (1988) believe there are two kinds of imagery: spatial and visual. These concepts are extremely complex and beyond the scope of this text. They are mentioned here to alert the reader that mental images are complex and can be different for different observers.

The results of the studies of imagery and QMD are less than conclusive, but they do illustrate the importance of vivid mental imagery in the observation of movement. In 1975 Hoffman and Sembiante studied whether a person qualitatively assessing human movement holds a mental image of the movement in memory. They concluded that the ability to formulate and hold a vivid mental image was related to QMD ability. In a second study (Hoffman and Armstrong, 1975), visual imagery did not appear to be a discriminating factor in QMD ability. The factors under examination here relate directly to the question of how a person observes and evaluates movement. Although the jury is still out, imagery would intuitively seem to be an important factor in matching our perception of performance with a prototypic, cognitive model of expected performance. Although research on imagery in QMD has been quiet, this area deserves further research to help us understand how visual perceptions of movement are evaluated.

Spatial Ability and QMD

People's ability to deal with spatial information in QMD has been only superficially investigated. It seems obvious that the more easily we can process visual–spatial information, the more easily we can observe movement, since movement is essentially change of position in space.

Researchers have assessed **spatial ability** by examining perceptual style (**field dependence** and **field independence**)—whether or not subjects can use background cues to extract information from the environment. A tennis player who tends to be field independent does not rely heavily on reference points such as court lines and the net or other environmental information to play. This type of player can gather sufficient information from the opponent to decide on appropriate responses. An observer who can separate performers from background factors is less susceptible to a disorganized sensory background and may not need to spend much time manipulating the observational situation. By contrast, an observer who has difficulty observing movement without reference to floor lines or background markers tends to have a field-dependent perceptual style. This person must make sure that the QMD environment is well organized to support the visual perception of movement.

Five studies (Gangstead, Cashel, and Beveridge, 1987; Knudson and Morrison, 2000; Morrison and Frederick, 1998; Morrison and Reeve, 1989, 1992) have looked at spatial ability from a verbal and a nonverbal (nonanalog and analog) point of view and its effects on QMD. A test of spatial ability can sometimes be solved using verbal clues, while other spatial tests resist verbal help. The Group Embedded Figures Test (Oltman, Raskin, Witkin, and Karp, 1971) is one in which verbal self-talk appears to aid in the solution of the visual problems. The Mental Rotation Test (Vandenberg, 1971) and the portable rod and frame test (Witkin, 1954) seem to resist verbal clues. Gangstead, Cashel, and Beveridge (1987) found a weak ($R = 0.27$) relationship between QMD and field independence. They used the rod and frame test (Whitkin, 1954) to determine analog spatial ability. Morrison and Reeve (1989, 1992) and Morrison and Frederick (1998) used the verbal–spatial Group Embedded Figures Test (Witkin et al., 1971) to see if spatial ability was important to QMD. Morrison and Reeve (1989, 1992) found that those tending toward a field-independent style of observation were better at QMD than those who were more field dependent. Morrison and Frederick (1998) uncovered a weak relationship ($R = -0.33$) between disembedding (separating an object from the visual background) and absolute error in QMD.

One study (Knudson and Morrison, 2000) added the Vandenberg Mental Rotation Test (1971) to the Group Embedded Figures Test (Whitkin et al., 1971) to assess mental rotation ability and disembedding in the assessment of angles and range of motion in the vertical jump. The authors reported no significant associations between disembedding and mental rotations on either linear or angular measures of movement in a sample of 43 untrained undergraduates.

The studies of the effects of spatial ability on QMD are not conclusive but tend to indicate that spatial ability may have little effect on QMD compared to other factors (knowledge, experience, and so on). Some people appear to use a combination of analog spatial ability and verbal spatial ability to help organize movement information. Although more research is needed, present data suggest a need for explicit verbal skill descriptions as well as practice with movement examples to develop QMD.

Development of Spatial Perceptual Ability

Studies of spatial perceptual ability could lead to enhancement of QMD ability. It has been demonstrated that visual–spatial ability can be enhanced by training. Since spatial ability appears to moderately correlate to QMD ability, development of spatial ability could underlie enhancement of QMD ability. Pinheiro and Cai (1999) found that undergraduate students who used an observational model similar to Dunham's (1986, 1994) were superior in their QMD to those who did not use an observational model. It is possible that observational models organized around spatial factors might help some observers through enhanced spatial perception of motion.

So far, only one perceptual training system appears to have been shown to improve perceptual spatial abilities (Secrist and Hartman, 1993). This system is available only to military fighter pilots. Essentially, it is a rapid-fire presentation of visual situations normally encountered by pilots on combat missions. Short bursts of combat sequences (lasting 5 to 6 seconds) are presented, and subjects are required to make nearly instantaneous decisions (in 0.33 to 0.67 seconds). All challenges are adapted to a subject's current level of performance. A similar system could be developed for training visual observation of movement. To some degree, systems like this already exist in video and virtual reality presentations, but they are expensive and are not purposefully matched to the live environment.

Good and Bad Examples of Performance

A number of authors have used good and bad examples of skill performances to teach QMD (Beveridge and Gangstead, 1984; Gangstead, 1984; Kelly, Walkley, and Tarrant, 1988; Morrison, 1994; Morrison and Reeve, 1989). Only Gangstead (1984), Morrison and Reeve (1989), and Morrison (1994) attempted to differentiate between the two types of information and QMD ability. Gangstead found that for QMD instruction, examples of good and bad performances were superior to good examples only. The study by Morrison and Reeve and the one by Morrison showed no significant difference in the effect of type of training information on QMD ability.

These research projects were initiated to establish the value of more and contrasting instructional information to the development of QMD ability. The authors felt that this information would be valuable in the cognitive interpretation and the comparison of the observed movement with the expected movement. Despite the conflicting results for types of examples, this book and the associated online features present both good and poor technique for examples and tutorials in QMD.

Expert Versus Novice Studies

An area of research that may have relevant information for QMD is expert versus novice differences in anticipation of movements (Abernethy, 1989, 1993; Abernethy and Russell, 1987; Abernethy and Zawi, 2007; Buckolz, Prapavesis, and Fairs, 1988; Davids, DePalmer, and Savelsbergh, 1989; Goulet et al., 1988). The essence of this research is this question: How much information does it take for an expert, as opposed to a novice, to anticipate an opponent's shot in racket sports or a punch in boxing? These studies used direction of shot and different amounts of movement information to examine these questions. Generally speaking, experts are better than novices at anticipating the end results of opponents' shots. They appear to acquire this information earlier in the stroke and need information from fewer body parts to make these decisions. This type of research appears to hold promise for QMD. Examination of the difference between expert and novice observers could produce information that indicates which body parts are most important for good QMD and when this information is acquired.

Research on Perception of Biological Motion

Recent psychological research has tied the idea of the *what* of focal vision and the *where* of ambient vision to the observation of biological motion (Heptulla-Chaterjee, Freyd, and Shiffrar, 1996; Shiffrar, 1994; Shiffrar and Freyd, 1993; Thornton, Pinto, and Shiffrar, 1998). Shiffrar (1994) indicated that these two visual systems converge in the higher centers of the brain and are used in the cognitive interpretation of biological motion. The two systems she mentions are the ventral (the *what*) and the dorsal (the *where*). She does not call these two systems the *focal* and the *ambient*, although she does appear to be referring to these two systems. In her experiment Shiffrar demonstrated that even when no movement path was given for apparent biological motion, the visual system constructed paths for the apparent motion consistent with biomechanical limitations. It appears in this case that the mind is working at the identification level of processing. That is, presented with enough time, the mind fills in the path of biological motion even though no real motion occurred. We cannot dismiss how active our perception is in the observation of human movement.

In 1990, Shiffrar and Freyd reported that when looking at apparent body motions (motions that were inferred from two still pictures but that never actually occurred), an observer interprets motion as occurring in the shortest path for fast motions regardless of how impossible that motion is. They also found that the slower the stimulus motion, the more anatomically possible the motion path observed by the subject. In faster motions, hands and arms appeared to move through the body, whereas at slower speeds the mind perceived the body parts as taking the anatomically correct way around the body. Without sufficient time, the mind notes the different positions of the body parts and seems to be operating at the detection level of movement. That is, the brain simply notes that a change has occurred without adding extra information such as the biomechanically possible path of the body part.

The visual perceptual system may also be seen to function as a hierarchy (Thornton, Pinto, and Shiffrar, 1998): a lower-level system with brief temporal and spatial limits and a higher-level system with longer temporal and larger spatial information. Although these levels are difficult to define, local information seems to be derived from the level of the joint and adjacent limbs; the global level acquires information from about half of the human body. Temporally, the local works within about 50 milliseconds or less, while the global works over longer periods of time.

Another interesting concept emerging from psychological research on biological motion is the idea that the recognition of biological motion may be tied to the motor aspect of the given movement (Thornton, Pinto, and Shiffrar, 1998). Psychologists have linked the visual and motor systems in the perception of human motion by noting significant effects for the recognition of human motion in conditions of relative masking. Masking is an experimental technique in which vision is cut off or masked at different times to determine the effect on visual perception. These researchers believe that the ability to recognize a type of motion is related to people's ability to use the motion they are asked to recognize. This preliminary evidence conflicts with the findings from kinesiology (Armstrong and Hoffman, 1979; Girardin and Hanson, 1967; Osborne and Gordon, 1972). Much more research is needed to see if this hypothesis has merit and could be extended to real-world QMD.

Perceptual research has implications for the two main approaches to organizing observational strategies: a less structured or gestalt approach (Dunham, 1986, 1994) or a more structured or localized approach (Gangstead and Beveridge, 1984). Both of these approaches are discussed in detail in the chapter on observation. These two approaches correspond to styles of cognition and have a common connection. The commonality between the two approaches (what and where, local and global information) appears to be temporal demands for processing sensory information.

The practical implications for these studies are that both approaches to observing human movement take time. The reason is that they use either local or global information as well as ambient and focal vision.

QMD **Choices** (?)

You have to observe an unfamiliar movement and assess it qualitatively. You are not sure about where to look for information during the movement. You decided beforehand that the most important information will come from the relationship of the limbs to each other and to the body. Will you take time to focus on each body part separately? Will you focus on the left or right side, or the top or bottom? How will your perception of the movement be influenced by this targeted approach to observation?

»KEY POINT 3.9

Research linking visual
perception to QMD has
been limited but has
focused on perceptual
styles, eye tracking, and
imagery. Much needs to
be learned about the interaction of visual perception
and cognitive processing
of movement that is essential to the observation task
of QMD.

Whatever observational strategy is used for QMD, the professional should allow time for information processing and not rush to judgment. This admonition speaks strongly to the ideas suggested throughout this text, namely, that good QMD relies on careful attention to all four tasks of the process. Professionals should not rush to judgment in QMD unless a serious safety issue is involved.

A GESTALT

Perhaps a good place to end our brief discussion of cognitive influences on perception of motion is at a place where perception starts, with common suggestions to novice teachers and coaches on how to observe human movement. Different people have varying approaches on how to look for meaning in movement—for example, "Just watch the arms in this skill" or "Watch for the follow-through in this movement." A volleyball coach may believe she can tell how well a spike was performed by watching the follow-through. While this suggestion may be helpful to some, it may be confusing or ineffective for most novice coaches.

One way for people to begin observing movement for qualitative assessment is to build a gestalt representation of the movement to be diagnosed (Chen, 1982; Treisman, 1986). A gestalt (named after a major branch of psychological theory) representation is a picture whose totality is greater than the sum of its parts. For instance, when observing batting practice, a coach might first look at the overall aspects of the hitter's swing; when observing a stroke patient's gait, a physical therapist might initially observe the whole body without any preconceived body segments in mind.

The mind tries to extract information from the stimuli presented by organizing them into patterns that have meaning. Ever wonder why we see patterns or groupings that we interpret as objects or surfaces from sight even though we may have never seen or interacted with them before? This study of grouping to make sense of visual perception was a strong stimulus for the development of gestalt psychology. Four basic principles (Anderson, 1990; Yantis, 1992) used to group objects are proximity, similarity, good continuation, and closure. Figure 3.3 from Palmer (1992) demonstrates these organizing principles that our mind wants to impose on images we see.

Notice in figure 3.3a that the dots with equal spacing tend to be perceived in pairs or groups, illustrating the principle of proximity. This effect is even greater in figure 3.3b. The principle of similarity (figure 3.3c) is illustrated by the pairing of dots with similar shading or lightness. The lines in figure 3.3d are usually perceived as two intersecting lines, rather than two "c" shapes, because of continuation (tendency to perceive the lines as continuous or connected). Notice that the same shapes are perceived as separate right–left objects when the ends are closed, an illustration of the grouping principle of closure (figure 3.3e).

Recent research and theoretical interpretation of a gestalt have formulated the idea of uniform connectedness or continuation as the foundation of information processing. That is, uniform connectedness (regions of homogeneous properties, such as lightness, color, or texture) is the first perceptual factor in a gestalt (Palmer and Rock, 1994). Continuation is often at work in the perception of object motion. For instance, when observing a fast sport, an observer perceives the motion that seems most continuous from one position to another. After uniform connectedness come proximity, similarity, continuation, and closure. What motion or holistic impression

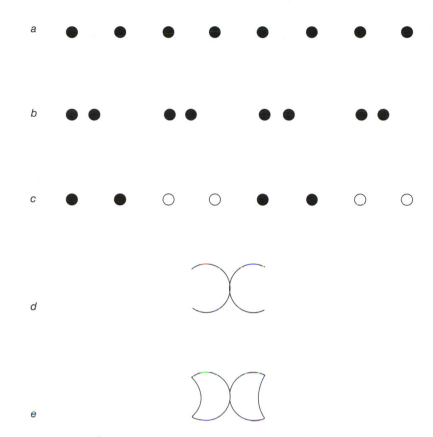

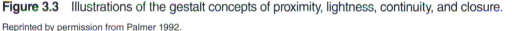

Figure 3.3 Illustrations of the gestalt concepts of proximity, lightness, continuity, and closure.

Reprinted by permission from Palmer 1992.

we perceive from unguided observation can be different from that of other observers given the interaction of these principles of grouping.

Although no QMD studies have directly addressed grouping, Petrakis (1986, 1987; Petrakis and Romjue, 1990) may have indirectly demonstrated grouping of stimuli by uniform connectedness as a factor in QMD. Her research examined eye tracking patterns to see if experienced observers looked at skills differently than inexperienced observers; she found that this was the case. The research further showed that the experts' visual search patterns were more systematic than the novices' and may have been based on the uniform connectedness of sections of the body. The experts tended to look for information concerning the movement being examined from the same regions of the body, while the novices searched wildly over the body for meaningful information. The experts seemed to group information by areas that appeared to be connected in a uniform fashion; novices did not. These results may further support the ideas of focal–ambient vision or what–where perception (Schmidt and Wrisberg, 2008; Shiffrar, 1994). The eyes appeared to be focused on a central point but were actually gathering information from ambient vision as to where the limbs were in relation to the body and each other. Focal vision could not have gathered information about these relationships.

Rather than watch for one specific aspect of a skill, observers should base their initial response on an overall impression or feeling about the quality of the performance. This may be especially true in continuous activities such as swimming, running, and rowing. Since we often get to see several repetitions of a skill performance when we are teaching or coaching, we can go back and look for specific skill components

»KEY POINT 3.10

The idea of a gestalt could be important in QMD. It provides a less structured approach to gathering information in observation that could be more natural for some people.

after making an overall decision about the quality of the performance. Then we can focus on the critical features that are organized (grouped) by knowledge and experience, or on the previous observations. Even if we do not get to see several performances of the skill, our short-term memory (Simon, 1979) may allow us to reformulate the performance in our minds so that we remember the individual parts of the skill.

Perception is a highly complex area of study. It can, at best, be hypothesized by indirect measures of brain function and observed behaviors. Although we do not have extensive hard evidence on how perceptual processes occur, perception is crucial to QMD and needs to be examined and understood if progress is going to continue in this area of our discipline.

PRACTICAL APPLICATIONS
Gestalt Observation

A teacher assessing a skill using a gestalt (holistic) approach to observation in QMD illustrates the cognitive power integrating information from all the senses. The whole performance is examined, and an overall impression of the quality of the performance is developed. The teacher picks up many visual, auditory, kinesthetic, and tactile stimuli. If the stimuli cannot be resolved into a coherent picture, then the process continues or the observer may try to resolve the problem by focusing on a particular part of the skill (the arms or the legs). Recall that another approach to observation is the use of a more structured, observational model. This back-and-forth sensory integration and interpretation is never ending and is fundamental to the interaction of senses and cognition and the tasks of QMD.

If one attentional focus is not useful, then we can switch to another and access the acquired information about that focus remaining in our memory. When processing sensory information, we must try to get the best and most complete information about movement performance. Teachers and coaches should be aware of the cognitive aspects and different approaches to skilled observation of movement.

SUMMARY

All of the senses are extremely important in QMD. Too often, professionals disregard the importance of the various senses and how these senses work together with cognition to provide information. Vision is generally the sense that predominates in the QMD process, but the best observation incorporates auditory, haptic, and kinesthetic information as well. Different senses process information at different speeds, and different senses may have primacy when dealing with certain types of information. The processing of input would be confused if we did not know what stimuli to focus on in different situations.

We process cognitive information on four levels (detection, discrimination, recognition, identification) based on how much information is available and what kinds of responses we need to provide. Sensory information is first organized, and then meaning is assigned to it. Research indicates that perception measured by disembedding and mental rotations and other perceptual factors may have an effect on observation within QMD. There is variation in how people make sense of sensory information, so some observers might prefer a structured plan of observation while others might prefer a more open, gestalt approach to observing movement for QMD.

Discussion Questions

❶ What visual limitations are most strongly related to a blown call by a sport official? For a given sporting event, who is more likely to make a visual mistake: an athlete, an official, or a spectator? Why?

❷ For movements you assess qualitatively, which sense provides the most relevant information? What are the limitations of this sense?

❸ Go to a driving range and compare the distance the golf ball travels to kinesthetic (feel of swing and impact), auditory (impact sound), and visual (initial flight) information. What sense seems most accurate?

❹ Experiment with a metronome or weights to determine the accuracy of your sense of audition or kinesthesis. Have a partner give you several "tests" to qualitatively evaluate the tempo or weight of various settings or weights. For example, your partner might ask you to rank six metronome tempos or rank, while blindfolded, six weights. Which sense do you think is more accurate or sensitive? Do you think this sensitivity would correspond to listening and spotting a movement in QMD?

❺ How are the senses related to cognition? Is it possible to see something and not understand it? Think of some optical illusions you have encountered.

❻ What types of research have been done on perceptual factors that may be related to observation within QMD?

❼ What is meant by a gestalt approach to observation? How is this approach useful for people involved in QMD? What would be an opposing approach to observing movement?

part

Four Tasks in the Diagnosis and Improvement of Human Movement

We have seen that the many subdisciplinary views of qualitative movement diagnosis (QMD) can be summarized in an interdisciplinary model with four important tasks. Chapter 4 outlines the knowledge that must be gathered in the preparation task. Several approaches to the second task of observation are illustrated in chapter 5. The third task, which may be the most important, has two parts: evaluation and diagnosis. Chapter 6 presents these two parts. Part II concludes in chapter 7 with a review of the many kinds of intervention that can be used to improve human movement.

chapter 4

Preparation: Gathering Relevant Knowledge

The booster club just asked you to judge the slam-dunk contest at the local junior college's basketball tournament. Since you are the local movement expert, the boosters want you to develop a rating scale and train the other judges before the contest. What are the key elements of a dunk in basketball? What aspects of a dunk should the judges look for and rate? Should the judges consider the height of the hand above the rim or the height of the jump? What is the range of correctness for a successful dunk? Could the judges make a fair and consistent rating without knowing the physical requirements and key elements of difficult dunks? How would you qualitatively assess the quality and difficulty of a dunk? Think about how different a qualitative movement diagnosis would be if you were going to teach safe dunking versus judging a dunking contest with skilled athletes.

Chapter Objectives

1. Identify areas of prerequisite knowledge that are important in the preparation task of QMD.

2. Define the critical features and explain how they are identified in the preparation task of QMD.

3. Explain how preparation in QMD is related to effective teaching and systematic observation.

4. Explain how preparing for QMD can be integrated with planning for teaching.

The first task in the qualitative movement diagnosis (QMD) of technique is the continuous process of building a prerequisite knowledge base. In our integrated model of QMD, this task is called preparation. Other scholars have called this the *preplanning step* (Philipp and Wilkerson, 1990) or the *preobservation phase* (Arend and Higgins, 1976; McPherson, 1990), or have simply noted that prerequisite information is needed before observation and evaluation can begin (Hay and Reid, 1988). Professionals must continually read and research to build a knowledge base and must think critically about the practice of their profession. For example, physical educators should keep up with current research in kinesiology. Medical and educational organizations have begun to formalize this weighting of knowledge into evidence-based practice procedures. The preparation task of QMD can certainly be based on this standard grading of evidence, but it can also be based on an individual professional's efforts to remain up to date on current standards of practice and research.

This chapter reviews important areas of prerequisite knowledge that professionals interested in human movement should be aware of in preparing for effective QMD. Good preparation involves weighing evidence from many subdisciplines of kinesiology. The three major areas of this prerequisite knowledge are

1. knowledge about the activity or movement,
2. knowledge about the performer(s), and
3. knowledge about effective instruction.

>> *KEY POINT 4.1*

The first task of QMD is preparation: gathering knowledge about the activity and performers. Professionals should continuously gather detailed prerequisite knowledge in order to be good qualitative assessors of movement, and they should strive to integrate their QMD with instruction to maximize their effectiveness.

The chapter concludes with a brief acknowledgement that another area of critical knowledge relates to the next task of QMD, observation. Professionals need to have the knowledge to develop and practice a systematic observational strategy for the movements they observe in QMD.

The philosophy of this approach to QMD is that knowledge is transient. Knowledge is different from mere information. **Knowledge** comprises systematic, data-supported ideas that make the best current explanation of reality for a particular academic discipline. Our state of knowledge and the standards of professional practice are dynamic. Professionals must realize that their career involves a never-ending search for the latest knowledge, our best approximation of the truth that can be applied to practice taking into account the individual and situation. For QMD to be most effective, the professional needs to maintain an up-to-date knowledge base. Otherwise, qualitative assessments may be based on erroneous or invalid information. A coach who is unaware of the rapid changes in sport equipment, for example, may be teaching inappropriate technique for the equipment her athletes are using.

KNOWLEDGE OF ACTIVITY

An extensive knowledge base about an activity is essential to a good QMD of that activity. The knowledge base for any activity comes from all the subdisciplines of kinesiology (Vickers, 1989). An elementary physical education teacher needs current information about motor skill development and the fundamental movement patterns related to those skills. This includes validated developmental sequences, ages of typical stages, rates of advancement, and changes in outcome measures of performance. The subdiscipline of motor learning contains important knowledge on practice schedules and stages of motor learning that affect all kinesiology professionals attempting to teach someone a new movement. Professionals must seek out the information they need from a variety of sources.

In secondary physical education, teachers and coaches need to know about the skills, strategy, and physical requirements of sports. They should update detailed knowledge of the individual skills and techniques. The goal or purpose of each sport skill needs to be determined (Gentile, 1972; James and Dufek, 1993). If the goal of a skill can be precisely defined, then the technique factors that lead to success in that skill can be more clearly identified for QMD. Recall that many models of QMD systems begin by defining the purpose of the movement (Arend and Higgins, 1976; Broer, 1960; Hay and Reid, 1982; Hoffman, 1983; McPherson, 1990).

A good example of a situational or strategic change in the goal of a movement involves the tennis serve. The first serve generally emphasizes placement, speed, or spin in order to put the opponent at a disadvantage, while the second serve's goal shifts toward greater accuracy to prevent a double fault. The importance of technique versus tactical errors changes depending on which serve is being attempted. In planning to qualitatively assess serve technique, tennis coaches use their knowledge of the sport to plan to observe both kinds of serves in practice and match play.

Sources of Knowledge

Three main sources of knowledge contribute to the prerequisite knowledge of an activity: experience, expert opinion, and scientific research. All three are important sources to consider in developing a prerequisite knowledge base for QMD. Two difficulties confront kinesiology professionals in this area: gathering the data from sometimes fragmented sources and weighing the evidence from each source.

Experience

Experience in any profession is invaluable, as evidenced by improvements in employability and salary with increasing years of experience. Most professionals develop positions on issues based on experiences with several patients or clients. This professional experience has the advantage of being population specific and environmentally relevant. Thoughtful coaches are likely to make valid generalizations from experience if their players are relatively homogeneous.

The weaknesses of experience as a source of valid information are that it is anecdotal and that it can be influenced by personal bias. Also, experience cannot control all the factors that may affect a particular issue, so professionals should search other sources to verify or refute their own conclusions.

Expert Opinion

The opinions of experts carry a lot of weight in many professions. People with a wealth of experience deserve the attention of their peers. The places to seek out this expert opinion are professional periodicals and professional meetings. In physical education, journals like *Strategies* and *Journal of Physical Education, Recreation and Dance (JOPERD)* feature articles written for professional practice. There are also professional magazines, newsletters, and websites (www.pecentral.org) that publish teaching tips and expert opinions. Sport-specific professional organizations also usually produce professional magazines (the Professional Tennis Registry [PTR] and U.S. Professional Tennis Association [USPTA]: *TennisPro, Addvantage*; USA Gymnastics: *Techniques*). Sport-specific popular publications (for example, *Tennis* or *Golf*) can be useful, often featuring interviews with successful coaches or teachers. In addition, professionals can seek out expert opinions personally at professional meetings or over the Internet or by consistently attending professional meetings and conferences where experts are often invited to speak.

The weaknesses of expert opinions are that they may conflict (figure 4.1), that like experience they are subject to bias, and that they often change over time. Teaching motor skills is frequently affected by the technique of a current champion or the agenda of the most prominent expert at that particular point in time (Hay and Reid, 1982). The latest rehabilitation protocol or technique in vogue may or may not be an improvement and may still be far from the ideal. For example, chance might bring to a coach a series of athletes with certain characteristics that influence the coach's professional opinion about the importance of a kind of training for all athletes.

Figure 4.1 Expert advice is often conflicting. This makes the task of building correct prerequisite knowledge for QMD difficult.

Both expert coaches and athletes have held opinions about performance in a particular skill that research has shown to be false. The history of sport is full of examples of incorrect notions about key points in many skills. Interested readers can find examples from track and field (Hay, 1993; Hay and Reid, 1982) and golf (Torrey, 1985). Skilled performers commonly have mistaken ideas about what is going on when they are performing. It is important to remember that championship athletic performance does not require expert knowledge about the kinesiology or QMD of the sport on the part of the athlete. The hallmark of a champion is to perform consistently and effectively, with little conscious attention to the considerable complexity of human movement.

Scientific Research

Another source of prerequisite knowledge for QMD is scientific research. Research in all the subdisciplines of kinesiology provides the most valid and accurate information available for basing decisions in QMD. Research can be either descriptive or experimental. For example, descriptive studies of injury rates of various populations of athletes or of modes of training would be relevant to the critical features selected for QMD. Experimental research on how different styles of teaching or teacher feedback affect learning also provides knowledge relevant to professionals planning QMD. Unfortunately, the controls needed in experimental research often limit its real-world application or ecological validity. It is also difficult for practitioners to gather relevant knowledge from research, for many reasons. Often research is published in journals with difficult and technical terminology. Reading and interpreting research may be difficult because of the abstract nature of the topics, the technical demands of some measurements, and the complex experimental or statistical designs. Advanced degrees and constant updating may be needed to accurately interpret research.

Some periodicals have the mission of bridging the gap between research and practice (Boyer, 1990). One of the first publications to attempt this was *Motor Skills: Theory into Practice*, but it ceased publication in 1985. Examples of good periodicals with scholarly research-based articles for practitioners are *Sports Coach, Track and Field Coaches Review, JOPERD, ACSM's Health and Fitness Journal*, and *Strength and Conditioning Journal*. In physical therapy, *Journal of Orthopedic and Sports Physical Therapy* has a reputation for readable and practitioner-friendly research. In addition, many newsletters associated with professional organizations or interest groups provide current information. Some examples are *Physical Activity Today* (Research Consortium, AAHPERD), *Physical Activity and Fitness Research Digest* (President's Council on Physical Fitness and Sports), and *Olympic Coach* (United States Olympic Committee Division of Coaching Development).

Research also has the problem of accessibility for many professionals, although powerful computers, data storage, and communication technology have begun to make it easier to acquire sport science information. This access increases the impact of good information, but unfortunately, it also increases the impact of poor research and incorrect expert advice. Computer-compiled databases of articles and abstracts are available in print, can be searched online, and are stored on CD-ROM. Some of the best databases are the Sport and Leisure Index, SPORT (Sport Information Resource Center [SIRC], Ottawa, Canada), Medline (National Library of Medicine), and ERIC (from the U.S. Department of Education). The online version of SPORT Discus is available in many university libraries.

The explosion of articles, blogs, and discussions on the World Wide Web has made access to information on human movement faster and easier. Unfortunately, what a professional wants is not information, but the knowledge that comes from systematic gathering of data in scientific research. The important thing to remember about the web is that it usually provides fast access to information but not necessarily accurate information (knowledge or the truth). People have a variety of motivations for posting information on the web, so professionals need to weigh the quality of evidence presented and the motivation or purpose of the web page. Readers must beware and look at the language and tone (and citations and statistics) of the web page for cues as to its potential accuracy or bias. Some scholarly journals have embraced a trend toward open access, with all articles freely available to the public all of the time or after a delay (6 months to several years). The best search engine to use to focus on published books and journal articles is Scholar Google (http://scholar.google.com/). For a directory of journals that provide free, full-text access, log on to the Directory of Open Access Journals (www.doaj.org/).

Professionals in human movement must weigh all the evidence for or knowledge about activities based on their own experience, expert opinion, and research to establish the most valid background knowledge for QMD. Scientific research should

Choices

Suppose you are a volunteer coach for your daughter's softball team. As you discuss the upcoming tryouts and draft with the other coaches, you wonder which aspects of hitting, catching, and throwing to look for in drafting the best team. What sources of knowledge are most useful for selecting players: your experience, other veteran coaches, or coaching literature? What source is most useful for teaching and training? Selecting equipment? Reducing the risk of injury?

be given the greatest weight because of the control, greater objectivity, and validity of the observations. The most important issue is to think critically about professional practice and the various sources of information. Some kinesiology scholars have argued that our professions should borrow the principles of evidence-based practice from medicine to provide a structure for weighing the strength of evidence in order to make decisions about interventions with clients (Amonette, English, and Ottenbacher, 2010; Faulkner et al., 2006; Knudson, 2005; Knudson, Elliott, and Ackland, 2012).

Terminology

One approach to studying and organizing knowledge of an activity is to compare the movements to those of similar activities. Although the terminology varies among authors and disciplines, many experts suggest that the study of human movement be based on classifications usually called fundamental movement patterns (Broer, 1960; Cooper and Glassow, 1963; Daniels, 1984; Philipp and Wilkerson, 1990; Wickstrom, 1983). This book uses the terminology in figure 4.2.

Fundamental movement patterns are broad categories of movements for a general purpose. Examples are walking, running, throwing, kicking, jumping, catching, striking, or carrying. Fundamental movement patterns can be adapted for specific purposes or combined with other fundamental movement patterns to complete a specific task. Some experts break movements into even smaller categories. A motor **skill** is an adapted fundamental movement pattern for a specific activity or goal. Typically skills are related to specific sports. Some skills related to throwing and kicking are football passing, baseball pitching, punting, and placekicking.

Techniques are skills with even more specific purposes. The selection of appropriate technique varies with each

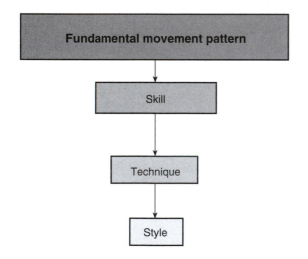

Figure 4.2 Hierarchy of terminology for describing human movement.

situation. A banana shot (a kind of placekick), for example, is appropriate in soccer, while the hang and hitch-kick techniques are appropriate in the skill of long jumping for maximum flight into the pit. Techniques are further divided into variations in **style**. Style aspects of movement are personal differences, rhythms, idiosyncrasies, or actions related to a specific performer. At this level, QMD is extremely difficult because the professional must decide whether minor variations in the movement detract from performance.

Whatever the level of qualitative assessment, critical features of the movement can be specified. An integrated QMD often uses critical features as the standards for observing, assessing, and improving movement. The term this book uses for short-term or long-term improvement in a motor skill is *performance*. In the subdiscipline of motor learning, performance often refers to short-term changes in motor skills;

learning refers to long-term or permanent improvements (Shea, Shebilske, and Worchel, 1993). This distinction between short- and long-term changes in a motor skill is important. Motor learning research has shown that long-term improvement often results from practice conditions that challenge the performer, sometimes making practice execution (short-term results) initially look worse (Schmidt and Wrisberg, 2008; Shea, Shebilske, and Worchel, 1993).

Critical Features

One of the most important areas of knowledge about an activity is establishment of its **critical features**. Critical features have been defined in many ways. Arend and Higgins (1976) define critical features as parts of a movement that can be least modified if the movement is to be successful. Critical features have also been defined as important aspects of performance that are related to the efficiency and effectiveness of the movement (Jones-Morton, 1990a). McPherson (1990) defined these features as statements describing specific body movements that are observable and that are then used to evaluate whether the key mechanical factors of the movement have been performed ideally. Critical features are also sometimes called *critical elements* or *critical performance elements*.

Critical features should be viewed as key features of a movement that are necessary for optimal performance. They are aspects of movement that are the most invariant across performers and are the least adaptable if the goal of the movement is to be achieved safely and efficiently. Critical features are the points defining good form that are used in teaching and should also be used to help determine the teacher's focus in the QMD of the skill. For example, the knee angle at deepest knee flexion (an indicator of the countermovement) in a standing vertical jump may be anywhere from 90 to 115 degrees because of range of motion, leverage, and muscle mechanical properties. Critical features for a conditioning exercise like the squat are knee angle, trunk lean, and neutral lumbar lordosis and are strongly related to training effects on muscle and risk of injury.

In other words, critical features are the most important aspects of a movement; they need to be performed in a certain way in order to be successful. We will see that it is useful for the movement professional to establish a range of correctness for the critical features and decide which are most important to performance. Integral to critical features is the idea of correct sequence. It does little good to have the correct critical features but the wrong sequence. (Imagine a softball batter starting the forward swing of the bat before the stride.) Chapter 5 shows how to organize an observational strategy to take advantage of the sequence or temporal organization of the critical features of the movement being assessed. Understanding of critical features and their order is important, as one suggested format for intervention is based on where in the movement sequence the error occurs.

Although it is helpful to express critical features in behavioral terms that can easily be evaluated visually, this may not always be possible. Some critical features of human movement are constructs or other abstract ideas rather than clearly defined

QMD **Technology 4.1**

Electronic databases of research articles are a critical technology supporting the preparation task of QMD. Accessing the most relevant research knowledge about a movement can be difficult given the huge volume of articles written. Professionals who can use electronic databases like SPORT Discus, Medline, or Scholar Google can quickly access important knowledge to plan for QMD. Being knowledgeable about professional terminology or common synonyms is also useful in searching within these databases. Good use of these words can help identify relevant sources among the ever-expanding body of knowledge.

biomechanical parameters. A teacher or coach may believe strongly in these ideas and may be struggling to find cues or ways to affect this aspect of performance. A good example related to several sports, among them baseball (batting) and pole vaulting, is fear. There are many ways coaches try to evaluate the attitude and confidence of athletes in situations in which fear of injury has a dramatic impact on performance.

Three issues can be used to justify the identification of critical features or desirable technique: safety or risk of injury to the performer, effectiveness in accomplishing the goal of the movement, and efficiency of goal attainment. Biomechanics is the primary sport science involved in identifying the quantitative underpinnings for the critical features of a movement. Biomechanical research provides kinematic (range of motion, body angles, length) knowledge on skilled and other performers. This kind of research can be of practical use because teachers and coaches may be able to observe some of these body or joint movement variables (Hudson, 1990a, 1990c).

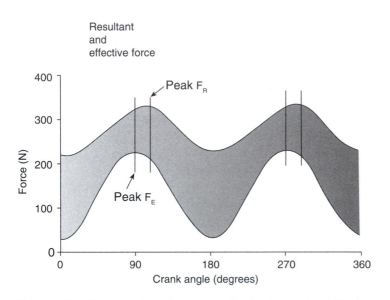

Figure 4.3 Mean resultant force and effective force applied to the pedals of a bicycle. The shaded region shows the amount of force that is not effective in rotating the pedals.

Reprinted by permission from Lafortune and Cavanagh 1983.

Another branch of biomechanical research is kinetics, the study of how forces and torques create the movement. Kinetic data determine which technique effectively applies force or provides smaller loads to tissues and may carry a lower risk of injury. For example, biomechanical research on the effectiveness of force application in cycling has found that only about 76% of pedal forces are used in driving the bike (Lafortune and Cavanagh, 1983). Figure 4.3 illustrates the total force on the pedal and the effective (rotary) force on the pedal. The shaded region shows the forces that are not effective in creating rotation. The complex interaction of muscle mechanical properties and musculoskeletal geometry may preclude the body from applying forces in a mechanically optimal direction in many movements. Even more important is using knowledge of kinetics to design exercises and training that provide an effective overload of body tissues without a high risk of injury.

Safety Rationale

A professional must decide if a particular action or technique is safe. Does it have a low probability of acute or chronic injury? These safety decisions depend on many factors, such as the age of the performer, level of conditioning, injury status, previous activity, and rest. For example, how much follow-through in pitching is desirable to help prevent shoulder injury? How much knee flexion is safe and effective when someone is landing from the vertical jump? Greater knee flexion clearly decreases the peak force from the ground on the body in landing, but as knee flexion increases, the shearing forces on the knee increase. So the amount of knee flexion needed to cushion landings could vary among individuals with knee injuries.

Effectiveness Rationale

Each human movement has an associated goal or outcome. The appropriateness of a person's form can be judged based on its effectiveness in achieving the movement goal. Principles of biomechanics can be used to evaluate whether a particular movement pattern or form is optimally effective in achieving a particular outcome. Does a particular body posture in a motor skill allow for desirable stability or the application of force in the direction of the target?

Many striking or throwing techniques employ weight shifts and linear motion that flatten the arc of the hand or implement toward the target, making the athlete more accurate. The linear momentum of the weight shift is transferred to the rotary motions (angular momentum) of the upper body. This weight shift is crucial for the effectiveness of these skills. Some movements have several goals that complicate the definition of effectiveness; consequently, there must be compromise between goals in defining critical features. For example, the weight shift in a golf swing is limited because of the greater requirements for accuracy in the swing to hit the ball. Not all goals are inversely related, but when they are, a compromise must be struck between the two competing performance objectives.

> **»KEY POINT 4.2**
>
> All human movements have critical features, the key factors that are necessary for optimal performance. Critical features are based on the safety, effectiveness, and efficiency of the movement. Their exact sequence or coordination is also very important.

Efficiency Rationale

The appropriateness of a movement technique can also be evaluated on its efficiency. Efficiency of human movement involves both minimizing the metabolic cost and maximizing the use of mechanical energies in achieving the movement goal. Some logical inefficiencies may be easy to spot. A large pause or hitch in a movement or the excessive up-and-down movement of a distance runner may be easily identified as wasted energy. Some small variations in technique may be exceedingly difficult to evaluate, however, particularly in terms of which is more efficient. Biomechanists have had difficulty establishing how to document the mechanical energy used to create a movement (Aleshinsky, 1986; Cavanagh, 1990; Cavanagh and Kram, 1985; McGibbon and Krebs, 1998). This level of complexity of documenting the work and power flows within a complex biomechanical system is well beyond the typical external work and power flow measurements commonly used in physiological and conditioning research. The implication for professionals is that they cannot currently expect explicit answers to important questions about optimal efficiency of technique when they plan for QMD.

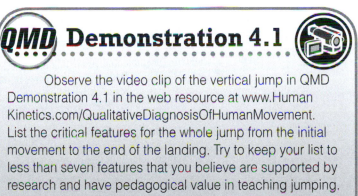

QMD Demonstration 4.1

Observe the video clip of the vertical jump in QMD Demonstration 4.1 in the web resource at www.Human Kinetics.com/QualitativeDiagnosisOfHumanMovement. List the critical features for the whole jump from the initial movement to the end of the landing. Try to keep your list to less than seven features that you believe are supported by research and have pedagogical value in teaching jumping.

Rationales in Action

Human movement is highly dependent on the environment or context, so the importance of safety, effectiveness, and efficiency can vary within and between techniques. For example, during preseason workouts, a cross-country coach noticed that an athlete appeared to have excessive rear foot pronation as he ran. Inspection of the athlete's shoes tended to confirm this diagnosis. The coach encouraged the athlete to change to shoes with more medial support and frequently change training routes to minimize the influence of a consistent slope of the terrain and streets he trained on.

The safety rationale, or reducing the risk of injury, was deemed most important in this case. The athlete may have felt that he was a better runner (more effective) with his old shoes, but the coach's knowledge of the rear foot motion and the deterioration of running shoes provided a powerful rationale for changing the athlete's equipment. Another safety example is changing one's walking gait in icy conditions. Even nonathletes shorten their stride lengths considerably to make up for the loss of horizontal friction forces.

Range of Correctness for Critical Features

One should define critical features in a skill as precisely as possible, bearing in mind that a range of correctness is needed to accommodate the variations inherent in people. Schleihauf (1983) hypothesized that the range of effective movement solutions varies with the nature of the movement. For example, the weight shifts are very different for a golf swing and a fencing lunge. Knowing the range of correctness of critical features makes evaluation and diagnostic decisions easier in QMD. Good examples are differences in stance and weight shift in throwing and striking. Weight shift is a critical feature in many skills because a weight shift followed by hip and trunk rotation is an efficient and effective way to generate speed in the upper extremity.

In baseball hitting, the emphasis is on bat accuracy because of the difficulty in hitting an unpredictable pitch. Baseball coaches should be aware that open, square, and closed stances may all be appropriate, but the step of the front foot should be a short distance (3 to 8 inches [8 to 20 centimeters]) toward the pitch. A forceful overarm throw, however, should have a square stance with a longer leg drive. The range of correctness that can be observed for QMD of high-speed throwing may be a forward step from half of standing height (Roberton and Halverson, 1984) to over 90% of standing height, typical in baseball pitchers (Atwater, 1979; Hay, 1993; Montgomery and Knudson, 2002). If research can establish desirable ranges of correctness that are observable, the reliability of QMD should improve.

There are two problems in defining the range of correctness: conflict between expert opinions and conflict between biomechanical theory and research. As already mentioned, experts in a particular area (sport, conditioning, therapy) often have conflicting opinions on the best practice. The field of biomechanics has only recently begun to attempt to define optimal performance in a particular task for a given environment, so the field is ripe for different theories and interpretations of research data.

There have been major limitations to the development of theories of optimal human movement. The complexity of the neuromuscular and musculoskeletal systems, experimental technique, computing power, optimization theory, psychological factors, and theoretical research have all limited the answers to the question of what is good or optimal form. Biomechanists continue to struggle to understand how movements are optimized.

Biomechanical research has shown that many optimization criteria (minimizing energy expenditure, muscle stress, or acceleration) can predict the overall patterns of electromyography and movement kinematics exhibited by subjects. Indeed, some would argue that the inherent variability of the body's physical abilities and of the learning process makes it impossible to establish one ideal form for a particular movement (Brisson and Alain, 1996; Duck, 1986; Gentile, 1972; Hay and Reid, 1982; Norman, 1975; Spaeth, 1972). What might be possible with biomechanical research is the documentation of a range of desirable form suited to human and environmental constraints (Schleihauf, 1983). Skills in a closed environment will likely have a narrower range of correctness than skills in an open environment (Higgins and Spaeth, 1972).

Most of the examples of critical features used so far have related to the movement of the body itself. We will see in chapter 7 that motor learning research calls this knowledge of performance (KP). It is important to understand, however, that critical features can also be related to the outcome of the movement, called knowledge of results (KR). For example, the initial trajectory of a typical basketball shot is between 49 and 55 degrees above the horizontal (Knudson, 1993). This angle of release is KR and may be a critical feature of shooting that coaches and teachers should plan to observe. These angles of release provide a range of correctness for the shot because they offer a good compromise between the angle of entry and the speed needed to reach the basket (Knudson, 1993). Many activities have distinct outcomes that may give an athlete an advantage and consequently should be evaluated in a QMD of the activity.

Critical features and their sequence can take the form of complex ideas, subtle points, common knowledge, professional jargon, or precise values based on modeling or research. Whatever the source or type of critical feature, the professional needs to gather teaching cues that correspond to it. Teaching cues translate critical features (usually expressed technically) into easily understood or highly descriptive language. For the professional to communicate effectively with the performer, critical features should be expressed in behavioral terms. This process involves the collection and organization of a wide variety of cue words and phrases. The professional then has a repertoire of cues to help communicate a point. This flexibility is important because people often interpret cues in different ways. Chapter 7 reviews the important points about providing appropriate intervention to improve performance.

Motor development knowledge is an important prerequisite to QMD of skill. Teachers should know the relevant developmental sequence people usually progress through and the typical ages for key milestones. Motor development knowledge is useful for knowing what to look for in early stages of motor learning (although many adults do not reach the mature level of many fundamental movement patterns). It also provides developmental milestones and changes that are useful in teaching adults new skills. Motor development knowledge, along with information on common errors, tells the professional some of what to look for in QMD.

> **» KEY POINT 4.3**
>
> QMD is easier and more reliable if the professional can establish a range of correctness for the critical features and technique points of a movement. This is often not easy because desirable and optimal movement patterns have not been established.

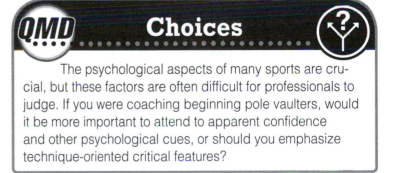

QMD Choices

The psychological aspects of many sports are crucial, but these factors are often difficult for professionals to judge. If you were coaching beginning pole vaulters, would it be more important to attend to apparent confidence and other psychological cues, or should you emphasize technique-oriented critical features?

Taxonomies and Common Errors

The physical education literature contains many books that provide basic skill information for teaching sport skills. Some books provide an overview of key skills for many sports; others provide detailed analysis of the skills of a particular sport (for example, the Human Kinetics *Steps to Success* activity series). In AAHPERD and the related district and state organizations, there are groups interested in basic instruction in physical activities.

Despite all this literature and interest, however, there is still a need for taxonomies of critical features of fundamental movement patterns and sport skills (Hoffman, 1974). We will see later that some books provide general technique points and cues for teaching motor skills, but often these are just recent versions of old opinion-based instructional books. It would be most desirable to get experts together with scholars from many subdisciplines of kinesiology and sport coaches to create universal or official taxonomies of critical features for the skills and fundamental movement patterns of specific sports. These taxonomies should also include common errors and an exhaustive list of cue words or phrases. Statistical techniques (Delphi study) exist to bring together common themes of professional opinion about issues and concepts.

QMD of novices is bound to be more effective if professionals are knowledgeable about the most common errors of the skills they teach. Teachers and coaches can also provide a variety of feedback if they are familiar with several of the most effective cue words or phrases for each critical feature and the associated **common errors**. In teaching softball batting, for example, it would be desirable for experts from teaching, coaching, and softball research to get together and identify the critical features of batting, the common errors, and the best teaching cues. Ideally, prospective research of teaching interventions using these critical features would provide validation of their utility with a wide variety of movers. Unfortunately, AAHPERD, sport governing bodies, and most coaching organizations have not tried to establish authoritative taxonomies of technique and common errors that would facilitate teaching and preparation for QMD.

KNOWLEDGE OF PERFORMERS

Extensive background knowledge about students, athletes, dancers, or clients is also needed to prepare for QMD. Performers come to an activity with a wide variety of abilities based on genetics, anthropometrics, age, gender, experience, training, and skill-related fitness components. The more knowledge the coach can gather about their mental and physical abilities, the better the coach can assess and evaluate performance.

For example, knowledge about the upper extremity strength limitations of young basketball players could be used to prescribe additional strength training, establish strength guidelines for initiating certain shooting techniques, or justify the purchase of developmentally appropriate equipment. Research on the forces and concussion risks of heading in soccer (figure 4.4) resulted in changes to the size and mass of the ball. The areas of motor development, anthropometrics, and kinanthropometrics all contribute valuable information on typical changes in the characteristics of humans in motion.

Knowledge about performers can serve as the basis for equipment and facility modifications that speed up the development of correct technique. Physical educators can select age-appropriate balls, while physical therapists with precise information on typical strength or flexibility of patients with a specific injury can select appropriate rehabilitative aids. Better knowledge of these functional capacities at various stages of rehabilitation can be used to improve diagnostic decisions in treatment. Potential sources of this information about clients include professional literature, clinic records, and communication with professional organizations or peers.

There are many good examples of dramatic physical and neuromuscular changes in people as a result of normal development. The cognitive development of children

Figure 4.4 Physical limitations of performers and appropriate equipment are important sources of information in QMD.

in early primary grades does not typically allow them to grasp and use abstract strategic information. So in the first few years of school, physical education teachers should probably not try to provide feedback related to a complex team tactic that fulfills a strategic game plan. One of the most dramatic changes in a short period of time is puberty. Over the course of a few months, adolescents may find it harder to coordinate their new longer and larger body segments. One adolescent may experience a minor improvement in strength, while another may have substantial increases in strength and stamina.

The science of motor development has begun to study components of movement (rate controllers) that influence the development and coordination of movements. Rate controllers may be the slowest-changing components in children, while aging adults may lose them more rapidly than other components of movement (Haywood and Getchell, 2009). Professionals must also adjust the kind of feedback they give and the practice they prescribe to each client's stage of motor development. Clearly movement professionals must be knowledgeable about the physical characteristics of the people they work with, because physical status can dramatically affect the level of performance that can be achieved.

> **» KEY POINT 4.4**
>
> Knowledge of the physical, cognitive, and emotional characteristics of performers is important in preparing for QMD.

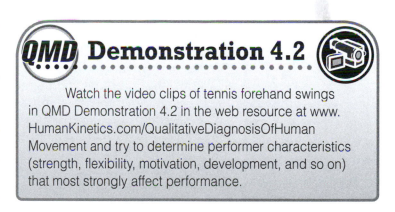

QMD Demonstration 4.2

Watch the video clips of tennis forehand swings in QMD Demonstration 4.2 in the web resource at www.HumanKinetics.com/QualitativeDiagnosisOfHuman Movement and try to determine performer characteristics (strength, flexibility, motivation, development, and so on) that most strongly affect performance.

KNOWLEDGE OF EFFECTIVE INSTRUCTION

All kinesiology professions involve some teaching of human movement. Teachers of motor skills need to be aware of pedagogical and motor learning research on appropriate and effective presentation of information. There is a natural connection between

PRACTICAL APPLICATIONS
Preparation Task of QMD

In the preparation task of QMD, coaches must maintain up-to-date knowledge of the performers they coach. In youth sports, this means a good working knowledge of motor development and exercise physiology. Imagine you are the coach watching a child swinging a bat that is obviously too large and heavy for her. Is the child's swinging with a bent arm a problem of technique, immature level of performance, or strength?

The science of motor development describes the typical stages and changes in the development of motor skills. Arm striking patterns tend to develop from  an overarm to a sidearm pattern. Are you familiar with the three major stages of sidearm striking proposed by Wickstrom (1983)? Does the child's performance look as though it was created by an arm-dominated action, a simultaneous body action, or a sequential strike? Motor development literature is extremely important to the understanding of the status of a person's movement, common actions, and typical changes with development.

When this knowledge of motor development is combined with an understanding of strength and training (exercise physiology), youth coaches have powerful tools for understanding the physical limitations of children and how to overcome them. The swing illustrated is not likely influenced by poor technique in the readiness and preparatory phases of the movement because she did not have to track a moving ball coordinated with the swing.

If the problem is strength, several intervention strategies could be effective. The child could choke up on the bat, use a lighter bat, and work on increasing upper body strength. Do you know about developmental changes in strength (Malina and Bouchard, 1991; Nelson, 1991; Roemmich and Rogol, 1995)? Did you know that prepubescent strength training results in strength gains related to neural factors rather than muscle hypertrophy (Kenney, Wilmore, and Costill, 2012) and that there are guidelines for youth weight training (Faigenbaum et al., 2009)? If you know that the child is strong enough, how should you try to change the technique? Should intervention focus on achieving the next developmental level, or should the child try to emulate adult form? The practical application of QMD is clearly an interdisciplinary process that must integrate many perspectives affecting a particular performer and performance. Consistent review of professional literature is essential to the preparation task of an integrated QMD.

presenting movement information and giving similar information as feedback that will be effective intervention. To communicate effectively, the professional and the performer must share a common vocabulary. To build this vocabulary and teach the movement skills, professionals must follow good instructional procedures. The two major factors in this teaching process are presenting appropriate information and presenting it effectively.

Presenting Appropriate Information

An essential task in presenting motor skill information for initial instruction and as feedback is translating the critical features into teaching cues. The teacher or coach must look for the most appropriate language to communicate the critical features. The performer's age, experience, and interest level may affect the choice of cue words or cue phrases to communicate skill information. Movement professionals should strive to collect and update a variety of teaching cue words and phrases to get through to the people they are serving. Because people have different backgrounds and make different inferences about words, it is helpful to be able to provide the same technique idea in several different ways. A good cue phrase for the preparatory weight shift in golf is "Shift your weight to your rear foot," while another golfer might better respond to "Rock your weight back" in the backswing. Yet another golfer might misinterpret this latter cue by interpreting "back" to mean toward the heels rather than opposite of the swing direction. "Straight" is a cue word a golfer could use to remember to keep the upper arm comfortably straight during the swing.

> **» KEY POINT 4.5**
>
> There is an important connection between teaching motor skills and later assessing them qualitatively to improve performance. A common language base and understanding of the critical features of a movement make QMD easier and more effective.

Authoritative taxonomies of cue words for critical features of sport skills would be invaluable in physical education and coaching. Unfortunately, kinesiology scholars and experts have not come to an agreement on the important critical features and cues for these skills. Good sources for ideas for customizing teaching cues for different clients include Dunham, Reeve, and Morrison (1989), Fronske (2012), Fronske and Dunn (1992), Fronske and Wilson (2002), Kovar and colleagues (1992), Landin (1994), and Masser (1985, 1993).

Teaching cues are among the best form of communication and feedback for performers. One key word, usually a verb (for example, "coil," "step," or "oppose"), can communicate the essence of the action of a critical feature. These words can be derived from the execution phase of Dunham's model (1994) if these phrases are written behaviorally. A cue phrase expresses the critical feature in behavioral terms. A figurative or descriptive phrase ("Arch your back as you block on the horse") is often better than a literal description of the action. Most diving coaches know that "Wrap your arms" is often a better cue phrase for a twisting dive than "Decrease your transverse plane moment of inertia," even though the latter is more accurate. Another example is the figurative cue to "scratch your back" with the tennis racket when you serve. Skilled tennis players do not literally scratch their backs with the racket, but they do drop the racket head behind their backs to make it easier to rotate the racket. Remember that cues need to be relevant to the performer. Small children may relate better to "Reach into the cookie jar" than to "Follow through high" as a cue for the follow-through in a basketball shot if they have a cookie jar in their home.

Good cue phrases communicate the essence of a critical feature or technique point concisely so that the performer can remember it during practice. Try to keep these behavioral descriptions to less than six words, the limit of most people's short-term memory. Remember that you can attach more information to the cue later and add finer points or details to build on the essentials. It is ironic that children can often teach each other a game or movement and get into action much more quickly than when taught by physical education teachers or coaches. Good teaching or coaching introduces the critical features of a skill quickly and in terms the performers can understand and relate to.

Housner and Griffey (1994) suggest that learning cues can be categorized based on verbal, visual, and kinesthetic/tactile information. These categories can help instructors select the most appropriate information for the needs of a specific learner. Verbal information presented too metaphorically may confuse a young person but can be quite helpful to adults. Tactile information and visual demonstrations may help young children more than verbal skill instruction. Auditory information is important in QMD because, many believe, it can help the performer. Information on the rhythm of a movement may be an important addition to more traditional verbal or visual cues in helping students understand movement.

Using cues or simple word labels as substitutes for a more complex description of a movement is verbal pretraining (Christina and Corcos, 1988). The phrases "hit and step" in baseball and "right left" for a basketball layup carry a great deal of information. Verbal pretraining is important for conveying complex information about human movement. The use of cue words or phrases in instruction can be helpful in the QMD process.

Some authors have begun to address miscues, common teaching cues that are truly incorrect or can be easily misinterpreted (Adrian and House, 1987a, 1987b). This is a tricky subject because sometimes cues that are incorrect still result in improvements or the correct movement. For example, one method of intervention is to exaggerate or overcorrect a problem area of performance, which often brings about smaller, desirable change in the person's technique. It is clear that some cues, although not literally accurate, have a history of success with performers. Two classic examples are "Run on your toes" for sprinting and "Throw by your ear" for throwing a baseball. More research is needed to determine how age, skill level, and aspects of cues interact to foster communication of movement information and aid learning of motor skills.

A variety of cues are also needed because of cognitive and perceptual style differences among performers of different ages. If a QMD of a novice in elementary school and an adult revealed the same problem, the cues that would best communicate to each learner would most likely be different. The small child with a short attention span would probably respond well to a figurative cue such as "Open a window when you toss the ball." The adult might respond equally well to a literal cue if he could attach the desired meaning to it. Until more research is done to identify the most effective cues, professionals should strive to develop and refine cues and to share them with others.

One of the best methods of writing cue phrases may be the method proposed by Morrison and Reeve (1993), based on Vickers' (1989) method of writing technical or qualitative objectives. The Morrison and Reeve approach uses the four parts of Vickers' behavioral objectives: action, content, qualifications, and special conditions. The action is a verb that expresses the desirable motion. The content points to what is doing the action. Qualifications specify how success can be gauged. And special conditions can be added to the teaching cues if more information is needed to evaluate the performance. This structure is illustrated in the cue phrase for overarm throwing "Swing your arm forward, level to the ground." "Swing" indicates the action; "arm" indicates the content; "forward" qualifies the action; and "level to ground" is a special condition defining the movement.

Relevant cues for movements can be created for each phase of a movement (preparation, execution, and follow-through), with variations for performers of different ages and levels of expertise (Vickers, 1989). This identification of special cues for various ability levels has also been emphasized by Strand (1988) and by Abendroth-

Smith, Kras, and Strand (1996). Future research should help professionals develop a variety of effective and developmentally appropriate cues.

Cues are highly effective as teaching and intervention tools. When kinesiology professionals know the best cues for the movements they are teaching, they can plan effective instruction that also sets the stage for effective QMD. Cues used in instruction are a common language that supports subsequent QMD of clients during practice activities. One relevant approach is the format for preparing teaching information in motor skills proposed by Dunham (1986, 1994). This format breaks movements into two phases, anatomical and motor, rather than the traditional three phases of preparation, execution, and follow-through. The anatomical phase occurs just as the skill performance starts. The motor phase consists of the major actions and the follow-through.

> **》KEY POINT 4.6**
>
> Cue words and phrases are an effective way to present information to performers. Cues need to be concise, accurate, and appropriate to the age and ability level of the performer.

Teaching cues used in the motor phase can be phrased behaviorally, as suggested by Morrison and Reeve (1993), to create a task sheet. Figure 4.5 shows such a task sheet for the soccer throw- in. Using task sheets for teaching physical education with examples from many activities has been published (Dunham, Reeve, and Morrison, 1989). The first word in the execution or motor phase can be used as the key cue word for feedback. These action words may also serve as cues or verbal pretraining for the performer (Christina and Corcos, 1988).

Task sheets tell performers exactly what to do for the execution of a movement. Additional details can be added to the teaching cues as they become necessary. As a part of their research, Pinheiro and Cai (1999) used a format that differs only slightly from the Dunham (1994) approach. Task sheets like those in figure 4.5 can be posted on the gym wall to augment instruction and provide a structure for observation (chapter 5). Which kind of task sheet used depends on the skill being evaluated and the observational strategy preference of the professional.

The quality of task sheets can be evaluated using a process proposed by Morrison and Reeve (1993). They developed a scale for evaluating the components of effective movement instruction materials or task sheets. The two most important content factors (behavioral cues and their sequence) are weighted the most in their system. Once instructors have used the score sheet a few times, they will not need to rely on explanations for scoring. A task sheet effectively provides the basis for a teaching loop. It is first used to present information to students; then it is used as the basis for movement evaluation, and finally it provides for feedback directly related to what was taught and evaluated. Thus, consistency between teaching, QMD, and the evaluation or grading of movement technique can be established in physical education (Pinheiro and Cai, 1999).

It is important to have clear teaching cues for instruction. Not only do they aid the learner in acquiring the skill; they also serve as the criteria against which the teacher judges the skill and the performer interprets feedback. Any discrepancy between what the professional teaches and the feedback can confuse the student. An example is the tennis instructor who taught her class

QMD Technology 4.2

If task sheets appeal to your preferences for movement instruction and your planned intervention in QMD, you might consider including photographic images on them. Simple images of key technique positions can be captured with a digital camera and used for task sheets and as feedback to athletes. Obtaining a good image of the exact instant or technique point of interest might take some time but could be worthwhile for facilitating your QMD and communicating with clients who have a preference for visual information.

Soccer

Observer's name: _____   Date: _____

Skills to be diagnosed: _Throw-in_____

Performer: _____

Illustration	Critical elements	Yes	No
Preparatory phase 	1. Face the target. 2. Feet shoulder-width apart or stride. 3. Feet **behind** the sideline. 4. Secure the ball **overhead** with both hands. 5. Fingers spread to form "**W**" on the ball.		
Execution phase 	1. Bring ball behind the head. 2. Body **arched** backward. 3. Body uncoils to release ball. 4. Ball released forward by both hands. 5. Release ball "**up in the sky**." 6. Feet **in contact** with the ground.		
Follow-through 	1. Feet **still in contact** with the ground. 2. Arms extended forward. 3. Body **follows direction** of throw.		

Figure 4.5   Observational model proposed by Pinheiro.

Reprinted by permission from Pinheiro 2000.

the forehand volley but forgot to tell students to keep their wrists stiff on contact. Her first correction was to keep the wrist stiff. How were the students to know that this critical feature was important if they were not taught it?

This poor preparation for instruction illustrates the connection between instruction and QMD. An explicit method of information preparation and presentation would have avoided or minimized the volley problem and allowed for greater learning. Also, a method of checking information presentation could have caught the omission. If the skill or task sheet for the forehand had been prepared properly, the process of instruction and QMD would have proceeded more smoothly. How does a professional take appropriate skill information and plan effective instruction? Although this text is not intended to explain teaching methods, the next section deals with the actual presentation of skill information to performers because it relates to QMD.

Effective Presentation of Information

Once appropriate instructional information is identified, the professional needs to present that information effectively. Pedagogical research has shown that highly effective teachers present information to students more efficiently than less effective teachers (Siedentop, 1991; Werner and Rink, 1987). The starting point for effective presentations is the preparation of teaching materials. If the skills to be taught can be organized in a systematic way and presented consistently, students can learn under any of a variety of teaching styles (Mosston and Ashworth, 1986; Siedentop, 1991).

This organization of information about motor skills for instruction is important because poorly taught skills will not be learned and will require greater QMD. Intervention will also become more difficult. Research on information presentation shows that certain components of instruction are essential for effective teaching. Most good teachers provide good demonstrations, explicit verbal explanation, and summary cues; they also direct student attention to important factors and have a way to check for student understanding (Brown, 1995; Graham, 1988; Graham et al., 1993; Harrison, 1999; Kwak, 1994; Masser, 1985, 1993).

Many pedagogy experts see the effective presentation of motor skill information and QMD as part of effective teaching or coaching. Graham (1988) tied the idea of good teaching presentation to the use of appropriate feedback and organization of the teaching environment. Kwak (1994) emphasized the importance of cognition in learning. Cognition is a cornerstone of understanding and using qualitative feedback. Masser (1985, 1993) provided evidence for the use of refining tasks in skill acquisition. This refining process relates instruction directly to suitable feedback. Reynolds (1992) describes monitoring, the process of following students carefully during class time so that feedback on performance can be tied directly to instruction. Finally, Harrison (1999), in a 10-year review of her research findings on teaching, supported these assertions on teaching for learning. She found that the more successful trials students had, the better they acquired a skill. She suggested that learning activities must be chosen carefully and that the learner must participate fully in the learning situation.

These results may seem obvious, but they explain why some teachers are more successful than others. Teachers and coaches can likely improve the effectiveness of QMD by planning for effective presentation of skill information and practice that set up effective QMD. All kinesiology professionals can improve their QMD of motor skills by studying pedagogical research on the effective presentation of motor skill information. For example, sports medicine professionals often need to teach new movement patterns to replace painful or dysfunctional techniques developed following injury. What good is advanced knowledge about human movement if you cannot communicate effectively to your clients?

PREPARING FOR THE NEXT TASK

In preparing for QMD, the professional should be aware that the next task is the observation of human movement. A key component of observation in QMD is the use of a systematic observational strategy (SOS), a plan to gather relevant information about a movement. Information from many subdisciplines of kinesiology affects the selection of the best SOS for a particular movement. In preparing for QMD, the professional may begin to plan a systematic observational strategy. If the movement is fast and complicated, a professional may practice the anticipated SOS to improve QMD on the job. This practice may help to create positive visual habits in the actual observation of live movement. Unfortunately, the question presented by Kretchmar, Sherman, and Mooney (1949)—namely, what are the best visual habits for observing human movement?—is still not answered. The next chapter provides a detailed review of important factors in the second task of QMD: observation.

SUMMARY

The first of the four tasks of QMD is preparation. Kinesiology professionals preparing for QMD must weigh evidence from the many subdisciplines on the activity or movement, the performers, and effective instruction. Professionals must continuously update their knowledge in all of these areas. Professionals weigh interdisciplinary evidence from three major sources: experience, research, and expert opinion. Since QMD is often part of the larger process of teaching motor skills, professionals must keep their knowledge about appropriate and effective presentation of information up to date. They often prepare for the observation of human movement, the next task of QMD.

Discussion Questions

1. In preparing for QMD, what factors should you weigh in evaluating the importance of various sources of information on human movement?

2. What sources of knowledge should be emphasized by professional teachers or coaches in preparing for QMD?

3. What factors are most important in selecting the *critical features* of a movement?

4. What subdisciplines of kinesiology are most important in selecting critical features of movements?

5. What factors are most important in establishing the range of correctness for the critical features of a movement?

6. In a world of rapidly changing knowledge and technology, how important are specialized publications that bridge the gap between theory and practice or that summarize the application potential of scholarly research?

7 How can professional organizations help their members integrate the many subdisciplines in preparing for QMD?

8 How can the integration of preparation with *effective pedagogy* (instructional strategies) make subsequent QMD more effective?

Observation: Developing a Systematic Observational Strategy

A friend has asked you to videotape her tennis match in the finals of a regional tournament to serve as a source of advice on her technique. What key strokes, player movements, and ball motions would be useful to capture on video for the tennis player? Should both players be in the field of view? What vantage points would provide the best view of the action, and what actions would be missed from those vantage points? These are some important questions to think about in planning to observe human movement within qualitative movement diagnosis. The videotape images, like the visual information that coaches collect in qualitative movement diagnosis, must be carefully selected to get as much relevant information about performance as possible.

> You can observe a lot just by watchin'.
>
> *Yogi Berra*

Chapter Objectives

❶ Explain how to compensate for perceptual limitations by planning a systematic observational strategy.

❷ Identify the key elements of a systematic observational strategy.

❸ Identify several effective systematic observational strategies.

❹ Explain how all the senses can be integrated to improve observation.

The observation of human movement is the second task of professional qualitative movement diagnosis (QMD). Visual observation is an important part of QMD, but there is much more to observation than is implied by the opening quote from Yogi Berra. The quote suggests that movement observation is an easy, natural task, but this is not true. Not only is there much more to it than "just watchin'," but observation also involves the use of the other senses that can contribute to the task of QMD. Once professionals have organized prerequisite information in the preparation task of QMD, they use this information to create a **systematic observational strategy (SOS)**: a plan to gather all the relevant information about a human movement. This chapter reviews several proposals for observational strategies, identifies key elements of an SOS, and discusses the integration of all the senses that contribute to the task of observation in QMD.

An SOS is necessary in QMD for many reasons. First, large amounts of information about the movement from many subdisciplines must be condensed into critical features that will be observed. Second, sensory and cognitive limitations (discussed in chapter 3) can be accommodated by a good SOS. Third, the knowledge and expectations of the observer strongly influence what is observed. Edgar Dale (1984: 58) pointed out the importance of prerequisite knowledge in observation: "We can only see in a picture what our experience permits us to see." In other words, professionals must have a background that allows them to know what to look for. A good SOS enables professionals to gather appropriate and unbiased multisensory information on a person's performance of a motor skill.

QMD Demonstration 5.1

Select one of the video clips of an activity that is unfamiliar to you (lacrosse or rugby) in QMD Demonstration 5.1 in the web resource at www.HumanKinetics.com/QualitativeDiagnosisOfHumanMovement. Observe the movement twice, and list what appears to you to be the performer's good and bad elements of technique. Was it easy or difficult to determine the performer's skill with unguided visual observation? Why?

The goal of an SOS is to provide a platform to gather relevant information on the status of a person's movement performance. All kinds of information must be attended to and apprehended. Remember that observation includes all sensory information a teacher or coach can garner about human movement. In the past, the predominant thinking has limited observation to identifying errors in performance based on some model of "good form" envisioned by the teacher or coach. Instead, the task of observation should be broad enough to encompass many modes of sensory perception but limited to the collection of this multifaceted information. The observation task of QMD includes only collecting and interpreting information, not evaluating or diagnosing quality.

This discussion separates the information gathering or perceptual task (observation) from the diagnostic task (evaluation and diagnosis in our model) to emphasize the two different processes. Radford (1989) reviewed the research on observation in kinesiology and concluded that observation and subsequent decision making must be conceptualized as separate. Some scholars argue that these two tasks are related and can occur at the same time (Pinheiro and Simon, 1992). But *observation* is defined here as the process of gathering, organizing, and giving meaning to sensory information about human motor performances. Most information-processing models consider this process separate even if they use direct mapping from the perceptual components to the decision areas.

»KEY POINT 5.1

The second task of QMD is observation. In this task, the observer gathers information from all the senses about a movement using a systematic observational strategy (SOS). There are several ways to organize an SOS for the diagnosis of human movement.

A simple view of observation of human movement essentially involves two main decisions: *what* (focus) to observe and *how* (a plan to observe). The preparation task of QMD identified the critical features of the movement. It also provided a format for observation. A discussion later in this chapter uses the Gangstead and Beveridge (1984) and Dunham (1994) observational models, which were presented in chapters 2 and 4. Use of models such as these may improve observation of human movement (Pinheiro and Cai, 1999). Critical features are the focus of the SOS, so observing them is crucial to gathering useful information. The next section summarizes key proposals from the kinesiology literature on how to observe human movement. The *how* of an SOS is more complicated than the *what* and may be different for different professionals.

This chapter began with a quote from Yogi Berra. The following quote better illustrates that perception is a large part of the observation process and that it is subject to change depending on many factors. A change in our perception of an event can enhance or diminish our powers of observation.

> Whilst part of what we perceive comes through our senses
> from the object before us, another part (and it may be the larger part)
> always comes out of our own mind.
>
> *William James*

Before we continue with our discussion of observation, it is important to reinforce the idea that observation relies strongly on the perceptual abilities of the observer. Perception is an active process, and we bring a great deal of our own informational and organizational structure to the process. This is one reason why, in many cases, we see things differently than others do. Figure 5.1 illustrates the Necker cube (for Swiss researcher Louis Albert Necker), a visual illusion based on a simple ambiguous drawing, which also shows how people can perceive things differently.

Look at figure 5.1 and note how your brain perceives the 12 straight lines as a three-dimensional object. Note what imaginary surface of the cube appears nearest to you. See if you can see another interpretation of the cube by focusing on different portions of the image. Why do you think one interpretation is easier for you to see than another? The Necker cube alerts us to the fact that visual perception is an active process and can lead to different interpretations of reality based on what the observer brings to the event.

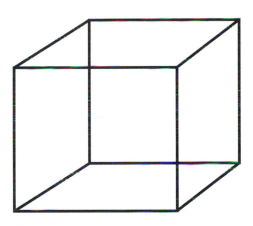

Figure 5.1 The Necker cube illustrates how ambiguous visual information can be interpreted differently by the observer.

Reprinted by permission from Bradley and Petry 1977.

PROPOSALS FOR OBSERVATIONAL STRATEGIES

Several scholars have proposed guidelines for developing skill in observing human movement. They offer different approaches for an SOS. Observation is only one task within QMD, and several observational strategies can be effective. Different plans of observation may even be needed to accommodate the perceptual differences among observers or to deal with challenging observational situations.

Barrett's System

Barrett (1977) identified three key tasks in the development of observational skill in physical education: analysis, planning, and positioning. She suggested that the lack of professional preparation for observational skill may have been due to trends in education that had emphasized the importance of reliability and outcome scores rather than human movement process variables. Barrett (1979c) later published a review paper on the growing body of literature that addressed the task of observation within the QMD process.

Her ideas on how to apply this research in planning observations were nicely summarized in some of her later papers (Barrett, 1979a, 1979b). According to Barrett, three components are needed in planning an observational strategy: deciding what to observe, planning how to observe it, and knowing what factors influence the ability to observe. In 1983 Barrett presented a model of observation and argued that observation was a key skill in teaching motor skills. A subsequent paper based on her work suggested that the first task, deciding what performance variables to observe, is the most important in planning movement observation (Allison, 1985b).

Radford's System

A paper by Radford (1989) reviewed the literature on observation in kinesiology and proposed a theoretical framework for movement observation. Observation was seen as three independent subprocesses: attention, template formation, and motivation. Attention is the process of limiting sensory information and can be controlled from the top down or the bottom up. For example, viewing the overall action draws attention to extraneous movements (bottom-up processing), while top-down processing is a conscious decision to direct observation. Viewing the overall performance would develop a gestalt of the skill, while an observational model such as the one presented by Gangstead and Beveridge (1984) could be used for top-down processing. As mentioned in earlier chapters, a gestalt is an overall feeling of the quality of a movement. The Dunham model (1994) employs a gestalt process. The Gangstead and Beveridge model (1984) is a temporal and spatial model designed to draw your attention to specific parts of a movement. These models are presented in this text to allow for exploration of two different approaches to movement observation: one that deals with a specific part of a movement and one that involves a general overall assessment.

Template formation (analogous to deciding on critical features and their acceptable ranges) is the cognitive, abstract, symbolic representation of a model of human movement. Templates of human movement are multilayered, plastic (shapeable), and generalizable to many performers. A template is not an image of an ideal movement but a generalized paradigm that accommodates differences within a definable range. This idea is similar to range of correctness presented in the previous chapter. The last subprocess, motivation, is involved throughout the whole process because good observation requires persistence, effort, practice, and the subsequent elaboration of movement templates (Radford, 1989: 23).

James and Dufek's System

James and Dufek (1993) proposed seven steps for the observation of movement. The focus of their paper is similar to QMD in this book. The steps in their observational strategy are to classify the skills to be assessed, divide the movement into phases, observe several times in order to evaluate each phase critically, and focus attention on four major areas.

James and Dufek deem the first step, classifying the fundamental movement pattern or the mechanical objective that is the focus of the skill, important to planning observation. Knowing the number and location of movement phases may assist in planning observations to see the important events in each phase. The last two steps are the guidelines for planning observation: focusing attention either on phases or on four different areas. The authors suggest that the plan for observation first focus on the total body (the rhythm or continuity of the whole body). Next, focus on the pelvis and trunk (the center of gravity) and the large muscles, which initiate most movements. The third focus is the base of support and how it changes, because it is often the source of balance and reaction forces that drive the movement. Fourth, focus on specific actions of the extremities. These movements are often fast and difficult to see, so the authors advocate concentrating on joint actions rather than specific segments. This model starts with a gestalt and concludes with a system of observation (temporal and spatial) advocated by other authors (Beveridge and Gangstead, 1984, 1988; Gangstead, 1984; Gangstead and Beveridge, 1984).

PRACTICAL APPLICATIONS
What's the Call?

The stolen-base leader in the conference is taking a big lead off first base. You are the field umpire of a two-person umpire crew. You crouch into position, anticipating the pitch and the runner's break to second base. Out of the stretch, the pitcher rockets a throw to the first baseman, who applies the tag as the runner dives back to the base. What's the call?

Your observation of this situation strongly depends on your attention and your ability to get into the correct position and systematically observe the movement of the ball and the runner. Were you attending to the pitcher at the time of release? Were your body, head, and eyes positioned at a 90-degree angle to the path of the runner? How close were you to the action? Where did you focus your eyes during the flight of the ball and when the tag was applied? Did your observation focus on the determining factor in the judgment of whether the runner was out or safe? Did the defense have to qualitatively prove the runner out, or did the runner have to prove he was safe?

A baseball umpire's systematic observation is similar to observation within other QMD. Choosing what to focus attention on prepares the professional for observation, but it may bias the observer to see some events and be less sensitive to others. How can professionals prepare to observe systematically to increase their accuracy, yet limit the effect of observer expectations in biasing information gathering for QMD?

KEY ELEMENTS IN A SYSTEMATIC OBSERVATIONAL STRATEGY

A critical review of the observation literature in kinesiology can be summarized in terms of five major areas that professionals should consider in developing an SOS. They should (1) plan to focus attention on critical features to aid in the assessment; (2) exercise as much control over the observational situation (teaching, coaching, or therapy environment) as possible; (3) plan the angle of view, or vantage points, from which to view the performance; and (4) plan the number of observations they expect to need. Finally, (5) they may sometimes need to include plans for extended observation. The primary sense involved is often vision, although the other senses are also used to gather information about the movement.

Focus of Observation

The first task of QMD, preparation, outlines the critical features of the movement to be diagnosed. These critical features and any other variables that may be related to the performer and the situation become focal points for the SOS to follow. Barrett (1979a) calls this idea of planning the foci of attention in observation a **scanning strategy**. A scanning strategy is a plan to define what to look for, when to look, and how long to look. Planning for a more open, gestalt approach is another observational strategy. Using a gestalt or observing with a specific observational model can achieve the goal of a scanning strategy.

Besides critical features, there are many variables a professional may need to observe in QMD. Aspects of the movement that relate to the rules of the sport need to be observed. For example, a badminton coach needs to focus on the racket head in serves because the shuttle must be contacted below the waist. Other observational variables to plan for might be environmental factors that constrain the appropriate technique in open motor skills, deviations in movement rhythm, extraneous motions, and cues indicating that fatigue or psychological stress is affecting the performer. For example, a beginning swimmer's fear may prevent proper execution of many swimming skills.

With all these aspects of performance to observe, how does one organize the scanning strategy? The QMD literature has proposed four approaches. Observation can be organized to follow (1) the phases of the movement, (2) balance, (3) the most important features, or (4) the path from general impressions to specific actions. While there have been numerous studies on the visual foci of athletes in a variety of sports (Vickers, 2011), only a couple of these have addressed what coaches visually focus on in observing sport skills (Moreno et al., 2002, 2006; Petrakis, 1993). Since there are limited data as to what scanning strategy or observational foci might be best, professionals are encouraged to use the observational strategy they are most comfortable with.

Observation by Phases of Movement

The most common scanning strategy in QMD may be to observe critical features within the normal order or phases of the movement (Gangstead and Beveridge, 1984; James and Dufek, 1993; Philipp and Wilkerson, 1990; Pinheiro, 1994). This observational strategy decreases the perceptual overload by focusing attention on the three primary phases of most movements: preparation, execution, and follow-through. It is often not clear what phase should be observed first, although it is assumed that observation follows the sequence of the movement. This observational strategy may be most appropriate for open motor skills. Since open skills are highly dependent on the environment, a strategy to observe the environment and how preparatory actions match the environment is very important.

This observational strategy is exemplified in the work of Gangstead and Beveridge (Beveridge and Gangstead, 1984, 1988; Gangstead, 1984; Gangstead and Beveridge, 1984). The focus is primarily on knowledge of performance for various parts of the body during three phases of the movement (figure 5.2). The movement itself is of primary concern, although it is important to remember that knowledge of results can be relevant to improving performance too.

Body components	TEMPORAL PHASING		
	Preparation	**Action**	**Follow-through**
Path of hub	Over base of support	Shift forward to target	Continue movement to target
Body weight	Over base of support	Shift forward to target	On front foot closest to target
Trunk action	Nonthrowing side to target	Rotate open to target	Follow arm to target
Head action	Face target	Eyes on target	Eyes on target
Leg action	Apart, weight on back leg	Step to target with closest leg	Bring back leg up to front leg
Arm action	Throwing arm extended back	Bring throwing arm forward	Throwing arm across body
Impact/release		Snap wrist	

Figure 5.2 Observational model proposed by Gangstead and Beveridge with overarm throwing cues. An example of observation by phases of the movement.

Adapted by permission from Gangstead and Beveridge 1984.

If this observational strategy is used without a gestalt, then a body component of interest is observed through the temporal phases of the movement. For example, the path of the hub (belt buckle or slowest-moving part of the body) is observed in the preparation, action, and follow-through phases. Any other body components are observed in a serial fashion. Descriptive phrases, like those about throwing illustrated in figure 5.2, can be inserted in the observational model to help guide the professional.

If a professional prefers a gestalt observational strategy, either the Gangstead and Beveridge (1984) or the Dunham (1994) model can be easily adapted to focus on specific aspects of performance. Recall that in a gestalt, the professional observes for an overall impression of the movement. If the professional suspects a problem in a particular part of the movement from the overall impression, the observational model will help the professional focus on a specific weakness in performance based on the technique factors on the task sheets. For example, you could use the technique points from the temporal and spatial model (Gangstead and Beveridge, 1984) of the overarm throw in the Dunham model (1994) to assess the movements in the QMD demonstrations. Figure 5.3 is a gestalt (Dunham, 1994) model that could be used in this approach to observation.

» **KEY POINT 5.2**

A good systematic observational strategy should include what critical features to focus attention on, how to control the situation, the vantage points of observation, the number of observations needed, and a decision on whether extended observation will be needed.

QMD Choices

You have two new students in class. You have never seen these students perform any of the skills you are teaching. You need to make a quick assessment of their abilities so that you can decide how to work with them. Which scanning strategy do you think would give you the most information in a quick observation?

Dunham Model

Body orientation: _____

Preparation: _____

Feet: _____

Knees: _____

Hips: _____

Trunk: _____

Shoulders: _____

Arms: _____

Hands: _____

Head: _____

Execution: _____

1. _____

2. _____

3. _____

4. _____

5. _____

6. _____

Figure 5.3 Task sheet for the QMD of the overarm throw using the Dunham (1994) model.

Observation of Balance

There is a saying in construction that you can't build a cathedral on the foundation of a house. The second way to organize an observational strategy is rooted in this principle. A scanning strategy can be based on the concept of balance and the origins of the movement. Many coaches like to focus on the base of support and the initial movements of the lower extremities for some skills because they believe that balance and the actions of the legs (or arms) strongly affect the actions of subsequent segments. Observing movement from the base of support and initial movements is especially appropriate for gymnastics and any other activity in which balance and base of support can dramatically affect subsequent actions. Movement in many sports requiring great accuracy, like baseball pitching, is strongly affected by variations in balance. DeRenne and House (1987) deem balance one of the four most important aspects of baseball pitching.

Observation Based on Importance

The third type of observational strategy organization is based on a ranking of the importance of the critical features identified earlier. This approach is favored in research by Morrison (Morrison, 1994; Morrison and Harrison, 1985; Morrison and Reeve, 1986, 1988b, 1989, 1992). The approach evolved from an earlier study by Harrison (1973). Obviously, if a professional believes that a particular critical feature is most important for safe movement, that feature should be observed first. The logic behind this observational strategy is similar to that for the origins-of-movement approach, because some critical features may influence other aspects of the movement. Examples of this approach can be seen in typical feedback from cross-country skiing coaches and ice hockey coaches. It is not uncommon for a skate-skiing coach to say to a performer, "You can only be on one ski at a time," meaning that the weight must be squarely over the gliding ski and is often considered by coaches the most important factor in successful skate-skiing technique. The hockey coach may say, "Sit down while skating," meaning that the player must balance over the skate blades. Again, coaches frequently consider this factor the most important for success in hockey skating.

Generally, the sequence aspect of this approach drives this system of observation. Professionals who have thoroughly researched a movement and reflected on their practice to identify critical features may have definite opinions on what aspects of the movement deserve the most attention. Biomechanical models of QMD tend to emphasize this approach by selecting variables for observation that are related to the goal or primary mechanical purpose of the movement.

Observation From General to Specific

The fourth approach is to move from the general to the specific. This has been proposed by several authors (Brown, 1982; Hay and Reid, 1982; James and Dufek, 1993; McPherson, 1990). It is what Radford (1989) called *bottom-up attentional processing* and what chapter 3 refers to as a *gestalt observational model* (Dunham, 1986, 1994). This approach is also similar to the gestalt talked about in chapter 3, in which the professional considers all the parts of a movement and develops an overall impression of the quality of the movement. The whole, complete skill is greater than the sum of its parts. If the

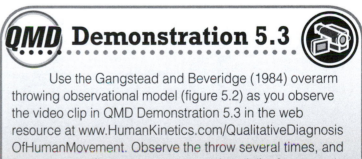

QMD Demonstration 5.2

Use the Gangstead and Beveridge (1984) overarm throwing observational model (figure 5.2) as you observe the video clip in QMD Demonstration 5.2 in the web resource at www.HumanKinetics.com/Qualitative DiagnosisOfHumanMovement. Observe the throw three times, and focus on one temporal phase of the movement each time. Does the timing of the phases seem correct? Does the initiation of arm movement follow the beginning of the rotation of the hips and trunk? Does the throwing elbow get in front of the hand before it moves forward?

QMD Demonstration 5.3

Use the Gangstead and Beveridge (1984) overarm throwing observational model (figure 5.2) as you observe the video clip in QMD Demonstration 5.3 in the web resource at www.HumanKinetics.com/QualitativeDiagnosis OfHumanMovement. Observe the throw several times, and focus on specific body components (spatial)—for example, path of hub, body weight, and trunk action. Which approach to organizing observation (temporal or spatial) was easier? What parts of the movement were difficult to see?

Use the Dunham (1994) task sheet (figure 5.3) to rank the most important technique points within all the phases of the overarm throw, as you observe the video clip in QMD Demonstration 5.4 in the web resource at www. HumanKinetics.com/QualitativeDiagnosisOfHuman Movement. Observe the throw, focusing on the most important critical features in each phase. Does your judgment of the most important critical features predispose you to observing some technique points and missing others? Do you end up using results similar to those with the Gangstead and Beveridge (1984) observational strategy?

You are a physical education teacher who has to teach several different skills in several different units. One of your units involves fundamental gymnastics skills. Another presents fundamental volleyball skills. As part of your SOS, would you observe skills in each unit differently? If so, why?

professional feels there is something wrong in the skill, he can pinpoint the deficiencies by looking at the phases of the movement or individual body parts, or a combination of phases and body parts.

Whatever the approach used to organize the observational strategy, some experts advocate written plans for observation. Written plans can take many forms, including

- assessment rubrics (Satern, 2011),
- checklists (Adrian and Cooper, 1989; Bayless, 1981; Davis, 1980; Frederick, 1977; Hoffman, 1977a; McPherson, 1990; Pinheiro, 1994),
- diagrams (McPherson, 1990),
- rating scales (Hensley, 1983; Hensley, Morrow, and East, 1990; Rose, Heath, and Megale, 1990), or
- task sheets (Dunham 1986, 1994; Klesius and Bowers, 1990; Morrison and Reeve, 1993; Reeve and Morrison, 1986).

Situation for Observation

The exact nature of the movement task and the environment in which the task is performed should be controlled as much as possible by the professional. Yet the task performance must be as realistic as possible for the QMD to be most effective. The Balan and Davis (1993) model (chapters 2 and 7) could provide an answer to the problem of how to teach in a way that improves QMD while successfully structuring an environment that is friendly to both QMD and practice. Unfortunately, QMD is often performed in situations in which the lack of environmental control either minimizes or exaggerates relevant technique problems. The environment should be carefully planned so that modifying for speed, competition, distractions, or psychological pressure will elicit realistic performances.

For example, most QMD of tennis ground strokes should be conducted during normal play and practice rallies or with a ball machine that can project balls in an inconsistent fashion. A player's forehand or backhand strokes are often dramatically affected by the environment. At the advanced level of open motor skills, the athlete's movement and tactics may be of primary interest. Coaches need to plan situations for their observations of open skills that mimic the competitive environment—for example, a point guard dribbling the ball up the floor who must adjust his dribbling to the defense.

Since closed skills are performed in a relatively stable environment, they do not need adjustment as they proceed. But even these can be made more realistic with psychological pressure. An excellent situation for the QMD of free-throw shooting in basketball is a free-throw competition at the end of practice. The combination of fatigue and psychological pressure to perform (to win or avoid penalty) creates a situation in which the coach can evaluate good and bad habits that may affect the outcome of a game. Remember, the very fact that the instructor or coach is watching has an effect on performers. Some thrive on the attention and pressure; others perform worse. Professionals must take this factor into account when setting up observations and later when evaluating and diagnosing performance.

The speed and timing of movements for observation should be matched as closely as possible to the situation in which the movement occurs in competition or public performance. For example, a novice tennis player usually encounters slower ground strokes from opponents and should not consistently be observed in a time-stressed position of returning shots with great speed or spin. Early skill practice and systematic observation by instructors should take place in closed environments. Unfortunately, practice routines in sports are often organized for convenience rather than for effective QMD. For example, coaching or teaching a movement with players in a shuttle position (fielding a ground ball, throwing to first, and returning to the back of the line) does not allow the professional to observe several trials. And multiple observations are essential to evaluation and diagnosis.

Once the play or movement situation is as realistic as possible, the professional may have control over the background only from her own vantage point (Brown, 1982). An effort should be made to set up the subject's movement to allow for a stable background. Setting up observation of a basketball player shooting with the wall or bleachers in the background is likely better than using a busy game or a drill if the coach is facing into the center of a large gym. Outdoors, similar stable background issues can be created by manipulating performer motion and position of the observer.

Vantage Points for Observation

The professional should specify the optimal positions for observation for a particular movement. A specific vantage point may be crucial to seeing the critical features identified for the QMD. In most cases, the best vantage point for observing a particular movement or critical feature is at a right angle to the plane of motion. This often means that several vantage points are required. Figure 5.4 illustrates split-screen video images of the address position of a beginning golfer. What different aspects of his technique are observable from the different vantage points?

Because most human movement involves important motions in all three cardinal planes, observation often requires several vantage points to obtain undistorted views of key actions. An example is the apparent excessive elbow flexion in preparing to throw a ball as illustrated in figure 9.7 (4th image). The knowledgeable observer knows that this view can exaggerate the appearance of elbow flexion due to the alignment of the arm with the line of sight in this phase of the throw.

Brown (1982) suggested that vantage points should have stable backgrounds, without distractions or moving objects. A uniform background with a contrasting color relative to the subject would be ideal. Horizontal or vertical references in the background make the visual estimation of motion and angles easier. The practical matter of getting to and from desirable vantage points may also affect the fine-tuning of what to observe in the SOS.

The observer's distance from the movement is another important factor in selecting vantage points for observation. It is clear from the research on vision (chapter 3) that this distance affects the angular motion of the observer's eyes as they track the movement. Therefore, the distance from the movement should be as large as

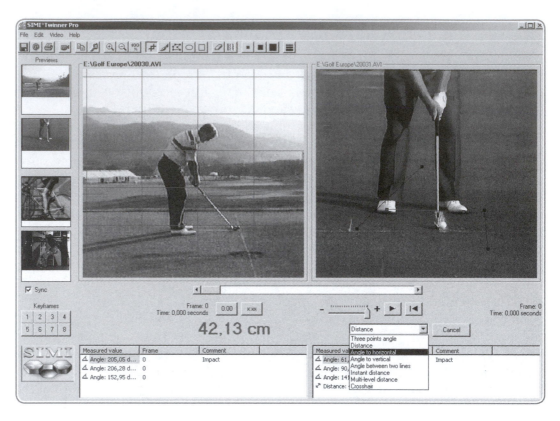

Figure 5.4 Split-screen video images of the address position of a golfer from two vantage points. Which aspects of performance are visible from only one vantage point?

Courtesy of SIMI, Unterschleissheim, Germany.

possible while still allowing for observation of important details. Sufficient distance provides a background on which to have the movement superimposed and reduces the tracking demand on the eyes.

How much distance should there be between the observer and the movement? There is no universal answer because the speed and complexity of human movements vary, the critical features of interest vary, and the environment may limit the selection of a viewing distance. The faster the movement, the greater the viewing distance should be to limit the demands on the observer's eyes. Hay and Reid (1982) advocated viewing distances of 10 to 15 meters for movements that have limited ground speed or that take place in small areas and 20 to 40 meters for movements that are fast or that cover a large distance.

But viewing distances beyond 10 meters are much larger than those many physical therapists or physical educators can use. The lab or gym space may not allow these large distances. These distances would also require the observer to travel or yell to the performer to provide feedback or intervention. Five to 10 meters seems to be a good rule of thumb for the minimum distance for observation of most human movements. If the movement or event is very small and slow, the professional may want to observe from closer vantage points in order to consistently see the action of interest. For example, a golf coach may not be getting enough information on ball spin from the flight of the ball and may choose to stand close to the golfer to examine the location of divots relative to ball placement. This provides a clue about the path of the club through impact and, consequently, the spin on the ball.

The evidence suggests that the best rule when using visual observation is to observe from as far away as pragmatically possible. If the distance is too great, however, it may limit the gathering of auditory, tactile, and kinesthetic information. Particularly when using the latter senses, the observer must be able to touch the performer.

Peripheral vision should also be taken into account as part of the vantage point. If clothing (hats, visors, and so on), people, or objects limit peripheral vision, understanding of skill performance may be inhibited. Peripheral vision is important in the development of cognitive maps. This concept is explained in chapter 3. Peripheral vision may be seen as ambient vision (Schmidt and Wrisberg, 2008) and provides us with the *where* of movement (Shiffrar, 1994). The *where* of movement allows us to understand the relationship of different body parts to each other.

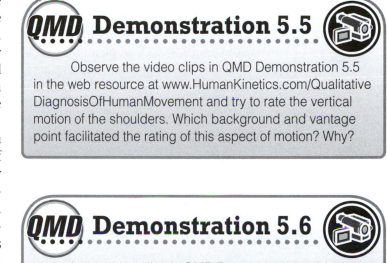

QMD Demonstration 5.5

Observe the video clips in QMD Demonstration 5.5 in the web resource at www.HumanKinetics.com/Qualitative DiagnosisOfHumanMovement and try to rate the vertical motion of the shoulders. Which background and vantage point facilitated the rating of this aspect of motion? Why?

QMD Demonstration 5.6

In the video clips in QMD Demonstration 5.6 in the web resource at www.HumanKinetics.com/Qualitative DiagnosisOfHumanMovement, how much elbow flexion do the runners have? Which distance and vantage point is most likely to result in a good estimate?

Number of Observations

The professional needs to plan the number of observations that should be necessary to gather enough information for diagnosis and intervention. Clients need to repeat the movement because of the observer's perceptual limitations and because the consistency of good or bad technique points is an important issue in diagnosis and intervention. Clark, Stamm, and Urquia (1979) found that the observation of six trials of a balancing task for children was sufficient to provide a reliable relative estimate of performance.

The problem is that many teaching, coaching, or clinical situations do not allow for large blocks of time for individual attention and observation of many trials. Unlike what happens at the elite level, where most coaching is one-on-one, typically coaches must divide their attention among several athletes. Therefore, they must compromise between gathering enough information for a good evaluation and diagnosis of the performance and bowing to the time constraints of the situation.

There are few guidelines for planning the number of observations needed for a QMD. The exact number is probably best determined by the individual doing the observation for a particular situation. It can also vary between observers, depending on the complexity of the skill and the ease of observation. Logan and McKinney (1970) recommended observing a minimum of eight trials, while Hay and Reid (1982) suggest 15 trials as a guideline. Based on QMD instruction studies by Morrison (Morrison and Harrison, 1985; Morrison and Reeve, 1989, 1992), some sport skills require only five repetitions for consistent assessment of technique.

In Morrison's original videotaped QMD skill test, the children performed each skill five times in sequence, and then that sequence was repeated. The latter two of the studies just cited did not use the second set of performances because those viewing the test analysis tapes (in intervening projects) felt they had decided on the merits of the performances during the first sequence. Initial scores in the latter studies with similar subjects were in the same range as scores from the original study. Reliability studies (Mosher and Schutz, 1983; Painter, 1990; Ulrich, Ulrich, and Branta, 1988) also suggest that five observations are usually enough.

Clearly it is best if multiple trials are observed systematically before any intervention is decided on. Based on this discussion and the reliability studies reviewed in chapter 2, a reasonable rule of thumb for the number of trials observed systematically in most QMD situations is between five and eight. Although some simple and slow body actions can be reliably observed in one trial, it is important to observe more because of the variability of performance. More observations also let the observer focus on information from all the senses. Unfortunately, sport officials and gymnastics judges must base their judgments on the observation of a single event.

> **» KEY POINT 5.3**
>
> Observation for QMD should normally be made from several vantage points and should include five to eight repetitions of the movement.

Perceptual limitations and the variability of human performance suggest that multiple trials are needed in QMD. It is not appropriate to correct a performer after observing only one attempt (unless the situation or competition involves only one observable performance). The consistency of a performer's strengths and weaknesses is a key issue in the evaluation and diagnosis task of QMD (chapter 6). The number of trials observed may also vary according to the skill of the performer, the skill of the professional, and other aspects of the situation. For example, novices to a motor skill often exhibit inconsistent performance and errors, so they should be observed more times than skilled performers.

Extended Observation

Extended observation is a plan for gathering more information on a movement than is usually observable. Two good examples of extended observation are using multiple observers and recording a performance on videotape. Observational power is greatly increased with the use of **freeze-frame**, frame-advance, and **slow-motion** replay features. A detailed discussion of the use of videotape to assist in QMD is presented in chapter 8. The use of senses besides vision is another example of extended observation. A coach or therapist may use the sounds created by a performer to gather information on the rhythm or to identify a weak phase of a movement.

If a critical feature of interest in a QMD is difficult to see, extended observation is called for. For example, a gymnastics coach who thinks the athlete has adequate height on a stunt but still fails to rotate enough for consistently good landings may choose to spot several attempts. The sense of touch during spotting gives the coach a feel for the level of performance and what intervention should work best.

QMD Demonstration 5.7

Observe one replay of the regular speed video clip of the vertical jump in QMD Demonstration 5.7 in the web resource at www.HumanKinetics.com/QualitativeDiagnosis OfHumanMovement and rate the amount of knee flexion at the bottom of the countermovement. Observe the video clip several more times and estimate the deepest knee flexion again. Discuss how many observations you think were needed for you to get an accurate estimate of or feel for the amount of knee flexion this subject uses. Compare your impressions to those of other students. Now observe the same jump using the slow-motion clip. Does this slower speed and use of the pause feature make your visual assessment easier?

If videotape is not available and the senses are not enough, Hay and Reid (1982) suggest a visual trick that may help observers get a look at fast movements. They suggest the eye-close technique, in which observers close their eyes at the instant of an important event of the movement to fix a temporary image of it. This trick relies on the short-term sensory memory or short-term memory (or both) described by Pinheiro and Simon (1992). Short-term sensory memory allows you to hold a sound, vision, or haptic sensation in memory for a few seconds after the senses have been stimulated. Short-term memory allows you to remember a limited amount of information about an event for a slightly longer time. The idea can be transferred to other senses as long as no competing stimuli are allowed to compromise the most recently gathered information.

INTEGRATED USE OF ALL SENSES

How can professionals extend their observational power by using all the relevant senses? Several authors have suggested that all the senses work together in the observation task of QMD (Hay and Reid, 1982; Hoffman, 1983; Radford, 1989). Hay and Reid stated that aural, tactile, and kinesthetic observation can supplement visual information. Chapter 3 summarizes the research that supports this holistic view of observation.

Many sports generate distinct sounds that can be used to provide information on an outcome or the movement itself. The sounds of preparatory footwork in the tennis forehand or the rhythm of the final steps in the long jump, for example, supply valuable information about how the athlete performed the skill. The sound produced by the slice technique in tennis is quite different from that of the flat serve. These sounds of impact provide important cues on the amounts of spin and speed applied to the ball.

The sense of touch can also be used to increase observational power in QMD. Teachers who spot gymnastics, diving, or other skills can sense the athlete's ability to generate the forces and torques required. Teachers or coaches with good physical skills may compete with performers to simulate various styles of play. This provides a great deal of information on the strengths and weaknesses of an athlete. Smart coaches who use the sense of touch can quickly check the appropriateness of equipment or the strength of a performer.

Hay and Reid (1982) use the term *kinesthetic observation* to refer to the assessment of a performer's sensation of movement. They suggest that good QMD is a cooperative effort between performer and professional. As performers increase their skill level, they usually develop a greater kinesthetic sense and feel for their performance. Skilled athletes can often tell the coach exactly what position their body was in or what mistake they made in a particular trial. Wise coaches and teachers of skilled performers use the performers' observations to supplement their QMD. Baseball batters or tennis players may be asked whether they hit the "sweet spot" on particular trials. Players can assist the coach and learn to evaluate shots by the sound of the impact and the vibration they feel. Good communication with performers

> **≫KEY POINT 5.4**
>
> Observation for QMD should integrate the use of all the professional's senses. Often the client's perceptions of the movement can provide important supplementary information for QMD.

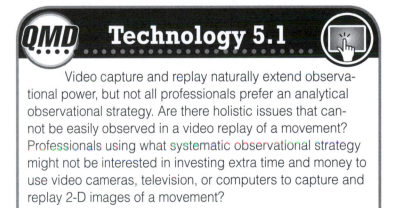

QMD Technology 5.1

Video capture and replay naturally extend observational power, but not all professionals prefer an analytical observational strategy. Are there holistic issues that cannot be easily observed in a video replay of a movement? Professionals using what systematic observational strategy might not be interested in investing extra time and money to use video cameras, television, or computers to capture and replay 2-D images of a movement?

is essential and may help motivate them to implement the corrections prescribed. Sports medicine professionals also use quite a bit of information from their patients or clients. The sensation of pain or fatigue a client feels during an exercise is important information to consider in QMD by athletic trainers, physical therapists, and strength coaches.

SUMMARY

The second task of QMD is observation. Good observation of human movement is based on a systematic observational strategy (SOS) to gather information about the critical features of a movement. An SOS can be organized based on the phases or sequence of the movement, by balance or base of support, by the importance of critical features, or from a general impression to specific aspects of performance. The key elements of an SOS are to focus attention, control the situation, plan vantage points, plan the number of observations, and extend observational power if needed. Professionals can extend observational power by getting information from all their senses, getting information from the client or mover, or recording the performances on videotape. The information gleaned from observation will be used in the next task of QMD, the evaluation and diagnosis of performance.

The following is a review of important points in an SOS:

- Observation is based on knowledge of the activity, the performer, effective instruction, and a systematic observational strategy.
- Observation is based on a variety of sensory information and the interaction of all the senses, not just vision.
- Because attention is an important component of observation, attentional focus is necessary in an SOS.
- An SOS can be organized by the phases of the movement, by balance, by ranking of critical features, or from the general to the specific.
- Observers should control the situation to optimize observation and the subject's performance.
- It is important to select appropriate vantage points, viewing distances, and numbers of observations.
- Coaches and teachers of skilled performers should integrate their observations with the performer's perceptions to supplement their QMD.
- Tools such as slow-motion video replay can extend observational power considerably.

Discussion Questions

① What factors make the use of a *systematic observational strategy* necessary in QMD?

② How do differences in professional's psychological and cognitive styles affect the choice of observational strategy?

③ What kinds of motor skills are best suited for visual observation? Auditory observation? Tactile observation?

4 Does specific knowledge about the critical features of a movement and about a performer's physical limitations improve or bias the observation? In what ways?

5 Do specific observational strategies bias professionals toward certain critical features? Explain.

6 How much practice in using a specific observational strategy is needed to become proficient and consistent in identifying characteristics of movement?

7 Explain why different critical features might require different observation strategies.

Evaluation and Diagnosis: Professional Critical Thinking

A young person who has several obvious faults in his bowling technique asks for your help. After viewing a few performances, you tell him the four major technique weaknesses you noticed. He has a bounding and inconsistent approach, an exaggerated backswing, and an inside-out downswing. He responds, "Yes, but what can I do to knock down more pins?" This young person has put your assessment of his technique to an important test. How do you know which fault is most strongly related to pin count? How will you select the intervention that will help this bowler the most? Which is more important, short-term or long-term score improvement? These questions point to the higher-order knowledge and thinking expected from movement experts who are kinesiology professionals.

Chapter Objectives

1 Explain why evaluation of performance errors is necessary for QMD.

2 Discuss the four major difficulties in evaluating strengths and weaknesses of performance.

3 Discuss six strategies for prioritizing strengths and weaknesses that professionals can use in diagnosing performance.

After observing a performance, the professional must identify desirable and undesirable aspects of that performance. A critical evaluation of these aspects and a diagnosis of the performance lead to a ranking of priorities for the intervention that the professional will provide the performer. Evaluation and diagnosis are the two parts of the third task of an integrated qualitative movement diagnosis (QMD). This chapter reviews the various theories or rationales for prioritizing possible interventions in improving performance. Because of the many interrelated factors involved in human movement, the evaluation and diagnosis of performance may be the most difficult task in QMD.

The systematic observation of human movement results in a large amount of information about a person's performance. This information must be processed in the observer's mind. Essential skills in this third task of QMD are the ability to evaluate the strengths and weaknesses of performance and to diagnose the implications of those strengths and weaknesses for performance. Diagnosis identifies the weaknesses that directly limit performance so that they can be corrected in the intervention task of QMD. Not all "errors" or differences in movement technique are related to performance, and providing too much or incorrect intervention will have a negative effect on performance.

In evaluating and diagnosing performance, the teacher, coach, trainer, or therapist essentially acts like a human movement detective. The diagnostic tasks of deciding "whatdunit" or what caused the problem are difficult. These diagnoses may have far-reaching consequences too because sometimes the intervention selected can do harm rather than good. A professional who focuses the performer's attention on minor or symptomatic errors at the expense of more important problems may indirectly contribute to an injury or poor performance. A coach who focuses practice on errors symptomatic of another problem is also wasting valuable practice time. Evaluation of performance is the important first step in making sense out of the information gathered in the systematic observation of human movement.

EVALUATION

The terms *evaluation* and *diagnosis* are used to emphasize the essential processes in the third task of QMD. Evaluation typically refers to a judgment of quality to ascertain the value or amount of something. This is important because the professional often must establish the good points of the performance as well as the errors or weaknesses.

Much of the early literature on QMD focused only on identification of errors or faults in performance. Early in its development, QMD was often referred to as **error detection**. This is understandable, because the primary method of QMD involves comparing the observed performance to a model of good form, with the goal of identifying differences or errors. Some motor development scholars argue that it makes little sense to judge performance as right or wrong; movement should be interpreted with reference to a continuum of development (Painter, 1990).

An integrated QMD is interested in more than identifying errors. The evaluation of a performance's strengths affects the diagnosis of the weaknesses and how the professional chooses intervention. A good evaluation should take into account the critical features of the movement that are within a desirable range, noting these as strengths of the performance. Those critical features not within that range are the

> **》KEY POINT 6.1**
>
> The third task of an integrated QMD has two distinct processes: (1) evaluation of the strengths and weaknesses of performance and (2) diagnosis to select the most appropriate intervention to improve performance.

weaknesses or errors. The Arend and Higgins (1976) model of QMD used the term *evaluation of performance,* noting that evaluation should focus on the efficiency of the movement and its appropriateness to the environment.

Process of Evaluation

It has often been assumed that professional experience and knowledge allow professionals to observe and evaluate human movement. There has been little research on the process of evaluation in QMD, and only a few authors have hypothesized as to what occurs in this process. It is likely that several cognitive methods are used to identify strengths and weaknesses of critical features. What is known about the accuracy of these methods is reviewed in the validity section of chapter 2.

The kinesiology literature about QMD suggests that the evaluation of performance can occur in two distinct ways. Hay and Reid (1988) call these the *sequential method* and the *mechanical method.* The sequential method involves comparing mental images of expected body positions throughout each phase of the movement with the visual perceptions of body positions and motion. Most coaches compare the actual performance with this mind's-eye image of the desirable actions and phases of a movement. Figure 6.1 illustrates how this visual comparison might look in the evaluation of a volleyball bump or pass. This focusing on the difference between a model of good form and the actual movement has been hypothesized to be the primary method of evaluation in QMD (Arend and Higgins, 1976; Hoffman, 1983; Pinheiro and Simon, 1992). This likely contributed to the overemphasis on error detection.

The mechanical method, illustrated in chapter 2 (figure 2.1), uses a model of the mechanical factors that affect performance. The evaluation task of QMD then becomes a process of deciding to what extent each mechanical variable was achieved. QMD models based on this approach suggest that knowledge of a few biomechanical principles can be used to evaluate a variety of motor skills and that these principles are directly related to corrections that can improve performance (Hay and Reid, 1988;

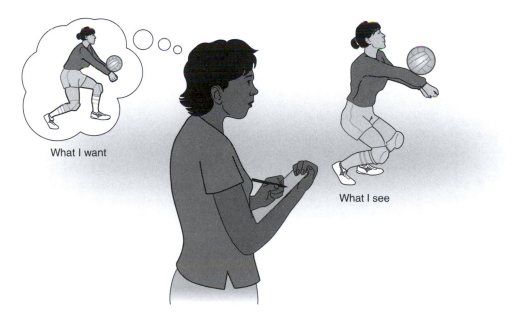

What I want

What I see

Figure 6.1   The evaluation of performance in QMD typically compares the observed performance with a mental image of desired performance. Evaluation is more than just identifying errors.

Experts in a particular sport may find the sequential method of evaluation appealing for QMD of that sport. Should they try to improve their QMD of their sport by blending biomechanical principles into their evaluation (mechanical method) strategy? If new techniques develop in their sport, should they rely more on sequential or mechanical evaluation in the new variations in their sport?

Hudson, 1995; Knudson, 2007b; Norman, 1975). Examples of critical features using mechanical principles are the concepts of sequential coordination, inertia, optimal projection, and range of motion. Like different observational strategies, different approaches to evaluation can be effective in the diagnosis of human movement.

The process of evaluation within an integrated QMD can benefit from using either the sequential or the mechanical method. The majority of observers are probably most comfortable using mental images of desirable form for critical features of the movements they observe and evaluate qualitatively. To keep the task manageable, evaluation of performance should be based on a few (about four to eight) critical features. So the quality of the critical features of the movement becomes the focus of the evaluation task. This evaluation is not viewed as dualistic; the performer's technique is not judged as either correct or incorrect (error).

If a performer consistently exhibits technique outside the range of correctness for a critical feature, this must be evaluated as a weakness or performance error. These errors should eventually be corrected because of their negative effect on performance or risk of injury. This is why it is important to specify as completely as possible in the preparation task the range of correctness of all critical features of a movement to be evaluated. The range of correctness can be defined or quantified in the professional's mind (for example, step with the left foot 3 to 7 inches), but correctness is usually expressed behaviorally to performers. A baseball coach may teach hitters to "step over a bat" to achieve the small, controlled weight shift that fosters the quick and accurate swing required in baseball.

The evaluation of critical features in an integrated QMD should result in an unbiased, accurate assessment of the critical features relevant to performance. Two approaches to evaluation using the sequential or the mechanical method have the potential to be accurate and reliable: rating performance on a three-point ordinal scale and rating on a visual analog scale (figure 6.2). The ordinal scale uses three categories, rating critical features into one of three levels: inadequate, within the desirable range, or excessive. The major advantage of this approach is that the categories lead directly to diagnostic decision making and intervention. This approach increases reliability by decreasing the chance of disagreement due to artificial precision.

The second approach evaluates the quality of the movement with a more holistic impression using essentially a visual analog scale. Professionals attach some personal meaning to the ends of a continuum and use their intuitive feel for the quality of performance of a critical feature to place a mark or rating on the scale. If the ends of the scale represent too much and too little of the critical feature (the middle of the scale usually meaning good performance), using the visual analog scale is much like the ordinal approach to evaluation.

»KEY POINT 6.2

Evaluation in QMD is concerned with identifying the strengths and weaknesses of performance. Effective approaches include rating critical features into one of three categories (inadequate, within the desirable range, or excessive) or using a visual analog scale.

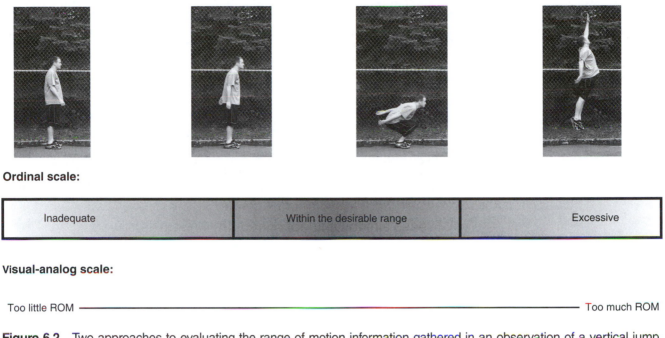

Ordinal scale:

Inadequate	Within the desirable range	Excessive

Visual-analog scale:

Too little ROM ————————————————————————————————— Too much ROM

Figure 6.2 Two approaches to evaluating the range of motion information gathered in an observation of a vertical jump. The vertical jump illustrated can be rated on the visual analog scale or an ordinal scale. A balance must be struck between accuracy and potential inconsistency.

Difficulties in Evaluation

There may also be deviations from prototypic form that are difficult to judge as in or out of the range of correctness of critical features or desirable form. The variety of environmental constraints and differences in performer anatomy and physiology may lead to differences in technique that cannot easily be identified as errors or bad technique. Some important difficulties in evaluating performance are variability, kinds of errors, differences between critical features and ideal form, and bias.

Performance Variability

A major problem in establishing whether a critical feature is within the desirable range of correctness or is a weakness is the issue of performance consistency or variability. This is why the systematic observational strategy should include at least five and up to eight observations. The professional should evaluate an inadequate level of a critical feature differently if it occurs once or across several trials. Early in motor development, performers exhibit a wide variety of errors, and an error in one trial may not be significant. The more advanced the performer, the more consistent performance becomes. In advanced performers the strengths and weakness are often more subtle, but they tend to be more consistent over repeated trials. Whatever the performer's level, use repeated observations to help you evaluate the critical features of a movement.

Unfortunately, movement consistency is not the only problem in the evaluation of performance. Other problems are finding agreement on the critical features, their range of correctness, and their order of importance to performance. Chapter 5 discusses some of the areas of disagreement on what is important to performance. This is a problem since there is no universal agreement about which features are critical for a movement, much less which critical features are more important than others. Professionals must use their own knowledge and experience to establish critical features and their range of correctness and must determine how to deal with movement consistency.

Kinds of Movement Errors

Even if evaluation were limited to error detection from an agreed-on set of critical features, it would still not be a simple task. Several authors have discussed how problems or errors in performance can have several sources. Hoffman (1983) proposed that errors could be related to critical abilities, skill, or psychosocial factors. A schematic of these classifications of errors is presented in figure 6.3. Note that even when errors are isolated as skill related, they may still have three different causes: technique, perception, or decision.

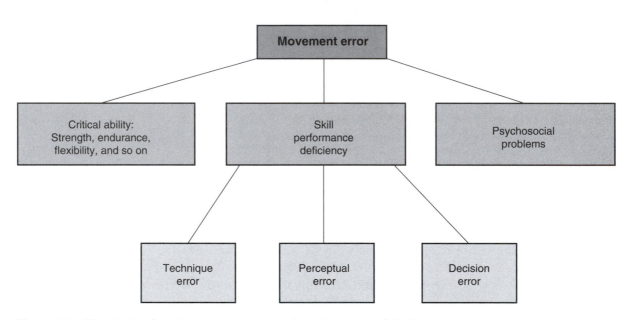

Figure 6.3 The kinds of performance errors that form the basis of Hoffman's diagnostic problem-solving approach to QMD (1983).

A similar approach was proposed by Philipp and Wilkerson (1990), who classified errors as biomechanical, physiological, perceptual, or psychological. Biomechanical errors relate to technique problems in body position or timing. Physiological errors are deficiencies in physical capacities such as strength, endurance, or flexibility. Perceptual errors are misunderstandings of technique or mistakes in evaluating environmental cues. Psychological errors are motivational and attitude problems that interfere with performance. These error classifications are built into the Philipp and Wilkerson model of QMD and are linked to specific interventions to improve performance. The problem with this approach is that a performance problem may be caused by the interaction of several of these errors, and there is no diagnostic process to select the best intervention to improve performance.

These factors may arise from a particular deviation in prototypical performance, making evaluation and diagnosis of performance difficult. For example, a professional would want to determine whether a volleyball player is missing passes to the setter because of bumping technique, fatigue, or perceptual problems in tracking the serve.

Critical Features Versus Ideal Form

The underlying assumption in the teaching of most motor skills is that the teacher knows what is the best or most appropriate movement response. The belief that a certain technique or form is best has been at the core of motor skills instruction for years. Brown (1982: 21) illustrates this belief: "While some performers may compensate for the lack of 'good form' through excessive practice, strength and speed of movement, improper form usually limits the ultimate level of performance." Most motor skill instructors would agree with this premise, but most have also encountered disagreement or conflicting expert advice similar to that illustrated in figure 4.1. There is a large difference between maintaining current knowledge to refine critical features and knowing what optimal form is for a specific person and situation.

If optimal form and the critical features for a movement are easy to establish, why have some skills undergone radical changes in technique independent of changes in equipment or rules? Why do many sports go through cyclic changes in predominant technique based on the best-known athlete or teacher at a given time? Why do experts in biomechanics disagree on the causes of particular movements? Why do some scholars argue (Gentile, 1972) the inappropriateness of assuming that the teacher can specify one ideal form that would help all learners succeed?

Clearly the evaluation of human movement is a difficult task complicated by a lack of consensus and knowledge about optimal human movement. For example, there are two highly effective techniques for the basketball free throw, the underhanded technique of Rick Barry (a professional basketball player who led the league in free-throw percentage) and the traditional set shot technique used by current pros. Is there an ideal form for a free throw? Even if there was, the free throw is a closed motor skill, while shooting in game conditions is an open motor skill. Skills in closed environments are more likely to have a tighter range of correctness approaching the idea of an ideal form. Skills in an open environment require greater flexibility to accomplish the goal and have fewer restrictions on effective technique.

The critical features of any skill are dynamic and interact with a multitude of other factors that affect performance. This interaction of critical features, performer traits, movement environment, and other factors makes it difficult to establish one ideal form for a particular movement. Kinesiology professionals should make a concerted effort to summarize the body of literature by inviting scholars from many subdisciplines and practitioners to begin to develop taxonomies and position papers on motor skills. Defining the critical features of motor skills and their range of correctness would make the evaluation of performance in QMD easier.

Observer Bias

Whatever the approach to evaluation, an unconscious tendency or bias may creep into evaluation in QMD. The reliability study by Ulrich and colleagues (1988) discussed in chapter 2 found statistically significant

QMD Demonstration 6.1

An important critical feature of overarm throwing is the amount of abduction of the shoulder. The upper arm should be held about 90 degrees to the spine (like an extension from the line of the shoulders) to maximize the effect of trunk rotation and protect some interior shoulder structures. Evaluate the video clips in QMD Demonstration 6.1 in the web resource at www.HumanKinetics.com/QualitativeDiagnosisOfHumanMovement and determine if this critical feature is a strength or weakness of these performers. Use a visual analog scale or an ordinal scale to evaluate the performances. Is one method of evaluation better than another? Do you think one method would be better for different types of skills (open vs. closed, discrete vs. continuous)?

interactions between observers and performers. The age of the performer tended to color the professional's evaluations. Young performers' skill levels tended to be underrated, while older performers' levels tended to be overrated. Two studies on bias in gymnastics judging (Ste-Marie and Lee, 1991; Ste-Marie and Valiquette, 1996) showed that prior knowledge of a performer could also influence current judgments of performance. Professionals should strive to use consistent evaluation standards in QMD, always asking themselves if they might be evaluating a mover more or less favorably based on their expectations for the person's age, gender, or position in a series of clients.

Other Considerations

Morrison (2000) suggested that we consider some other factors when trying to reduce the difficulties in evaluating movement. Along with presenting some of the ideas discussed earlier, he indicated that a systematic way of defining movement (possibly an observational model) was essential to aid in evaluation. He also suggested that we should be aware that everyone sees things differently due to different perceptual abilities, experiences, and perceptual illusions.

Remember that observational assessments of human movement have not been as reliable as quantitative measurements. Some scholars believe that the strong measurement and reliability emphasis of the physical education profession has hurt the development of QMD (Barrett, 1979c). The difficulty of achieving highly consistent, unbiased evaluation of human movement technique is not an argument against using QMD. QMD can be reliable and unbiased when observers plan and conduct their evaluations carefully.

The more specific the range of correctness of a critical feature can be made, the less sensitive the observer will be to any unconscious tendency to rate a performer differently from the standards. Too much specificity, however, reduces reliability by decreasing the potential agreement among different professionals. Professionals should not assume that their mark of a particular point on a 10-point rating scale or on the continuum of a visual analog scale is as precise as the assessment appears.

Evaluating human movement within QMD involves judging the strengths and weaknesses of a person's performance. Critical features in a movement should be evaluated only after an adequate systematic observation has been performed to gather relevant information. A few (four to eight) critical features should be evaluated using a three-category ordinal scale (inadequate, within the desirable range, excessive) or a visual analog scale. These evaluation approaches are not complicated and, if there is a problem, point logically to the intervention appropriate for the given critical feature.

DIAGNOSIS

Once the characteristics of performance have been evaluated, the professional must diagnose the situation to establish what specific intervention is likely best for the performer. Diagnosis usually refers to critical scrutiny and judgment in differentiating a problem from its symptoms. The many subdisciplines of kinesiology have often used other terminology for these steps or included them within the process of observation. It is not surprising that several subdisciplines have slightly different names or definitions for this process.

What is surprising is that with well over 130 years of history in kinesiology in the United States, there is no consistent rationale for the diagnosis of movement technique. This is analogous to medical schools teaching anatomy, physiology, and the variability of these parameters but forgoing courses in the diagnosis of diseases and the clinical rotations used to teach this skill. What would a patient think of a physician who had no basis for diagnosing disease and prescribing a remedy? Physician to nurse: "I'm not sure what's wrong with Mrs. Smith. She's the fifth patient today with a sore throat. Have her take a sample of medication X home. I'm not sure why or if it will work, but it's what my old doctor always used." (Substitute "coach" and movement details to make this story more familiar.)

One problem that may account for the lack of a theoretical basis for diagnosing movement is the difficulty of establishing critical features discussed in chapter 4 and the previous section of this chapter. Another potential barrier to the development of a theoretical basis of diagnosis is the erroneous belief that QMD is easy and that practitioners will develop a talent for it with experience alone. Remember that in the early stages of motor learning and development, errors are often obvious deviations from good form. The difficulty is in the frequency and variability of these errors. Despite the controversies, one aspect of diagnosis appears to be agreed on. It should narrow the strengths and weaknesses in performance to focus the performer and professional on the single most important intervention (Arend and Higgins, 1976; Hay and Reid, 1982; Hoffman, 1983; McPherson, 1990).

Prioritizing Intervention

Focusing on one intervention in QMD is important because the research in psychology and motor learning suggests that most learners can focus on only one correction at a time during practice (Christina and Corcos, 1988; Schmidt and Wrisberg, 2008). To prevent paralysis by analysis, the professional must prioritize intervention and select one solution as the best (Arend and Higgins, 1976; Hay and Reid, 1982; Hoffman, 1983; McPherson, 1990). Due to the lack of guidelines for the diagnosis of performance to select the best intervention, this is not an easy task. Diagnosis is also difficult because the cause of a particular problem may be far removed from its observable effect(s) (Hay, 1993; Luttgens and Wells, 1982).

The subdiscipline of kinesiology traditionally associated with diagnosis of motor skills is biomechanics. Hoffman (1974: 6) stated, "Diagnosis appears to rely on the observer's ability to use biomechanical concepts to accurately interpret the visual data at hand. The nature of the task suggests that a thorough understanding of principles in biomechanics, including the structural and functional relationships of joints and skeletal segments, is a prerequisite for diagnosis." Biomechanics is the science of how forces and torques move the human body.

While the science of biomechanics forms the vast majority of knowledge for prioritizing intervention on movement technique, the professional should also use knowledge integrated from many subdisciplines of human movement (Hoffman, 1983). Even if an optimal form could be established for a particular movement task and context, the professional would still have to decide whether the performer was skilled, flexible, and strong enough to use that form. If a performer is several developmental stages from desirable or mature form, is it wise to try to emulate this form if it is beyond the person's physical abilities? Biomechanics, motor development, exercise physiology, motor learning, pedagogy, and psychology all affect the answer to this question.

Does kinesiology research support any theories or guidelines for prioritizing movement errors into appropriate intervention? Do we know what is the best feedback or intervention to help a person move from, say, an immature level of throwing to a mature overarm throwing pattern? Unfortunately, the answer to both questions is no. There has been limited research to test which kinds of diagnostic decisions result in faster short-term and long-term improvement. The government and private foundations have not been willing to invest in these prospective-type studies of optimizing improvement in motor skills.

The kinesiology literature has only begun to stress that diagnosis is a critical task in QMD (Hoffman, 1974, 1983; Hay and Reid, 1982; Knudson, 2000; McPherson, 1990; Pinheiro and Simon, 1992). Pinheiro and Simon (1992) proposed a three-stage model of motor skill diagnosis based on their research on novice and expert coaches of the shot put and the medical diagnosis literature. They found that expert coaches were better at the observation task of QMD, called *cue acquisition* in their diagnosis model. The second stage was the interpretation of cues, or making connections between the cues and the cognitive representation (schema) of the skill. This finding of meaning in the information observed is what we call the process of evaluation. The last

> **»KEY POINT 6.3**
>
> Diagnosis within QMD involves a judgment that identifies the underlying causes of poor performance from the observed strengths and weaknesses. Diagnosis is used to set priorities for possible intervention.

stage of the model was considering the causes of the errors observed and making a decision. Pinheiro and Simon suggested that experts in QMD simultaneously use these three stages in their thinking when they evaluate and diagnose movement.

Few QMD scholars have tried to hypothesize about how the diagnosis process occurs. The papers by Hoffman (1983) and Pinheiro and Simon (1992) are among the few that have addressed this problem. Most QMD models do not elaborate on how the diagnosis of performance occurs. Since there has been little systematic research on which approaches to prioritizing corrections result in faster and longer-term improvement, we can only review some logical approaches and discuss their merits based on research that has focused on other kinesiology issues.

Rationales for Prioritizing Intervention

In prioritizing possible intervention, it is very important to keep in mind the goal or purpose of the movement being diagnosed in the ecological context of performance. Just as a physician's diagnosis is based on knowledge from the patient and knowledge about the many possible outcomes of a disease, diagnosis within QMD must be based on knowledge of the situation and the desired outcome of the movement. The diagnosis of a person's overarm throw would differ depending on its purpose. The purposes of a throw from the outfield, from the catcher to second base, and from the pitcher to home plate are different. Clearly the professional must understand the context and purpose of the movement and the goals of the performer. This knowledge of goals and outcomes will shape the priorities of corrections.

A review of the kinesiology literature on diagnosis within QMD supports the view that prioritizing possible intervention or corrections is important. Christina and Corcos (1988) warn professionals against rushing to provide feedback if they are uncertain about the diagnosis, because inappropriate feedback may frustrate the performer and damage the professional's credibility.

A review of the variety of QMD literature revealed six logical rationales that scholars have proposed for prioritizing intervention in the diagnosis of performance. These rationales may be summed up by the following phrases: relationship to previous actions, maximizing improvement, order of difficulty, correct sequence, base of support, and critical features first.

Kinesiologists may unconsciously be using one or more of these criteria when they select specific feedback or corrections over other possible interventions. Or they may just have strong opinions on what types of movement errors must be corrected first. No one rationale for prioritizing intervention should be considered the best, because research has not directly compared the various rationales and because the best rationale may be specific to the person or the motor skill. For gymnastics skills that are strongly determined by balance, corrections should likely be prioritized from the base of support up. A coach's experience and reading may lead her to emphasize execution actions over preparatory or follow-through actions in a particular sport skill. The variety of successful approaches is like the different strategies for the systematic observation of movement, which can all be used to gather information about performance.

Relationship to Previous Actions

The first rationale for prioritizing intervention is to relate actions to previous actions in a movement (Hay and Reid, 1982, 1988). It may be possible to identify errors or weaknesses that are only symptoms because they are caused by another problem. This rationale is similar to Hoffman's (1983) idea of primary and secondary errors. Some deviations from the desirable form, or other style differences, may be less important or may be mere symptoms of more important problems. If a missed ball (secondary error) can be related to a previous action (primary error) like improper preparation, vision, or directed attention, it would be foolish to correct the swing mechanics that are merely symptomatic of the real problem.

Knowledge needed to relate an action to a previous action comes from biomechanical principles, research, and practical experience in QMD. For example, if the goal of an overarm throw is distance, it is important to know the critical features of throwing that relate to the ball's height, angle, and speed of release. These three factors determine the distance of the throw. The throwing actions that relate to these parameters will help the professional diagnose throwing performance. This is one of the strengths of QMD models based on the principles of biomechanics. The evaluation and diagnosis of performance are focused on key movement production issues that lead directly to potential intervention (Norman, 1975).

Teachers, coaches, and athletes often have good hunches or opinions on what action relates to another. If the majority of coaches and professional athletes in a sport have the same opinion, it is most likely correct. It should not be assumed, however, that the opinions of athletes and coaches are always correct; in many instances they are wrong. Researchers have a difficult time relating actions to previous actions even in controlled biomechanical research and computer modeling. The two biggest problems in relating biomechanical actions to other biomechanical actions are the interaction of the segments and the effects of muscle actions at joints they do not even cross (Neptune, Zajac, and Kautz, 2004; Zajac, 2002; Zajac and Gordon, 1989). These and other biomechanical research issues make it difficult to establish exactly what actions are related to what other actions.

Good examples of relating an action to an earlier action can be found in both open and closed motor skills. Suppose a gymnast overrotates in a vault and stumbles in landing. Biomechanics tells the coach that two things can result in too much rotation: (a) the conditions at takeoff that determine angular momentum, or (b) the timing and changes of body position that manipulate the resistance to rotation (moment of inertia) that control the athlete's rotation in flight. If the professional's evaluation of the takeoff was good (trying to relate to a previous action), he may focus corrective efforts to the flight phase of the vault.

An open motor skill like baseball batting might have several factors related to an awkward action in the follow-through of the swing. The pitch may have been in a bad position, the batter may have been fooled by the pitch, or a loss of balance could have accounted for the poor follow-through. Diagnosis of these issues would help the baseball coach provide good intervention to the hitter. Notice in this example that generic feedback about a more "correct" follow-through would do nothing to deal with any of the three possible causes.

Another example relevant to many overarm patterns involves sequential coordination. The powerful trunk rotation in an overarm pattern (throw, spike, tennis serve) results in the arm lagging behind because of its inertia. This humeral horizontal abduction and the external rotation create an eccentric stretch of important muscles (pectoralis major, subscapularis) that contribute to the last 50 milliseconds of the movement. For performers to have advanced arm actions in overarm patterns, they must have good leg drive and trunk rotation. Cues to keep the arm back will not lead directly to the coordination and timing needed in elite-level overarm skills. Students interested in coordination in overarm patterns can examine the video clips from QMD Practice 9.9 and QMD Exploration 9.6 for overarm throwing and the tennis serve.

Attentional, psychological, or motivational factors may create performance errors. The awkward baseball swing could be related to nervousness or just a lack of attention and concentrated effort. A professional may notice that one performer is rarely enthusiastic and is very attentive to other students' reactions to his performance. If this performer has confidence and self-esteem problems, the best diagnosis may be to emphasize success and praise. Future work with this performer could get him to focus more on the movement and less on the result. Relating actions to previous actions in the diagnosis of motor skills must be based on principles from all the subdisciplines of kinesiology.

The science of biomechanics studies how muscles and external forces create motion of the human body. Biomechanical research on throwing technique and exercise provides knowledge on which to base a diagnosis of the throwing problem of this baseball player. What muscle groups would you emphasize in weight training or conditioning exercises to help improve throwing? Most professionals would likely emphasize the triceps brachii. It might surprise you to learn that biomechanical research suggests humerus internal rotation as one of the most important joint actions (Adrian and Cooper, 1995; Hong et al., 2001). The high angular velocities of elbow extension appear to be related to the coordination and interaction of the arm segments (Atwater, 1979). Roberts (1971) reported on a study by Dobbins that used a radial nerve block to paralyze the triceps brachii . The subject could create about 82% of initial throwing velocity after several practice trials. Triceps brachii do appear to contribute to throwing speed, but probably not to the extent that many coaches expect. Coaches should review research on testing and conditioning for overarm throwing (DeRenne, Ho, and Blitzblau, 1990; DeRenne, Ho, and Murphy, 2001; DeRenne and Szymanski, 2009; Hermassi et al., 2010; Janda and Loubert, 1991; Jones, 1987; Lachowetz, Evon, and Pastiglione, 1998; van den Tillar, 2004) to optimize performance and, more importantly, to minimize the risk of overuse injuries.

If you already have a good conditioning program, this athlete may improve if you prescribe appropriate practice to improve the sequential coordination of the throw. Unfortunately, biomechanical research is not conclusive on why sequential coordination is best in all kinds of throwing of light objects. Biomechanists have not even agreed on a way to define *coordination* using biomechanical variables. There are various approaches to documenting the kinetics or mechanical causes of movement. The biomechanical research on what is commonly called the kinetic chain or link principle (vague transfer of energy in multisegment high-speed movements like throwing and striking) is not conclusive. Biomechanics research over decades has tried to clarify the kinetic sources of complex 3-D motions. Distal musculature is simply too small and weak to create the large angular accelerations just before release in throwing, so passive dynamics (transfer of force or power across joints) clearly contributes to multisegment motion. What is controversial is whether the slowing proximal segment speeds up the distal segment or the acceleration of the distal segment slows the proximal segment (Feltner, 1989; Hirashima et al., 2008; Hong et al., 2001; Naito and Maruyama, 2008; Phillips, Roberts, and Huang, 1983; Putnam, 1991). The most recent work supports the theory of proximal to distal transfer of torques across joints (Hirashima et al., 2008; Naito and Maruyama, 2008) as the mechanism of passive dynamics of high-speed movements. Currently biomechanics research is not in a position to provide clear guidance for coaches trying to improve coordination in throwing and other sequentially coordinated skills.

QMD Demonstration 6.2

Observe the video clips of tennis ground strokes in QMD Demonstration 6.2 in the web resource at www.HumanKinetics.com/QualitativeDiagnosisOfHuman Movement. Use a gestalt observational strategy to get an overall feel for the movement and identify the most obvious movement weaknesses and strengths. If you are not familiar with tennis, you may need to review the critical features of ground strokes. Observe the performances again and check to see if any preceding actions contributed to the weakness you observed. Compare your results to another student's diagnosis.

PRACTICAL APPLICATIONS
Deciding on Focus

One of your best infielders is having trouble creating speed on her overarm throw. Fast batters and long throws to first base will be problems unless you can improve the speed of release. This player has good technique on the major critical features of the overarm throw. This leads you to believe that improvement will be difficult and will require improved coordination or conditioning. (See chapter 9 for the critical features of the overarm throw.) How do you decide if conditioning or throwing technique should be the focus of intervention?

Maximizing Improvement

Another rationale for prioritizing intervention is to select intervention that can be expected to maximize improvement (Hay and Reid, 1982, 1988). In 1988 Hay and Reid proposed that diagnosing performance is a two-step process of excluding faults that appear to be effects of other faults and prioritizing the faults that are left based on the improvement that can be expected in the time available. On the surface this seems like a logical approach to selecting corrections. The problem, however, is that it is not clear how to judge which correction leads to the most improvement and what time frame should be used.

Prioritizing to maximize improvement is probably a good approach, but research is needed to determine what factors are most significant in both short-term and long-term improvement. One technique change could create a lot of initial improvement but in the long run make it difficult to achieve advanced levels of performance. On

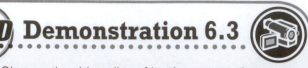

Observe the video clips of beginners trapping a soccer ball and catching in QMD Demonstration 6.3 in the web resource at www.HumanKinetics.com/QualitativeDiagnosis OfHumanMovement and diagnose performance based on intervention for a weakness that you believe would result in the most long-term improvement. Why would this intervention create the most improvement? Can you suggest another intervention that might work as well? Does the type of intervention depend more on the performer, the activity, or the skill level of the performer?

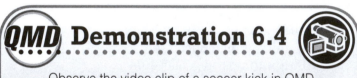

Observe the video clip of a soccer kick in QMD Demonstration 6.4 in the web resource at www.Human Kinetics.com/QualitativeDiagnosisOfHumanMovement, and diagnose performance based on the intervention that would be the easiest for the performer to implement. Why do you think this would be easiest for the performer? Does skill level affect the ease with which a performer can implement an intervention strategy?

the other hand, motor learning research has shown that more randomly assigned practice conditions (high contextual interference) create lower initial performance but better long-term performance (Schmidt and Wrisberg, 2008).

Order of Difficulty

Research in psychology, pedagogy, and motor learning has shown the importance of positive reinforcement and success in motivating practice and the learning of motor skills. These disciplines tend to recommend that the easiest corrections be made first if movement errors seem unrelated and cannot be ranked in importance (Christina and Corcos, 1988). The explanation is logical, since easy technique changes lead to the performer's perceived success, improvement, and greater motivation to continue practice. Selecting intervention in order of difficulty may produce small but consistent increases in skill. Again, though, the question is what is the best way to prioritize intervention to get the most improvement. A correction that is easy to communicate to a performer may not lead to improvement, especially if the action in question is related to another action. Like the rationale for prioritizing intervention, selecting intervention in order of difficulty for the performer is logical, but there is no clear research showing that it is most effective in improving performance.

Correct Sequence

Another rationale for prioritizing intervention is to correct in sequence, or provide intervention in the sequence of the actions in the motor skill. First correcting actions in the preparation phase of a movement may have effects on the execution and follow-through phases. This approach has been used by Morrison and Harrison (1985) in teaching QMD to classroom teachers with no background in kinesiology. This order of priority may also be implied by the many QMD models that break the movement into preparatory, execution, and follow-through phases.

There is little scientific evidence to support this "domino theory" of technique. It is logical that some skills could be highly sequential; actions in preparation might strongly influence later actions. The correcting-in-sequence rationale may be a good approach for making tough decisions between two very similar corrections, or for professionals without a strong background in biomechanics and other subdisciplines. A volunteer youth sport coach or early childhood specialist at a day care center might be able to help children improve a variety of motor skills by providing corrections in sequence. Fast sport skills in open environments are highly dependent on preparatory movements and might benefit from prioritizing intervention in sequence. The sequence in a slow movement may also be important. A strength coach teaching a complex pulley exercise might be very concerned with initial stance and posture position before final arm pull to maximize safety and sport specificity of the exercise.

Base of Support

In many activities, coaches choose to provide intervention to improve performance from the base of support up. For activities requiring balance or the control of large forces generated by the strong muscles of the lower extremities, this approach may be logical. Often gymnasts must support their body weight with their upper extremities and very small bases of support. Your author is aware of a golf pro who bases teaching and QMD of the golf swing (a closed motor skill) from the stance, or base of support, up. This PGA professional, who has a master's degree in kinesiology, has come to the conclusion that most errors later in the golf swing are a result of actions in the setup and backswing. The golf swing requires great precision to strike the ball with the correct speed and path for a specific shot.

A target shooting analogy illustrates the importance of stance and balance in accuracy sports. Tell performers to compare the accuracy and stability of an Olympic marksman to the accuracy they can expect with their current lower extremity technique in an accuracy skill such as a basketball free throw. Sports medicine professionals often diagnose performance of fundamental movements around base of support because a weak or injured person may be in physical danger from falling when walking or standing from a chair. As with the other approaches to diagnosis, research is needed to see whether balance or precision motor skills improve the most with intervention directed to the base of support and balance compared to other kinds of intervention.

> ## QMD Demonstration 6.5
>
> Observe the video clip of a soccer kick in QMD Demonstration 6.5 in the web resource at www.Human Kinetics.com/QualitativeDiagnosisOfHumanMovement, and diagnose performance based on the intervention related to the base of support and performer balance. Do you think this is the best intervention in this performer? Why?

Critical Features First

The last rationale for prioritizing intervention is to improve critical features first, before minor variations in performance. The kinesiology literature is full of professional articles offering opinions on the most important aspects of motor skills. That is *not* what is meant here. By definition, critical features are the most important factors in determining the success of a movement. They are established by rigorous review of all the professional experience and research in the preparation task of QMD. If the right critical features have been established, correcting them before addressing other general points of good form or style should help the performer achieve the movement goal faster and with greater safety.

The problem with this approach to diagnosis is that most movements have several critical features, and we have seen that it is difficult to establish which are the most important. A professional may have strong convictions about several critical features but may have only educated guesses or beliefs regarding their relative importance.

When selecting the rationale to use for diagnosis of performance, kinesiology professionals have virtually no prospective evidence to show what rationale is best. Sport-specific research and experience can make a professional believe that the sequence of a skill is very important. That professional may then diagnose performance based on a combination rationale: the *sequence of the critical features* of that skill. A therapist might prefer to prioritize intervention in gait QMD by relating actions to previous actions and the base of support. Professionals should select the rationale they feel is most relevant to the movement, client, and situation.

So what rationale can one recommend without knowing specific details about movement, client, or situation? The best diagnosis of performance may normally use knowledge from all relevant subdisciplines of kinesiology to try to relate actions to previous actions in combination with another rationale relevant to the situation. Relating actions to previous actions might prioritize two possible interventions of relatively equal influence. The professional then might select the weakness to focus on based on either sequence or base of support. It would be ideal to have some prospective training studies where specific interventions are compared to see what kinds of practice or feedback creates the greatest improvements in performance. It may be that the interaction of the movement, client, and ecological conditions influence what in general would be the best rationale for diagnosis in QMD.

FURTHER PRACTICE

You can continue to practice evaluation and diagnosis of performance in any situation where you can observe movement. If you record a televised sporting event, the comments of color commentators can be good practice examples. Replay the movements they are referring to and evaluate the performance to see if your diagnosis results in similar judgments. Remember that television experts are usually former sport or activity stars who may or may not have extensive coaching experience and training in kinesiology.

Another way to use video replay to practice evaluation and diagnosis is to record practice sessions or competitions of local sports. Do your evaluation and diagnosis of the performances point to the same interventions selected by the coach? This idea could also be extended to exercise movements in conditioning or rehabilitation settings.

Sometimes movement problems or improvement can be difficult for an athlete. This complicates the diagnosis of performance. One way to improve your diagnostic powers is to consult with fellow professionals. See chapter 8 for details on shared use of video with Siliconcoach LIVE or DartfishTV. Free (Skype) Internet conferencing programs allow professionals to communicate about QMD case studies. It is important that you obtain permission from your clients to consult with others or to use their images in online or shared environments. Health Insurance Portability Accountability Act (HIPAA) regulations protect the sharing of individuals' private medical information.

SUMMARY

The third task of QMD requires skill in two processes, the evaluation and the diagnosis of performance. This may be the most difficult task of an integrated QMD. The professional must evaluate the strengths and weaknesses of the movement's critical features, which were identified in the observation task. The process of diagnosis involves prioritizing these strengths and weaknesses so that one intervention can be selected to improve performance. There are six rationales that may be used to prioritize intervention: relating actions to previous actions, maximizing improvement, making the easiest corrections first (working in order of difficulty), correcting in sequence, moving upward from the base of support, and fixing critical features first.

In the absence of prospective research to show what kind of intervention creates the most improvement in performance, it is up to the professional to decide what rationale or combination of rationales to use. One good way to diagnose performance is to combine the rationale of relating actions to previous actions with another rationale relevant to the client, movement, and situation. Once a critical feature or aspect of performance has been selected as most likely to improve performance, the professional is ready for the fourth task of QMD, intervention.

Discussion Questions

1 How do the processes of *evaluation* and *diagnosis* differ?

2 What factors in the evaluation of performance affect *accuracy*, *reliability*, and potential *bias* of the QMD?

3 What subdisciplines of kinesiology contribute knowledge necessary for prioritizing strengths and weaknesses identified by the evaluation of performance?

4 What important research questions need to be answered to improve the diagnoses of performance in QMD?

5 In what ways does accurate performance evaluation make diagnosis easier?

6 What rationale for prioritizing intervention do you think is the best approach to diagnosis in most QMD situations? Why?

7 If you were to combine several diagnostic rationales, which would you combine and why?

8 What would be some of the advantages if kinesiology established standard critical features and their range of correctness for human movements?

Intervention: Strategies for Improving Performance

A badminton player is having difficulty getting the desirable speed and downward trajectory of the smash. Your observation of several trials suggests that the player positions himself poorly, letting the shuttle get over or behind his head. Your feedback to the player is to stay back a little before hitting the shot. Because the player has been learning the smash with his body turned at a right angle to the net, he misunderstands your cue to mean that he should move back toward the sideline. Clearly feedback must be phrased carefully to be effective. What other intervention would help this player? Would general feedback to "hustle" be effective, or should a more specific cue relevant to this performer be used?

Chapter Objectives

1. Identify the variety of intervention strategies used in QMD to improve performance.
2. Identify research-supported guidelines for the provision of augmented verbal feedback.
3. List the functions of feedback as intervention in QMD.
4. Describe how to develop appropriate cue words and phrases.
5. Identify situations in which the intervention of exaggeration, modification of practice, manual or mechanical guidance, conditioning, or ecological intervention would be an appropriate intervention to improve performance.

nce a systematic observation has been conducted and the strengths and weaknesses of the movement have been evaluated and diagnosed, the last task of qualitative movement diagnosis (QMD) is intervention. Intervention is the professional's selection of feedback, corrections, or other changes in the practice environment to improve performance. This is a critical step in the QMD process in which the instructor communicates with the learner about the desired change and practice that will lead to improvement. Intervention requires the integrated use of knowledge from all kinesiology subdisciplines. Improper intervention can result in decreased performance.

In various models of QMD, the task of intervention has been given a variety of names. Some scholars call it *feedback* (Arend and Higgins, 1976), whereas biomechanists call it *remediation* (Knudson and Morrison, 1996; McPherson, 1990) or *instructions to performers* (Hay and Reid, 1988; McGinnis, 2005). The integrated model of QMD uses the more general term *intervention* because it encompasses all possible actions a kinesiology professional can take. Intervention is not limited to the various forms of feedback, instruction, or remediation in previous QMD models. Instructors can choose to give verbal feedback—or they can choose not to intervene and can positively reinforce good aspects of performance (technique or effort). They can also use modeling, provide physical guidance, modify practice, prescribe training, and adjust competition or equipment. This chapter discusses how to optimize the intervention to improve performance.

> **»KEY POINT 7.1**
>
> The fourth task of QMD is intervention, which may involve providing feedback to performers, making technique corrections, or prescribing practice in order to improve performance.

FEEDBACK

The predominant mode of intervention in teaching motor skills is verbal feedback from the teacher. Whenever a person executes a movement, information about the outcome is available immediately. This information is **intrinsic feedback**. The physical sensations of walking in deep sand, the kinesthetics of carrying a heavy object, and visual information on the path of a thrown ball are examples of intrinsic feedback.

The other major kind of feedback is extrinsic or **augmented feedback**. It comes from an external source after the movement has been completed. Augmented feedback is the primary mode of intervention in most QMDs in kinesiology. Examples of augmented feedback are praise, corrective instructions, and specific information about the completed movement. This section focuses on how professionals can use feedback as intervention to improve performance. A great deal of research has been conducted on feedback in learning motor skills. Later sections of this chapter summarize the research, providing principles for using feedback as intervention; in the next two sections we review the functions that feedback serves in learning movements and the various classifications of feedback.

> **»KEY POINT 7.2**
>
> Feedback used as intervention in QMD has three major functions: information, reinforcement, or motivation. These functions of feedback help the performer improve.

Functions of Feedback

Movement feedback has three major functions in helping people improve their performance: information, reinforcement, and motivation (Schmidt and Wrisberg, 2008). Each function of feedback can be used as a target of intervention in the QMD process. Arguably the most important function of augmented feedback is guidance, or information about how to perform the next practice trial.

Information

The information function of feedback is essentially instructing performers how to correct movement errors. Performers use the information to plan the next movement response in the process of practicing and learning the movement. A therapist who wants to improve the safety of a patient's gait with a cane can use feedback to teach the patient how to position the cane relative to the injured leg.

The primary power of good augmented feedback is the guidance it gives in shaping future responses. Motor learning scholar Charles Shea illustrates this idea by characterizing feedback as cognitive training wheels for performers. Skill feedback provides the most appropriate mental images that help the person shape the next response, similar to bicycle training wheels set to an appropriate height. For example, a dance instructor might ask a dancer to keep her trunk more upright or vertical during a particular move.

A classic example of failure to use the information function of feedback occurs when a coach provides poor feedback, often in frustration as a performer repeats the same error over and over. Teachers or coaches may get caught in a "correction complex" that is sometimes manifested by a stream of don'ts: "Johnny, don't step in the bucket, son." "You did it again; don't lift your head!" Feedback with a "don't" message does not directly inform performers what they *should* do. It also communicates a negative and discouraging message. More effective intervention would be to provide feedback that helps the performer get an idea of what to do. For example, the teacher could say, "Johnny, remember to take a small step toward first base for outside pitches." Or, "Remember to step about the width of a bat, Johnny."

A common error of novice golfers is to lift their head during the swing, which is often accompanied by effects on the trunk and the plane of the swing. It would be poor feedback to say to such a student, "Don't lift your head." A more creative coach might use the information function of feedback: "I would like you to really focus on the ball. Pick your head up and away from the ball only when I say 'lift!'" The student takes his practice swing, concentrating on "head down, eyes on the ball," and executes a nice shot. "Hey! You didn't say 'lift,'" he would say. His nice shot and the coach's silence would have made the point.

Reinforcement

The second function of feedback is **reinforcement**, which can be either positive to help encourage correct technique or negative to diminish the frequency of undesirable actions. Readers may remember Thorndike's law of effect (1927). When applied to learning motor skills, Thorndike's law would say that people tend to repeat responses that are rewarded and avoid responses that are punished. Feedback to a performer should begin with reinforcement of the strengths of performance identified during evaluation. This is why it is important for evaluation to identify both strengths and weaknesses. If good movement components cannot be found, reinforcement feedback cannot be given. A therapist might tell a patient, "Great job! I know that was hard, but those were your smoothest, most even steps so far." A

QMD Choices

You are coaching a junior high school wrestling team, and several boys have been causing trouble, wrestling dangerously in drills with boys in smaller weight classes. Safety is important, but so is an aggressive attitude in wrestling. Initially you use negative feedback and punishment, but how can you use positive feedback to reinforce safe but competitive wrestling?

strength coach might tell an athlete, "Nice job, Sheila! You kept your back straight through some very heavy squats."

It is sometimes appropriate to provide negative feedback. Behavior that is inappropriate or dangerous may warrant a swift, negative response from the instructor. But this type of feedback should be used sparingly and only in the right circumstances. Relying on negative feedback can result in an adversarial relationship with clients, which is neither productive nor pleasant. Psychologists also say that several positive reinforcements are needed to compensate for one negative reinforcement or criticism. Emphasizing positive reinforcement of appropriate behavior tends to decrease inappropriate behavior.

Motivation

The third function of feedback is similar to reinforcement, but it focuses on providing motivation to practice. Indeed, many psychologists argue that this type of feedback is of primary importance. Teachers and coaches should provide positive feedback that rewards consistent effort and tends to create a positive attitude and climate. For example, elementary physical educators should carefully provide positive feedback that reinforces an "I can do it" attitude in each child. Pedagogy and psychology research has shown that good teachers and exercise leaders provide a great deal of praise and positive feedback (Rink and Hall, 2008; Roberts and Treasure, 2012; Siedentop, 1991).

One study showed that specific, corrective feedback in a volleyball unit increased the number of successful practice trials in a junior high school physical education unit (Pellett, Henschel-Pellett, and Harrison 1994). Clearly, a good intervention approach is to provide feedback that lets the performer know what to do next. This feedback should be expressed in a positive, encouraging manner. In coaching athletics or other high levels of performance, the coach is responsible for motivating the intense practice and training needed to prepare for competition.

The higher the skill level of the performer, the more important the motivation function of feedback becomes as intervention. The muscle memory that is the hallmark of skilled performance becomes a problem when a professional tries to correct a weakness in performance. Intermediate and advanced performers often resist the intervention (Langley, 1993). They are often not motivated to make these difficult technique adjustments unless performance improves immediately, but performance usually suffers until the new motor pattern is learned.

Langley (1993) makes some useful suggestions for overcoming learners' resistance to intervention. Performance is improved in these clients by individual attention that focuses on the performance limitations of a technique problem and personal motivation to make changes. The next section summarizes the research on how to select feedback that is effective in improving performance.

Classifications of Feedback

The two major classifications of feedback used in motor learning research are important to understand so that the professional may select appropriate feedback as intervention. Feedback can be classified as either **knowledge of results (KR)** or **knowledge of performance (KP)**. KR is information about the outcome of the movement or the extent to which the goal of the movement was achieved. KR is easily observable in some activities, such as when a ball misses the basket or an arrow hits the bull's-eye. Other times KR may be unknown or less obvious to the performer. A sprinter may believe she won the race, but the photo at the finish and times generated will provide more precise KR.

KP is information about the movement process, the actual execution or movements of the body. For example, a basketball student shooting a free throw receives KP from a physical education teacher who comments on his good wrist action at release. KP feedback in softball hitting might be to "keep more weight on your back foot at the beginning of the swing."

In some human movements, the outcome or goal is to move the body in a "perfect" way or in a way that conforms to certain stylistic criteria. Judges at dance, diving, or gymnastics competitions have the difficult task of assessing movement qualitatively. Coaches and teachers must realize that KP and KR are one and the same in these situations. The outcome of interest is the actual movement of the body. But usually these two classifications of feedback are different and provide unique information to performers when used as feedback during QMD. When providing augmented feedback as intervention in QMD, professionals should remember that KP is often more powerful.

Since the main function of feedback is to guide the learner in the next practice response, information on the actual movement and how it should be changed is often more valuable than the outcome or result of the movement. Several studies have examined motor learning when KP is given as feedback. A classic study that provided both KR and KP showed the power of KP feedback (Hatze, 1976). A 10-kilogram mass was attached to the foot of a subject, who attempted to minimize movement time in a leg-raising task. For the first 120 practice trials the subject received KR as movement time, producing the classic negatively accelerating learning curve (figure 7.1). With the effects of KR at a plateau, KP was provided as velocity curves of the optimal and the subject's movement. A second learning curve occurred that approached optimal movement.

Studies have shown the superiority of KP over KR if KP is provided as a continuous graph of the performance (Howell, 1956; Newell, Sparrow, and Quinn, 1985; Newell et al., 1983). Biomechanical studies have shown how graphic presentation of KP can improve performance in cycling (Broker, Gregor, and Schmidt, 1989; Sanderson and Cavanagh, 1990). Research has demonstrated that allowing the learner to control when the instructor provides KP results in greater learning than fixed schedules of KP feedback (Janelle, Kim, and Singer, 1995; Janelle et al., 1997). But don't let all this evidence give you the impression that KP is always the most effective form of feedback; in many of these studies, the goal of the movement can be easily defined in biomechanical terms.

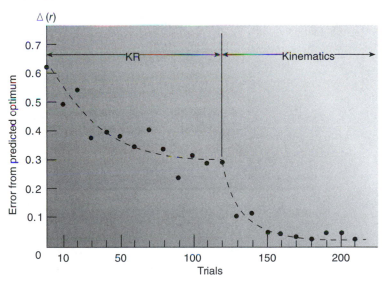

Figure 7.1 The guidance power of KP over KR in a kicking task.

Reprinted by permission from Hatze 1976.

In the real world of a clinic or gymnasium, much of this carefully controlled research is hard to apply. Gentile (1972) argued that too much emphasis can be placed on KP feedback in teaching motor skills. She proposed that KP is most appropriate when the movement itself is the goal of interest, as in a dive or a gymnastics routine. She also hypothesized that undue emphasis on KP, when the goal of the movement is some other outcome, could interfere with motor learning.

Gentile emphasized that kinesiology professionals may direct the learner's focus to form and teacher feedback, limiting attention on the outcome (KR) of the movement. In these more open environment movements, skill is essentially the ability to achieve the goal by reacting to changes in the environment. KR, therefore, is also important and powerful feedback. Remember that critical features can be related to the movement itself (KP) or a different outcome (KR). The trajectory of a basketball shot and the break of a baseball pitch are crucial aspects of performance and provide important information to the performer.

Another promising line of research involves the combined use of KP and KR feedback (McCullagh and Caird, 1990; McCullagh and Little, 1990). Kinesiology professionals performing an integrated QMD should select the mode of feedback that matches the movement and environment. When the professional is confident that a critical feature related to body motion needs improvement, the research suggests that KP feedback is more effective.

Principles for Providing Augmented Feedback

To summarize the kinesiology research on augmented feedback, seven guidelines are proposed that will help the professional shape the intervention chosen after the diagnosis phase. These feedback guidelines are based on the general consensus of the literature and are not intended as a scholarly review of the subject. Several good review articles deal with movement feedback (Annett, 1993; Bilodeau, 1969; Lee, Keh, and Magill, 1993; Magill, 1993, 1994; Newell, 1976; Newell, Morris, and Skully, 1985; Wulf and Shea, 2004; Wulf, Shea, and Lewthwaite, 2010). There are also good professional articles on providing feedback (Chen, 2001; Sharpe, 1993; Tobey, 1992). Tobey suggested that feedback strive to be specific, immediate, and positive.

These guidelines are not intended to define an optimal approach to feedback. They are presented here to allow you to consider factors you might include in your feedback. Kinesiology research has just begun to study what kinds of observational models (correct form, incorrect form, developing or learning form) given as feedback create the most learning (Martens, Burwitz, and Zuckerman, 1976; McCullagh and Caird, 1990; McCullagh, Stiehl, and Weiss, 1990). Other studies have tried to identify modeling and critical cues that improve performance best in children (Fronske, Abendroth-Smith, and Blakmore, 1995; Masser, 1993; Weiss, 1982). More such studies are needed. The best intervention for short-term versus long-term performance, for open or closed skills, and for the different stages of motor learning (Gentile, 1972) or motor development must also be determined. The best that can be offered at this time are some guidelines that have consistent support within the kinesiology literature.

1. Be judicious with feedback.
2. Be specific.
3. Give almost immediate feedback.
4. Keep it positive.
5. Provide frequent feedback, especially for novices.
6. Use cue words or phrases.
7. Use a variety of approaches.

Be Judicious With Feedback

A common mistake made by many professionals, especially novices, is to provide too much feedback to performers. Even if they understand the feedback, performers are often overloaded with several corrections, making it impossible for them to plan future executions to practice the movement. Coaches with a correction complex often rifle off a stream of corrections and feedback. This creates information overload (figure 7.2) for the performer and leads to paralysis by analysis. Remember that this problem is the reason there is an evaluation and diagnosis step in QMD. Diagnosis of performance results in the selection of one intervention to help the performer improve (Arend and Higgins, 1976; Hay and Reid, 1982; Hoffman, 1983; McPherson, 1990). Any feedback selected should also be directed at only one aspect of performance.

Figure 7.2 Too much augmented feedback as intervention can cause paralysis by analysis in performers. The subtle irony is that the important message of relaxation (essential in high-speed movements) is negated by the excessive corrections.

Practice and game performance can suffer from the psychological pressure of keeping too many things in mind. Too much information can be worse than no information. This overanalysis and extra feedback can create greater problems during competition. Psychological research has shown that in sport competition, the brain of a skilled right-handed performer has a highly active right hemisphere (visual, spatial functions) and the left hemisphere (analytical functions) is essentially turned off (Torrey, 1985). The phenomenon is parallel for left-handed players. When an athlete is so confident, focused, and relaxed that she achieves peak performance (flow theory), the experience is often referred to as playing *in the zone* (Chavez, 2008; Jackson, 2007). This experience may be accompanied by altered perception of effort, time, and pain. A coach who provides too much information, forcing the athlete to overanalyze what she is doing, may hinder performance.

Be Specific

For feedback to guide the performer's next practice trial, it should be as specific as possible. Specific feedback focuses on the exact element and how it needs to be changed (Christina and Corcos, 1988). The feedback should also be specific to the motor skill and at the student's level of understanding (Lee, Keh, and Magill, 1993).

A good example of specific feedback might be seen in a coach helping a Little League hitter. Saying "step" to remind players to stride in the direction of the throw could be too vague. It would be more helpful for the athlete if the coach said, "Sally, remember to step toward the target at least 2 feet." Using the cue "step" could then be a reinforcement for continued batting practice. In teaching and diagnosing weight training technique, "Take a wider grip on the bar" is less useful than "Line up your hands on the hash marks right here." Remember, if you prepare well for QMD and have developed task sheets and teaching cues, you already have a set of relevant cues for the movement. These cues are the action words in the execution phase of the movement description. One of the most important functions of cue words is building a common movement vocabulary between teacher and learner.

An excellent way to make feedback specific and motivational is to tailor the feedback to each individual. Many factors interact to determine what will be effective feedback for a particular person and situation. The performer's stress level, age, and personality are all factors in the choice of feedback. A good coach or therapist tailors (makes specific) to the individual the exact mode of feedback, the cue used, and the tone of voice. To do this, the professional needs a feel for the person's attitude, body language, and interaction with others.

Research in pedagogy has demonstrated that the pattern of instruction in physical education usually results in initial general feedback regarding a common error of many students (Siedentop, 1991). Later the teacher begins to provide more specific feedback focused on the individual. Unfortunately, the majority of feedback in physical education is general rather than specific (Siedentop, 1991). How can professionals create specific, individualized feedback as intervention?

Considerable research from psychology on personality and learning styles is helpful in individualizing feedback. In general, learners tend to be visual, auditory, or kinesthetic. Visual learners respond well to pictures or diagrams; thus, demonstrations or videotape replays are effective for them. Auditory learners tend to relate to and remember the words used by an instructor; cues are very helpful for them. Kinesthetic learners tend to use their bodies to help them learn. They could benefit from manual guidance or limited verbal feedback intended to focus their attention on a specific aspect of their movement during practice trials.

Perceptive professionals use their insights about their clients' learning styles and personality traits to devise specific feedback for each individual. A professional might choose to use a cue emphasizing proper technique with verbally oriented novices to help them get the basic coordination or **motor program**. For visual or kinesthetic learners, the professional would supplement traditional verbal feedback with other kinds of intervention like demonstrations and manual guidance.

Give Almost Immediate Feedback

Coaches' corrections or augmented feedback should be provided as soon as possible after the performance or trial because the feedback will be used to plan the next practice trial (Lee, Keh, and Magill, 1993). Immediate feedback helps learners make connections between that feedback and their kinesthetic sense and proprioceptive information (intrinsic KP) from the trial. Remember that the professional will be observing several trials, evaluating and diagnosing the movement before feedback is used as intervention. The learner must process the kinesthetic sense, proprioceptive information, and muscle memory of the movement before a professional provides augmented feedback (Chen, 2001). Part of the art of intervention is finding the minimal delay at which the intrinsic feedback can be combined with intervention and not lost.

The performer must focus on how the movement felt and relate this feeling to a judgment of its correctness. It takes time to think about the feedback and its association with the person's intrinsic KP. If feedback is nearly immediate (within a few seconds), the performer has time to compare it with her kinesthetic experience or intrinsic KP. At least five seconds should also be allowed for the performer to process and integrate these two sources of information before further practice (Schmidt and Wrisberg, 2008).

Research suggests that feedback need not be instantaneous and that *summary feedback* or *bandwidth feedback* is effective in learning motor skills (Chen, 2001; Schmidt and Wrisberg, 2008). Professionals can provide effective feedback that summarizes the performance of a block of several trials, although the potential link of the feedback with the performer's perceptions will be weaker. With bandwidth feedback, the instructor provides KP only when the performance is outside the range of correctness. The mover gets implied reinforcement if the professional provides no feedback after a performance. Many real-world settings are not like research settings, where one-to-one feedback can be given. Creating ideal feedback conditions may not be possible in most clinical and field situations.

Timing feedback so that it is close to the actual performance is another reason that the evaluation and diagnosis step of QMD is difficult. The professional must not only evaluate and diagnose the performance effectively but also attempt to do it in a timely fashion. Certainly the validity of the judgment is most important, but it is also desirable to maximize the guidance function of the feedback by responding as quickly as possible.

Keep It Positive

To be most effective, augmented feedback should be worded to instruct the performer in a positive way (Christina and Corcos, 1988; Lee, Keh, and Magill, 1993). Unfortunately, teachers of motor skills have not always followed this advice. Because motor skill teachers typically use visual models to identify errors in performance, their feedback may lapse into a series of don'ts. Selecting cues that are positive also tells the performer what to do, but without sending a negative message. Remember, a good QMD focuses on the qualities of a skill and not just on errors.

Research in typical physical education settings has shown that the majority of feedback is negative (Siedentop, 1991). Negative feedback is sometimes appropriate, but it should not be the primary mode of feedback. There is a big psychological difference between "Let's work on this" and "You still haven't done this right!"

In short, feedback should convey the message that the professional believes in the performer so that the performer is more likely to think, *I can do it.* When performers have problems with self-confidence and self-concept, positive feedback is very important. Feedback should encourage students and paint a positive picture of their potential (Ziegler, 1987).

James and Dufek (1993) suggest that intervention in QMD should refine strengths before correcting weaknesses in performance. Reinforcing or rewarding good performance characteristics helps athletes learn that part of the performance. It may also help motivate continued practice. This is a good intervention strategy to make sure that the feedback has a positive tone. A coach should suggest, "Step with your left foot over the blue line" rather than "You stepped with the wrong foot, Chet!"

So far we have focused on the tone of verbal feedback, but nonverbal communication should also be positive. Gestures, facial expressions, and other body language can strongly communicate positive or negative messages to performers. Professionals need to choose their intervention carefully and make sure their nonverbal communication is consistent with their positive verbal feedback. Patients and athletes will begin to doubt your credibility when they read differences between what you say and what your body language says.

QMD Demonstration 7.1

Observe the video clip in QMD Demonstration 7.1 in the web resource at www.HumanKinetics.com/Qualitative DiagnosisOfHumanMovement. Perform a QMD and propose two specific sets of augmented feedback for this performer. What principles of feedback seem most important in this case?

Provide Frequent Feedback, Especially for Novices

Early in motor learning it is important for the teacher to provide frequent feedback to guide the learner's subsequent practice trials. To help prevent confusion, provide some time for young learners to mentally process the feedback. In large classes, teachers do not usually have a chance to provide too much feedback to any one student. Thus, pedagogy research suggests that feedback rates should be high enough that every individual receives some feedback. In situations in which there is more one-on-one QMD, intervention should follow the motor learning principle of faded reinforcement. It is best to provide relevant feedback with every few trials initially, with decreasing feedback as practice proceeds or skill improves (Schmidt and Wrisberg, 2008).

As learners become more skilled, they need to rely more on their kinesthetic and proprioceptive intrinsic feedback than on the augmented feedback from the teacher. This is why professionals need to decrease the frequency of their feedback as performers become more skilled. In the advanced stages of motor skills, the kinesiology professional may question performers on what they feel were the good and bad points of their performances. Enlisting the athlete's help in the diagnosis can focus attention on the occasional correction.

Use Cue Words or Phrases

Kinesiology professionals have known for years that performers can remember and use information in practice better if it is presented as concise **cues**—cue words or phrases (Masser, 1993). Many books provide cues for teaching the skills of specific sports. There are also articles that suggest cues for teaching motor skills (for example, Fronske, Wilson, and Dunn, 1992; Kovar et al., 1992; Parson, 1998). Fronske has compiled several books of cues for teaching activities to students of various ages (Fronske, 2012; Fronske and Wilson, 2002). A particularly effective approach is to use metaphors to communicate the desirable technique (Gassner, 1999).

Good dance instructors are highly effective in their use of cues. Dance instructors must carefully time their cues to the music, use as few words as possible, and combine nonverbal cues to lead the group effectively (Shields, 1995). Simple cue words like *grapevine*, *box*, *cross*, and *plyo* can carry a great deal of information to a performer who has been

QMD Choices

You are teaching beginning, intermediate, and advanced tennis classes at the local junior college. Would the cues for the serve be different in these classes? If so, how might the cues be different in each of the classes? In which classes might students need more positive feedback with corrective cues to keep their confidence up?

taught these movements. A way to translate critical features into cue words or phrases is discussed in chapter 4.

Use a Variety of Approaches

A dictionary may lead people to believe that words in the English language have multiple meanings but that these meanings are stable. Teachers who have written test questions or experimented with various cue words, however, know that changes in the meanings of words may be quite dynamic. This is why kinesiology professionals should collect cue words or phrases for each critical feature of the motor skills they teach. They will then be able to provide several cues to communicate the essential idea of each critical feature they evaluate.

Professionals should also try to have a variety of modes to communicate cues as feedback. Some performers respond best to a verbal cue, others to a visual one, and still others to a kinesthetic one. Cues for the last step in a basketball layup might be verbal ("long step and jump"), visual (footprints on the floor), or kinesthetic ("Make it feel like stepping over a beach ball").

Professionals need to use age-appropriate cues when teaching small children and cues using current slang when teaching in elementary and secondary schools. Most veteran physical educators or coaches will have had an experience similar to that of Jeanne Jones, who was teaching a basketball unit to junior high students. A particular student was having trouble shooting along the appropriate trajectory. Jones had worked with this student over several class periods, providing cues such as "Shoot with a high arch" and "Shoot upward" to help correct the angle of release. Finally she found a cue that clicked with the student: "Shoot up through the top of a phone booth." When the student began shooting better, he said, "Why didn't you tell me this before?" Jones smiled, even though she wanted to say, "I've been telling you for the past week!" She knew that she needed a variety of cues to communicate with the culturally diverse students she taught. Nowadays instead of a "phone booth" coaches might use "top of the stairwell" or "elevator."

To determine if different feedback is needed, question students or clients on what feedback was given. A dance teacher could ask, "Jenny, what were we working on with your waltz last time?" If the student is not making the changes you want, repeat or rephrase your feedback. Questioning is a good technique when you can't tell whether the performer understands what you want her to work on specifically (Christina and Corcos, 1988).

Kinesiology research has shown that feedback is an effective intervention in QMD. Feedback should be as specific as possible, given as a cue about a single aspect of performance, and phrased in positive terms. Feedback is most effective when it is provided as KP with minimal delay. Professionals must carefully choose whether to focus feedback on the movement itself (KP) or the outcome of the movement (KR). They should also decrease the frequency of feedback with increasing skill level and be prepared to provide feedback in a variety of ways.

> **» KEY POINT 7.3**
>
> In general, verbal augmented feedback should be specific and should be expressed positively with cue words or phrases appropriate to the performer.

QMD Technology 7.1

Coming up with effective cue words and phrases can be difficult. Children and adolescents produce slang and attribute novel meanings to words, which can impede child–adult communication and thus interfere with intervention. Using the Internet to search for similar words is a good strategy to expand the power of movement cue words and phrases. Check out the following websites when you are revising cues: http://thesaurus.com or www.merriam-webster.com. What are other web sources for current terminology and slang that might be appropriate for intervention?

Imagine you are a physical education teacher in a typical gym with 50 fifth graders. Your basketball unit is under way, and you have the class at eight stations practicing shooting and ball-handling skills. Research in sport pedagogy suggests that to be effective, the teacher should move randomly throughout the gym, actively supervising practice, providing positive feedback to motivate students, and performing QMD of their performances. Motor learning research has shown that novice subjects like many of those in this class require somewhat frequent augmented feedback to improve. How are you going to work it all in?

Your attention is drawn to Hugo, who is having difficulty hitting the spot on the wall for the chest pass. Observation and evaluation and diagnosis suggest that his step is correct, but his arms do not appear to propel the ball with enough speed. How do you select appropriate intervention when you focus on one performer to qualitatively diagnose shooting? Your mind flies. *Quick! Hugo's throw, shooting, and push-up score are above average. Hugo is a great little person who always tries his best. I bet he needs a little more forearm pronation and coordination with the extending arm*. Your voice and smile turn on. "Nice pass, Hugo! I bet you could pass like Jason Kidd or Steve Nash if you used your hands more. Give me that upside-down five (hand slap) on this next pass. Great!"

Moving on to the next performer, in your gym voice you give another general prompt to hustle to the whole class: "Squad four looks great! Kim and Hugo almost knocked the targets off the wall!" Looking over to squad five, you think, *Bobby looks like he needs some praise. Let's see what looks good in his shot*. In 15 seconds you have weighed information from many subdisciplines of kinesiology. Pedagogy and psychology considerations made it important for you to praise the effort and good aspects of the children's performances. Knowledge of the children made diagnosis and intervention of their basketball skills more accurate. You were also sensitive to the body language of a child who might need a little attention and praise to feel good about his shooting ability. The intervention you used for Hugo included augmented verbal feedback about the forearm action of the pass. You linked the feedback to a previous cue (upside-down five) and used some manual guidance by putting your hand in the follow-through position for Hugo's next pass.

BEYOND TRADITIONAL FEEDBACK

There are some tricks of the trade beyond traditional augmented verbal feedback for providing intervention to improve motor skills. They are used effectively by teachers, coaches, and clinicians in all kinds of environments:

- Using visual models
- Exaggeration or overcompensation
- Modification of practice
- Manual or mechanical guidance
- Conditioning

- Attentional cueing
- Ecological intervention

While some of these could be viewed as a kind of feedback (i.e., manual or mechanical guidance, attention cueing), others are clearly not classic verbal feedback. Unfortunately, a narrow view of intervention (traditional feedback) can limit the effectiveness of professional's QMD. These forms of intervention, which typically do not qualify as traditional feedback, have not been studied from an interdisciplinary perspective to determine which are the most effective. The following sections describe some of these methods; professionals must make their own decisions about which methods are most appropriate.

Visual Models

There are several ways to provide visual feedback on the status of performance or a desirable correction to a performer. Demonstrations by the instructor may be effective because most people have a visual learning style. Years of motor learning research have shown that modeling or observational learning is the most effective way to convey information to people who are learning a new pattern of coordination or are establishing coordination in the early stages of learning a skill (McCullagh and Caird, 1990; Messier and Cirillo, 1989; Wood et al., 1992). In observational learning the professional provides visual information like a picture or demonstration to a learner.

Posters of key body positions in exercise are highly effective in the weight room. The walls of elementary school gyms could have key cues and corresponding pictures for teachers to use as intervention. Such visual aids are excellent for improving the atmosphere of the gym and decreasing the stress on a teacher's body resulting from demonstrations in many classes every day of the week. Figure 7.3 illustrates how a golf instructor can use a hoop to create a visual image of the swing plane of the golf swing.

Several scholarly reviews of modeling or **observational learning** research can assist the professional in using this form of intervention (Gould and Roberts, 1982; McCullagh, Weiss, and Ross, 1989). Ashford, Bennett, and Davids (2006) reported a meta-analysis of this large body of research and concluded that observational learning compared to regular practice has a greater effect in helping adults adapt coordination than on eliciting improvement in movement outcomes. Interestingly, children do not appear to share these coordination improvements with observational modeling, but do have small improvements in movement outcomes with observational models compared to regular practice (Ashford, Davids, and Bennett, 2007). Professionals who want to use observational learning as intervention can use the

Figure 7.3 An instructor can use the visual image of a hoop to illustrate the desirable swing plane of the golf swing.

following guidelines for providing good demonstrations. Multiple trials should be presented by a skilled model similar to the performers (McCullagh, 1986, 1987; Wiese-Bjornstal, 1993) with verbal guidance by the instructor (Williams, 1989a, 1989b, 1989c). The demonstration should normally also be presented from the performer's perspective (Ishikura and Inomata, 1995).

It is important, however, for professionals to understand that there are two situations in which observational learning has not been shown to be superior to other feedback as intervention. Observational learning is not very effective in creating improvement in motor control or for refining and customizing a movement to situational constraints. For example, watching a pitching coach demonstrate a pitching technique is not likely to immediately help a high school pitcher improve pitching (control or pitch location for the new technique) in different game conditions.

Another common form of visual intervention is the video replay. There has been a great deal of research on the effectiveness of motion pictures, loop films, and video replays in teaching motor skills. A classic review of video replay as intervention in learning motor skills was presented by Rothstein and Arnold (1976). The review examined 52 studies; the majority found no significant difference between video replay and normal teacher feedback. Additional research on video intervention has supported these results (Emmen et al., 1985; Miller and Gabbard, 1988; van Wieringen et al., 1989). Although little scientific evidence exists to show that video replay as intervention is superior to traditional teacher feedback, its use for QMD may be justified. This justification, however, is not that a mover viewing a video will receive unique feedback that will enhance performance beyond other kinds of feedback. Chapter 8 discusses how to maximize the effectiveness of video by the kinesiology professional in several tasks of QMD, not just intervention.

Exaggeration or Overcompensation

The muscle memory that is necessary for skilled movement is also sometimes our biggest obstacle in creating changes in a movement. Because even small changes in technique can be very difficult for players to make, many experienced teachers and coaches exaggerate the desired correction in feedback. Good examples of this are the serve in tennis and shooting in basketball.

The initial trajectory of the ball in the tennis serve of recreational players needs to hit nearly horizontal or slightly upward. A common error is to hit downward on the serve and to net a lot of serves. Players who have trouble getting the feeling of the upward action of the serve can be encouraged to try to hit the back fence with the serve. Often this results in the desired upward service action and a serve that lands deep in the service box.

Exaggeration or overcompensation can also be used to correct a common error in basketball shooting. Players often shoot on a low trajectory at the rim when they should have a higher angle of release. Telling players with this problem to shoot with a high arc is effective in improving shooting, although skilled basketball players do not really shoot with a high trajectory (Hay, 1993). This cue and the QMD of the jump shot are reviewed in chapter 8.

A professional using overcompensation as intervention should do so with care. Whether the exaggeration is effective or not, the performer should be informed later that the cue words were not literally accurate. They later can explain that the exaggeration was necessary to create the desired change, or just that it is a good description of what the desirable technique may feel like. Do not let misconceptions about performance persist in athletes. Eventually the exaggerated technique may develop, or the cue may be passed on to other performers when it is inappropriate. Remember that the performers of today are often the instructors of tomorrow, and they are likely to teach as they have been taught.

Modification of Practice

An option often overlooked by professionals is to change practice as an intervention to improve performance. The kind of practice used varies with the kind of motor skill being learned and the performer's skill level. Professionals must often make the practice easier for novice performers to accommodate deficits in strength and skill. Coaches working with advanced athletes should change practice tasks frequently to challenge the athletes and maintain their motivation to practice. Remember that advanced performers require a great deal of practice to improve a small amount.

> **» KEY POINT 7.4**
> Techniques of intervention beyond traditional verbal feedback in QMD include the use of visual models, exaggeration of corrections, modification of practice, manual and mechanical guidance, conditioning, attentional cueing, and ecological intervention.

Fortunately, there is a large body of motor learning research on the effectiveness of various forms of practice and practice schedules. For example, the first (cognitive) stage of motor learning focuses on learning the basic motor program for a particular task. Good intervention for a person having trouble in this stage might be a combination of feedback and modified practice. Breaking the task into parts, making it easier, or eliminating attention to outcome could all be effective ways to help learning. Schleihauf (1983) demonstrated that a successful way to modify breaststroke swimming practice was exaggerating the glide phase to give the athletes more time to cognitively process the coach's feedback.

Practice Environment

The large cognitive demands of learning a new motor skill usually require that any change in practice as intervention should occur in a closed environment, in which the conditions of the immediate environment do not change a great deal. The external factors of performance are reasonably consistent. Conversely, in an open environment, the conditions (opponents, obstacles, and so on) are changing.

Modifying practice in progression for a beginner learning to dribble a basketball would typically mean moving from a closed to a more open practice environment. The performer would dribble in a small area, dribble moving slowly in one direction, dribble around objects, dribble with others moving randomly, and finally dribble to avoid defensive pressure. In the early stages of learning, cognitive attention focuses on the movement. As the performer learns, attention can progressively be moved from the skill to the environment.

Practice Equipment

Another possible intervention in QMD is to change the equipment used by performers. Practice with different equipment can improve performance in several ways. Improvement in performance can be dramatic when appropriate equipment for a performer is identified, like the correct club length for a golfer's height or a child's tennis racket for a young learner. Equipment that is lighter or heavier than normal may be used in practice to provide an overload or training effect. Training with under- and overweight baseballs has been shown to improve throwing velocity more than normal throwing (DeRenne, Ho, and Blitzblau, 1990).

Another way to improve performance with the modification of practice equipment is to use teaching aids. This approach is common in golf, in which simple homemade aids such as hoops, stance or target line mats, and club face orientation attachments are often used (as in figure 7.3). A professional can modify equipment for practice or long-term use as an intervention strategy to improve movement.

Practice Schedule

The prescription of practice and practice schedules has been a major area of motor learning research. The classic practice scheme in athletics—repeated practice trials without rest—may not be the best way to learn motor skills. Practices may be organized (practice:rest schedule) along a continuum from massed (blocked) practice to distributed practice. Blocked practice involves many repetitions or trials of a task in a block before another practice task is introduced, with little or no rest between trials. Distributed (random) practice involves a number of different movement tasks with variation in order and rest intervals; trials alternate rapidly among different movements.

These two practice schedules have slightly different effects on immediate performance in practice and long-term performance or learning effects. For discrete motor skills (tennis serve, basketball shot), massed practice does not usually result in degradation of learning. For continuous motor skills where fatigue can be a factor, however, massed practice tends to decrease practice trial performance, but has a small effect on learning (Schmidt and Wrisberg, 2008).

The interesting paradox is that random practice tends to outperform blocked practice when one examines their long-term (learning) effects. Early in learning a new movement, blocked practice is effective in helping performers develop a basic motor program. But motor learning research suggests that blocked practice leads to good performance in practice and an exaggerated sense of skill, followed by poorer long-term learning. Repeating the same task over and over may initially lead to what looks like improvement (traditional behaviorist approach) but will not generalize to other movement situations as well as random practice does. Random practice, with trials alternating rapidly among different movements, results in poorer practice performance but better long-term learning (Schmidt and Wrisberg, 2008). This contextual interference effect (Schmidt and Wrisberg, 2008) means that practice as intervention for most movements should vary rather than repeat the same task. This emphasis has also been called behavior training (Vickers, 2011) or games-approach learning.

Planning practice with multiple skills or tasks then becomes a factor in modifying practice as intervention. When prescribing practice as intervention, carefully evaluate the performer's situation and goals. Modification of practice interacts with many other aspects of physical training and learning other motor skills.

Variations in Modification of Practice

If a complex movement can easily be broken down into phases, it might be a good idea to provide intervention by progression. Part–whole learning is generally not as meaningful or as easy to put together as whole–part learning, but it's a good remedial approach for someone who is having a problem. An example is a student having trouble with footwork in the basketball layup. It might be good intervention to modify the practice task by allowing the student to carry the ball and work on the approach footwork in isolation.

The consensus of research suggests that if practice is to be modified to help improve motor learning, the approach should depend on the level of the performer, the kind of movement, and the other movements being learned. The quality of practice is highly important; the time spent on a task is less important than was once thought. It is clear that frequent changes in practice conditions, tasks, and repetitions assist in motor learning. A challenging practice session may look worse in execution, but the retention and transfer of skill are better in the long run (Schmidt and Wrisberg, 2008).

Frequent changes in task and feedback are important if a performer is working on a difficult correction or change in a movement pattern. A good intervention strategy in situations like this is to change the practice task and then return to the observation task of QMD. Remember that in the real world, the professional can move from intervention back to observation. This immediate gathering of new information on performance and on the effectiveness of the intervention should speed up the improvement in the client's or student's performance.

The last thing to remember about modifying practice is that competition may be a big help in QMD. Most coaches know that you cannot use practice to predict performance in actual competition. Some people thrive on pressure, while others fold in competitive situations. Motor learning research shows that there is no simple association between practice performance and performance in retention trials (learning). Drills and practice should be changed frequently, with many activities involving competition and game-like situations. Making practice drills competitive can help nervous performers get used to playing under pressure. In an intermittent activity like tennis, practice can mimic match conditions with short, intense drills followed by a short rest. Coaches can use this intervention and can teach relaxation techniques for players to use during the rest breaks.

Manual and Mechanical Guidance

Sometimes a teacher physically moves or holds a performer's body in specific positions to give the athlete a feel for the position or action. This is **manual guidance**. Spotting in gymnastics and having athletes freeze on command so that the coach can manually change the athlete's body positions are examples of manual guidance. Giving performers this kinesthetic sense of the position or action can be highly effective, but the professional must be careful not to violate people's cultural or personal taboos against being touched. Kinesiology professionals need to be sensitive to these concerns of their clients.

Mechanical guidance involves using some aid or mechanical device to help the performer make the appropriate movements (Lockhart, 1966). For example, in golf, mechanical guidance may be provided by a swing aid, brace, or strap. The sports of golf and basketball seem to generate a continuous stream of new mechanical devices that are supposed to help players get the feel of or "groove" the desirable swing or shot.

One of the problems with both manual and mechanical guidance is transferring the new feeling into practice movements and unlearning the older muscle memory or faulty motor program. It is difficult for players to unlearn motor programs, especially if they have been using the movement for some time. Manual guidance should be used carefully; the professional needs to make sure the performer understands that it will take concentrated effort and a lot of perfect practice to make movement changes. Manual and mechanical guidance may be most effective when the performer is at a low skill level and is comfortable with instructor contact (Lockhart, 1966).

Another problem is that guidance can increase the risk of injury. An example of this risk was given by tennis pro Vic Braden (1983). A tennis pro wanted a player to get the feel of keeping his feet on the court during the serve to correct an early jump in his serve. The pro asked the player to serve with his foot in a shoe that had been fixed to a board, which had been nailed to the clay tennis court. The player served and got the feeling of an injured ankle! Helping the performer feel the correct movement can be effective, but it must be done with care. A tennis player dragging his toe and a golfer with a weight-shift problem can better feel their performances if they remove their shoes to focus their attention on the foot in question.

Conditioning

If the professional believes that a critical ability is lacking, appropriate intervention may be physical training or conditioning. A student who lacks strength to perform a skill might be helped by several intervention strategies: changing equipment, modifying practice, or prescribing physical training to increase strength. If a performer lacks flexibility to perform a skill optimally, an integrated QMD might prescribe a static stretching program in conjunction with modified skill practice to gradually increase range of motion. The increase in resistance to muscle stretch (dynamic flexibility) is different from the passive motion limits of joints (static flexibility); however, the best and safest conditioning for both properties is static stretching exercises (Knudson, 2008). If a lack of flexibility poses a risk of injury, intervention should focus exclusively on this issue until safe practice is possible.

Kinesiology professionals often try to verify their judgments of physical limitations with physical tests (40-yard dash, vertical jump, three-hop test). Advancements in motor development, exercise physiology, kinanthropometrics, and sports medicine research may help identify key quantitative measurements and fitness variables related to performance and potential injury.

Attentional Cueing

Attentional cueing refers to learners' ability to take themselves through a movement with specific cue words (Janelle et al., 1997; Minas, 1977; Morrison and Reeve, 1993). This process can involve the instructor's providing intervention as cue words to guide subsequent performance, or the cue words can be developed cooperatively with the performer. Cue words should focus on the actions of the movement and the most meaningful parts of the movement, like the Vickers teaching cues illustrated in chapter 4. The action words or the first words in the execution phase of the movement description become the cues to movement. Overarm throw examples include "turn," "step," "rotate," "swing," "snap," and "follow through." Cues for the soccer throw-in can be seen in figure 4.5. This intervention strategy can teach performers to become more aware of the movement as they are learning it and can be useful for self-instruction in the future.

Ecological Intervention

The ecological approach to skill acquisition is primarily founded on the notion of direct realism (Handford et al., 1997). This approach says that biological systems do not need a highly developed mental representation to use sensory information. Skilled movements are based principally on the relationships among the performer, the task, and the environment (Handford et al., 1997). There is a continual revolving relationship between perception and movement that shapes the movement form. The movement product is not based on a schema of the movement but is continually adaptable to the changing environment. This is seen in a hockey player who is adapting the backswing on a slap shot based on defensive conditions.

To enhance this interaction between perception and movement, the environment needs to be manipulated in relation to the responses the performer will make in varying game situations. This is vital if the performer is to enhance her particular movement responses within the constraints of a particular environment. Because the number of environments is nearly infinite, infinite responses are available. A professional sensitive to an ecological approach would provide interventions with as much variety in the environment as possible for the level of the performer. Recall that the environmental adjustments need to take into account the level of the learner. Appropriate environmental changes that add contextual interference can enhance learning.

The goal of an ecological approach is the performance of a particular movement. An objective for a movement is established, and attainment of that objective is sought. Take, for example, the objective of throwing a ball for distance as opposed to throwing a ball for accuracy. The objective of the movement is distance, while attaining the form to throw the ball that distance is the goal of the ecological instruction. How the movement proceeds is of paramount importance. The quality of the movement is the desired outcome of the ecological approach.

Balan and Davis (1993) have shown that an ecological approach to teaching physical education naturally aligns with ecological strategies of intervention in QMD. Their model has four main components: task goals, choices, manipulation, and instruction. Figure 7.4 illustrates this ecological instructional model. Each of the components is designed to allow the teacher to partner with students to collaboratively improve the quality of their movement. When considering the goal (a general movement objective like locomotion, object manipulation) of a movement to be learned, one should design the practice environment and intervention to elicit the desired movement goal. The second stage of the model allows the performer to move in a way that will achieve the goal of the task. For example, a physical educator teaching catching would set up a variety of activities in which students could experiment with techniques to catch different kinds of balls. Again, the observer evaluates the quality of the movement within the context of environment. Intervention (instruction in figure 7.4) may again focus on changing the environment to change the action of the performer. The teacher may challenge the student by modifying the task or environmental constraints to train the mover's technique or physical capacity. In the

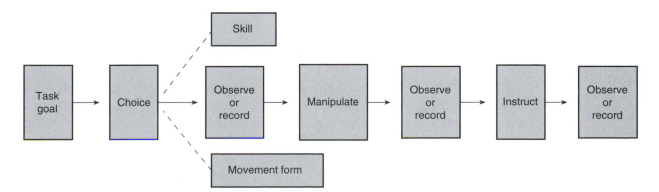

Figure 7.4 The Balan and Davis (1993) model for assessment and instruction in an ecological approach to teaching.

last component, the professional may actually select explicit instruction in technique in order to enhance the movement. Implicit in this overall approach is that students collaborate with the instructor to develop a movement pattern that is optimal for them and one that is closely matched to task and environmental constraints.

In an ecological approach to intervention, it is crucial that the professional understand the *range of correctness* concept advocated in this text. Ecological intervention is also consistent with the contextual interference effect of motor learning noted earlier. Other principles that seem to underlie this intervention strategy of movement enhancement include the following proposals by Handford and colleagues (1997):

- Practice should be variable.
- The majority of skills are seen as open skills.
- High cognitive and contextual interference enhances acquisition.
- Practice conditions should match competition conditions.

Biomechanical research may propose an apparent optimal movement pattern for cross-country skiing, but an interdisciplinary and ecological approach to QMD remains open to a range of correct techniques. Physiological, psychological, and anatomical along with environmental conditions might suggest that different movement patterns are optimal rather than the idealized form. The focus of the ecological approach is to get the performer to be as biomechanically efficient as possible within the environmental constraints. If professionals do not take these constraints into account, they will not be able to manipulate the environment in such a way as to optimize the learner's performance. The ecological approach to intervention in cross-country skiing may involve practicing on different types of snow—granular or powder. Skiers should also practice on well-groomed and ungroomed tracks as well as up and down slopes and on cambers where one ski is higher than the other. A coach might modify training in preparation for a competition where certain elevation and wind conditions are expected. All these variations should be mixed and combined to provide environments with different perceptual challenges, which should stimulate a variety of responses to enhance performance.

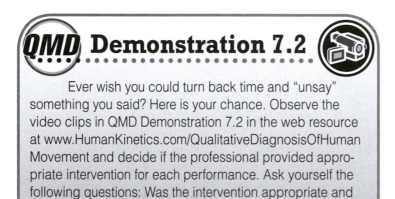

QMD Demonstration 7.2

Ever wish you could turn back time and "unsay" something you said? Here is your chance. Observe the video clips in QMD Demonstration 7.2 in the web resource at www.HumanKinetics.com/QualitativeDiagnosisOfHuman Movement and decide if the professional provided appropriate intervention for each performance. Ask yourself the following questions: Was the intervention appropriate and Why? If not, is there a better intervention?

FURTHER PRACTICE

For ethical reasons, it is not recommended that you try out or experiment with feedback and intervention on people without weighing evidence and client consent. A thoughtful professional always strives to give the most effective intervention and, following many interactions with clients, critically examines the effectiveness of past intervention. New kinds of practice or cues can be explored, but professionals should not experiment with clients unless there are good logical and empirical reasons to try the new intervention. Care should also be taken to obtain a client's consent to try new or experimental practice that is not supported by evidence. Further practice for intervention will have to happen over the long term, within a career-long vision of professional development.

SUMMARY

The fourth task of QMD is the provision of some intervention to help the performer improve. Intervention is not limited to traditional augmented verbal feedback. It includes other methods to teach and train a person to move better. A rich history of research on feedback suggests that augmented verbal feedback used as intervention should be limited to a cue word or phrase that is as specific as possible and is given with minimum delay. Augmented feedback can be highly effective if it is expressed in positive terms, as knowledge of performance (KP), and in language that is age appropriate or specific to the individual. Another consideration for those who wish to improve performance is the use of intervention techniques that are beyond verbal feedback. Simple changes in the environment can add to the effectiveness of the intervention.

Discussion Questions

1. What kinds of *augmented feedback* are best for a novice, an intermediate, and an expert performer? How do these kinds of feedback vary with different sports or activities?

2. What motor skills would benefit from a performer focusing attention on *intrinsic feedback* as intervention?

3. How does the age of the performer affect selection of the mode of intervention in QMD?

4. You are a high school soccer coach. What hand signals can you use to provide intervention to players who cannot hear you in competition?

5. What factors should one consider when selecting cue words for particular movements?

6. What rationale for intervention is best?

7. Should intervention focus on short-term or long-term performance? Why?

8. Which kind of intervention is most important, reinforcement or corrections? Explain.

part

III

Practical Applications of Human Movement Diagnosis

Part III is dedicated to the practical application of our integrated model of qualitative movement diagnosis (QMD). The best way to illustrate such a model is to describe the process in several examples. Chapter 8 presents examples of how digital video and computer technology may enhance QMD. Chapter 9 presents selected examples from a variety of human movements that are designed to illustrate the interdisciplinary nature and tasks of QMD. Video from real movement situations was captured for these tutorials. Chapter 10 poses hypothetical situations in which QMD could help solve a problem in human movement. These "theory-into-practice" situations are helpful for beginning discussions on QMD issues with other professionals.

Video Replay Within Qualitative Movement Diagnosis

You are a coach who has used video to record and qualitatively evaluate the technique of an athlete. You receive information on a new handheld computer with integrated video and wireless Internet access. Can use of this system improve your qualitative movement diagnosis? Video replay has often been used to extend observational power, but does this system share these advantages or does it have limitations? Which of the computer-enhanced features of the digital video clips are of real value? Which of the four tasks of qualitative movement diagnosis benefit the most from a system like this?

Chapter Objectives

❶ Describe the uses of video replay for extending observational power within QMD.

❷ Explain how to use video and computer replay to maximize QMD ability.

❸ Describe spatial and temporal limitations of commercial and consumer video equipment in documenting human movement.

An important tool in extending observational power within qualitative movement diagnosis (QMD) is the use of video replay, especially slow-motion replay. Video replay may be most useful in providing information to the professional that is unavailable to real-time observation. Video can capture fast elements of the movement that are unobservable by the naked eye. This greater movement detail and unlimited capacity for replay make video an important tool for extending the observational power of the kinesiology professional within QMD. This chapter reviews the factors that are important in using video to improve QMD of motor skills. These factors also apply to recent advances in computer grabbing, storage, and presentation of video clips that have been developed for QMD.

INTRODUCTION TO VIDEO REPLAY IN KINESIOLOGY

Many of the early studies of video in teaching motor skills focused on its use as visual feedback. Reviews of these early studies found that video replay used as feedback to performers was not significantly different from regular practice and teacher-augmented feedback in improving motor skills (Rothstein, 1980; Rothstein and Arnold, 1976). People seeing themselves perform on video replay do not appear to spontaneously perceive important aspects of their movement to improve performance. However, some studies have shown that video replay may have some benefits, due to the observational learning or modeling of the performer's behavior, beyond intrinsic feedback (Dowrick, 1991; Gould and Roberts, 1982). Recall the summary of observational learning research in kinesiology in chapter 7. Even though the knowledge of performance (KP) or modeling information in video replay is not any more useful to the performer than other inherent feedback, video replay can be an important supplement to QMD for the professional. We will see that the benefits are related to extending observational power for evaluation, for diagnosis, and for providing unique visual intervention through video replay. The section on using video replay for QMD reviews important aspects of this research in more detail.

> ## »KEY POINT 8.1
>
> Research has shown that video replay does not provide any special information to the mover beyond other forms of inherent feedback from movement practice. Video replay, however, can provide additional information on performance to the professional skilled in QMD using this technology.

If used correctly, video replay may have benefits for the QMD of motor skills (Franks and Maile, 1991; Rothstein, 1980; Trower and Kiely, 1983). A major advantage of video replay is that it shows high-speed details of the movement that are not perceivable in real-time observation. Video-recorded performances also have virtually unlimited and slow-motion replay potential to increase observational power. Computer programs can even extend and enhance these replay advantages of video. Video replay can be useful in working with clients needing some technique adjustments, in that the clients cannot argue with a picture that proves that they have not mastered a technique or critical feature.

Before any discussion of the use of video in QMD, teachers and coaches should know how video imaging works and should be aware of the limitations of the medium. The rest of this section describes the basics of video imaging and how these technical facts shape the professional's use of video within QMD.

Two-Dimensional Image

A normal photographic or video image from one camera is a 2-D representation of a 3-D scene. This means that only the motion of objects oriented at right angles to the field of view are represented accurately, in the two dimensions of the image. These two dimensions are usually the vertical and horizontal when the camera is aligned to these important real-world directions. Anything aligned toward or away from the camera (any plane not parallel to the image) will be distorted in the image (2-D representation) taken of the real world (3-D reality). For a simple demonstration, extend your first and second fingers and hold them vertically at arm's length from your eyes with your forearm pronated (so you are looking at the back side of your hand and fingers). Note the angle formed between the fingers and how it appears to shrink when you rotate your forearm so your thumb is now toward you.

Suppose three TV cameras were placed next to the runway of the gymnastics vault (figure 8.1). The coach believes that the angle between the gymnast's extended arms prior to the block on the horse should be a specific angle for this athlete. The coach can get an accurate estimate of the angle between the two arms of the gymnast only when the camera is at a right angle to the motion of the arms at the instant of interest (camera B in figure 8.1). The views from cameras A and C show an arm angle that is smaller than the actual angle between the gymnast's arms. This is why an athlete can look as if he is stepping on a boundary line in one camera view, while a camera view down the boundary line shows that the athlete is inside the line.

The point to remember is that video images are 2-D representations of a 3-D reality. Some predictable distortions of the image are caused by the position and orientation of the objects in the field of view.

> **» KEY POINT 8.2**
>
> The 2-D images created by normal photography or video often provide distorted representations of the 3-D reality. For example, objects not parallel to the camera can appear smaller than they actually are.

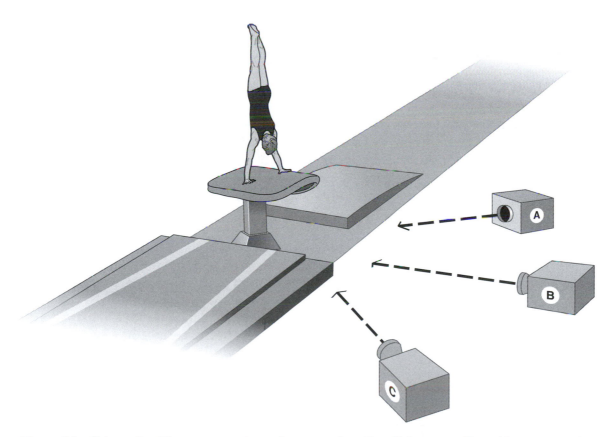

Figure 8.1 Schematic of three camera views of a gymnast vaulting. Only camera B provides an accurate image of the angle of the performer's arms as she blocks on the table.

People who use video for QMD should set up recording sessions to minimize these distortions and should know when these distortions may affect judgments about the performances. We will see that these distortions can also ruin many calculations that might be made with the digitizing and calculating capabilities of computer-enhanced video replay and presentation systems.

Video Image

Video images are also limited representations of reality because they are made up of a 2-D array of dots. These dots that compose each picture are called **pixels** (short for *picture elements*). Each picture element is given a shade on the gray scale (from black to white) for a black-and-white video picture. Color video is made up of pixels that have been given an intensity or mix of red, green, and blue light. A video picture is called a **frame** and is made up of two halves, or **fields**. One field is the odd-numbered horizontal lines of pixels, and the other field is the even-numbered horizontal lines of pixels. This is why normal video is called *interlaced* video. The more pixels and lines of pixels in a video frame, the higher the resolution and the better the quality of the image. Interlacing of video is a technology holdover from early television in which vertical resolution was sacrificed to balance image quality and smoothness of fast motions.

Resolution and Bit Depth

The number and color depth of the pixels in a video frame determine the quality of the image. Traditional television production video equipment can have 525 (National Television Standards Committee [NTSC] format) or 625 (Phase Alternation Line [PAL] format) horizontal lines of pixels (Feldman, 1988). The number of pixels for a given field of view determines the **resolution** of the video picture and thus the image quality. Resolution in all directions is a critical issue when video is used for actual measurements (quantitative analysis), but it is still somewhat important in QMD if the images are of poor quality or the subject is small in a large field of view.

Video picture format is determined by broadcasting conventions, which vary around the world. The NTSC standards are used in North America and Japan. The United Kingdom, Australia, and European countries use PAL video picture format, while the French and Russians use Séquentiel Couleur à Mémoire (SECAM) video. NTSC video has 30 frames or pictures each second (actually 29.97); PAL and SECAM video have 25 frames per second. Video equipment that can switch among these international formats is available.

When consumer electronics are used to video record and display human movement, the resolution of the video varies based on the equipment and system purchased. There is a difference between the resolution of one video component (camera) and the resolution of the whole system. For example, even a professional NTSC video signal may show up on a good TV as only 490 horizontal lines because it takes time to jump from the bottom of one frame to the top of the next. Camcorder video signals may have lower horizontal and vertical resolution than professional video. NTSC video produced by consumer camcorders, for example, generally creates images with 300 to 330 horizontal lines, with 235 to 245 pixels in each line. High-band consumer video formats (Super VHS [S-VHS] or 8 millimeter [Hi-8]) typically have 400 to 480 lines of horizontal resolution, with up to 640 pixels per line. A problem with higher-resolution video is that TVs and video cassette recorders (VCRs) sometimes use only 330 horizontal lines in a video frame.

Computer monitors and large-screen televisions are capable of taking advantage of high-definition (HD) video. HD video has about 720 to 1080 and 1280 to 1920 pixels in each line. Some recent HD video formats have even higher resolution (QSXGA: 2560 by 2048 pixels). HD video files, even when compressed by a codec, are quite large and only recently have been easily handled by fast computers and inexpensive digital data storage (chips, CDs, and flash drives). Codecs are compression or decompression routines that facilitate the storage and presentation of the large amounts of information in digital video files. Many digital video cameras now provide a variety of temporal and spatial resolution options, for example, of 60-, 210-, 420-, and 1000-Hz video. The horizontal and vertical resolutions of these formats are 480:360 at 60 Hz and 224:56 at 1000 Hz. To take advantage of the improvements in resolution of higher-resolution or HD video, one must factor in the added cost of tapes, digital chips or cards, computers, and monitors.

Digital video has been integrated into camcorders, still cameras, tablets, and cellular phones. Some older digital camcorders may use the same TV field (60 interlaced fields to create 30 frames per second of video) formatting to be compatible with analog equipment along with digital microcomputers. Note that the vertical resolution (the number of horizontal lines) is essentially fixed by the format of the video. The horizontal resolution (number of pixels in each horizontal line) varies according to how the analog video signal is sampled (digitized) or how the digital video is captured. When a video ad claims 400 lines of resolution, this typically means that the equipment can create 400 pixels in each horizontal line. Computer video systems can often handle larger horizontal resolutions because of differences in computer and TV displays. Added horizontal resolution depends on the video board or the digital camera that does the digitizing. Video files exported from cell phones or for web use are of lower temporal (typically 10 to 15 Hz) and spatial resolution (160 by 120 or 320 by 240 pixels) than other video formats to minimize the computer memory and transmission costs.

Bit depth is the factor that controls the number of colors that can be used in digital video files. Image quality increases with larger bit depths, but so too does the file size. An 8-bit video image can represent 256 levels of colors while a 24-bit has 16.7 million levels of colors.

Paused Video Image

The resolution of video, and consequently the quality of the picture, may become worse when a VCR, camcorder, or computer software implements freeze-frame or pause mode. When many consumer VCRs are put in pause mode, only one field is shown in order to prevent a flickering picture. Flicker is caused when motion occurs between the two halves of a video frame (1/60 or 1/50 of a second apart), making the image look blurry or seem to vibrate between the two positions on the screen. Some VCRs or frame grabbers used for computer-assisted video replay effectively present 60 or 50 half-pictures (fields) to look at when the frame-advance or slow-motion features are activated. Most digital camcorders in pause and frame advance show images 1/30 of a second apart. Unfortunately, most VCRs are similar, jumping from the first field of a frame to the first field of the next frame when the frame advance is pushed, giving 30 lower-resolution pictures per second.

Video pictures are physically bigger horizontally than vertically. The ratio of field width to height is the aspect ratio of the image. Video images have nowhere near the aspect ratio provided by the horizon and normal vision. Normal video has an aspect ratio of 4 to 3. This difference in field of view, combined with pixel resolution and lost resolution in freeze-frame mode, affects the quality of objects represented in these directions on almost all video (analog or digital) images.

When you use consumer video for slow-motion or frame-by-frame QMD, you are often looking at a video field with half the vertical resolution of a normal video frame. If the video images were created with poor lighting, a small image size for the performer, a noncontrasting background, an unfavorable camera location, or some other problem, the video would probably not provide a good medium for the observation of small details. For example, if a subject's image is small and her arm rotates out of a position parallel with the camera, it may be impossible to discern the position or angle of her arm.

Measurement From Video

Biomechanics research has been using video for kinematic measurements (quantitative analysis) for many years. Careful setup and procedural techniques (for example, scaling, camera alignment, shuttering, data smoothing) are needed to collect images, digitize, scale, and make accurate calculations from video data. Just the fact that a video presentation software program provides a distance or angle measuring tool does not mean that the measurement it provides will be accurate. The two major sources of error are perspective error, as a result of different distance of the object from the camera, and the motion or positions of objects not in a plane parallel to the field of view. Imagine the error in calculating elbow angle in a side view of a vertical jump when the jumper moves his arms (shoulder horizontal abduction) out of plane toward the camera. The angle would appear smaller just as with the finger demonstration described earlier.

Many biomechanical quantitative video systems are based on digital and commercial video equipment that breaks the video signal into each field (noninterlaced video) or may use sophisticated image processing to create subpixel resolution in measurements. Stereophotogrammetric techniques and special computer programs are needed to generate 3-D data from synchronized multiple camera images (Allard, Stokes, and Blanchi, 1995) or nonstationary (panning) camera setups. For example, traditional 3-D video systems are based on two to eight or more cameras that are genlocked or synchronized to take images at exactly the same time.

Just the fact that a QMD software program provides some calculation features also does not mean that the numbers generated will be accurate. Considerable data collection and processing expertise is required to do quantitative measurements from video. Most human movement professionals should use video primarily for QMD and should not attempt measurements from video unless they are trained in these techniques or biomechanical assistance is available. Making videos for QMD is less complicated than for quantitative analysis. The following sections discuss important information in video technology and procedures for using video to enhance QMD. A variety of computer software programs are available to augment video capture, assessment, and presentation.

VIDEO TECHNOLOGY

This section summarizes the basics of video technology relevant to QMD. The important characteristics of normal video that professionals need to be familiar with are presented: frame rate and features of camcorders and common playback devices.

> ## PRACTICAL APPLICATIONS
> ## Hindsight Is 20/20
>
> You are a first-year football coach at a high school, and the head coach assigns you to coach the punters. Since many of the important actions of kicking are too fast to observe reliably in real time, you bring your camcorder to practice and video the punters trying out for the team. You record side views of the punters, maximizing the size of the punter in the field of view so that the recordings can be reviewed in freeze-frame and slow-motion replay.
>
> As you review the video, you can identify differences in knowledge of performance (KP) like punting technique, impact position, and ball position on the foot. Several strengths and weaknesses can be identified, and one punter is close to having the desirable punting form that you expect. Unfortunately, you now realize the problem. You did not create a written or auditory record on the video of the distance and hang time of each punt. What kind of information are the distance, hang time, and accuracy of each punt? Slow-motion video replay can extend observational power of only some of the critical features of punting. If videos are made from a position above the press box, coaches can document distance, accuracy, and hang time of the punts. In short, video replay can extend observational power in QMD, but it is not a magic bullet.

Frame Rate

An analog or digital camcorder generates a video picture by using a CCD or a CMOS chip to assign the appropriate gray scale or color to the pixels of a video picture. The sensors that create video images are typically either charge-coupled devices (CCD) or complementary metal-oxide semiconductor (CMOS) chips. In the United States, the alternating current (AC) electrical power has a frequency of 60 Hz, so standard video is set up so that the phases of the current are used to scan each field of video. This gives NTSC video an effective frame rate of 30 frames per second (fps). Normal cinematography (movie film) frame rates are 24 fps. Compared to real-time visual observation, stop-action or slow-motion video dramatically increases the ability to see details of human movement. PAL and SECAM videos have an effective frame rate of 25 fps.

The **sampling rates** of human vision and normal video are important to understand because they help determine what aspects of human movement can be observed. For very high-speed movements like pitching, tennis serving, or the golf swing, normal video may sample at a high enough rate to consistently capture details of interest. A good example of distortions created by inadequate sampling is the motion of the wheels of a stagecoach or wagon recorded by the typical frame rates in old westerns. When you watch these old movies, it appears that the wagon wheels are rotating backward because the frame rate is too slow to represent their actual rotation. The wheels rotate too quickly, making the spoke position of sequential frames appear behind the initial position, when in reality the spoke has rotated all the way around to a position behind the previous image. Similar illusions can appear in modern films when vehicles that have spokes on wheels are filmed at normal sampling rates.

Be aware that some consumer camcorders or video cameras described as high-speed cameras in marketing materials are sometimes in fact *not* high-speed video.

They are really just regular video (e.g., 30 fps) but with high shutter speeds (very small exposure times for each picture, for example, 1/10,000 of a second). There is a marked difference in the kinematic information created between video with 30 fps and video with 200 or 1000 fps. Compare the golf club motion in the illustrations in figure 8.2 that were taken from high-definition, but regular digital, video (60 fields). Sixty images typically will not capture an image that contains impact (image *b* is just after impact). Notice the forward bend in the club through impact. Also notice how far the club can move when images are separated by 1/30th of a second (images *a-b*). The importance of the shutter speed to freezing this fast motion of the club and ball is discussed in the next section on camcorders.

a   *b*

Figure 8.2 Illustration of video sampling (30 fps) of a high-speed golf swing. Regular video rarely can identify impact (just before image *a*), an event that lasts only 0.005 second. The sampling of normal video does not show considerable detail of high-speed movements. Images *a* and *b* are 0.03 second apart.

Special **high-speed video** cameras that were once commercially available only at very high costs for research and manufacturing have now decreased in cost so that many people can afford them. Video cameras capable of frame rates of a thousand frames per second (fps) are not really necessary for most QMD. If a professional has skilled clients interested in high-speed movements and the computer memory and power to handle these files, using a high-speed video camera might be worth the investment. Several video cameras with high-speed capability are available now for under $400. High-speed video can show human movement to a very high temporal resolution, giving the professional and client the ability to see details of technique that are not captured by regular video. For example, if golf instructors want to know about the timing of the uncocking of the hands in a full golf swing, 60 Hz video will not be very effective. A high-speed video camera with a sampling rate higher than 180 Hz would be necessary to get accurate timing of the "release" of the club during full swings. Examples of high-speed video clips can be seen in QMD Demonstration 3.1 and QMD Technology 8.1.

Some companies market video equipment specifically for sport instructors and coaches. Special video cameras, editing equipment, and computer software have been state of the art in the imaging of football games for coaches to evaluate game plans and opponents. These game videos have replaced the old 16-millimeter films that were used to record games and to scout opponents. Many video cameras are marketed as appropriate video equipment for the QMD of sports. In fact, most camcorders or video cameras have automated focusing, white balance, and shutter features that make them suitable for obtaining video images for QMD. Let's look at the video camera features that are desirable for QMD.

Video Camera

The camcorder is the combination of two machines that were separate in the early versions of video technology: the video camera and the recorder or machine to code the video signal on storage media. This text uses the term *video camera* synonymously with *camcorder*. Factors that help determine a camcorder's suitability include the shutter, image quality, and format. As video capture capabilities are integrated into more and more consumer electronics like still cameras, cell phones, computers, and tablets, professionals need to stay informed about these factors.

Shutter

Perhaps the most important feature for imaging human movement for QMD is the **shutter**, which limits the exposure time during which each field is scanned by the chip. Capturing a video field in 1/500 or 1/1000 of a second ensures that moving objects do not move much as the image is being captured. This allows the imaging of high-speed movements without blurring. The term *high-speed shutter* is a misnomer, since the shutter prevents image blurring but does not allow the camera to capture more images per second. In other words, a shutter does not make a high-speed camera, but rather a shutter allows any speed camera to "freeze" motion, making crisp images of fast-moving objects in the fields of view.

The problem a shutter may create is a need for extra light when exposure times are very small. This is rarely a problem with use of a camcorder outside in natural light, but indoors, shuttered video may be underexposed unless extra light is used. Current camcorders have shutters that can generate exposure times as small as 1/10,000 of a second. These settings are rarely necessary in capturing human movement. In fact, video recording slower movements may not require the use of a shutter at all. Table 8.1 lists several common human movements and sports and the corresponding exposure times that may be needed to freeze the action.

It is also important to check how the camcorder operates in the stop mode. Some video cameras storing the video have a short rewind or roll-up of a couple of seconds after the recording is stopped. A good practice is to check if your camcorder has this delay and, if it does, to start recording a couple of seconds before the action of interest and stop a couple of seconds after. Even digital video cameras and recording media cannot instantly begin capture, and the uncertainty of human reaction usually

Table 8.1 Recommended Exposure Times for Video Recording Typical Human Movements

Activity	Exposure time (shutter setting)
Walking	1/60 (off)
Sit to stand	1/60 (off)
Bowling	1/60 (off)
Basketball	1/100
Vertical jump	1/100
Jogging	1/100 to 1/200
Sprinting	1/200 to 1/500
Baseball pitching	1/500 to 1/1000
Baseball hitting	1/500 to 1/1000
Soccer kicking	1/500 to 1/1000
Tennis	1/500 to 1/1000
Golf	1/1000 or smaller

means that a professional should plan to activate the camera just before the movement of interest.

Image Quality

Since a professional may not have complete control over the situation being recorded (weather, competition, logistics, and so on), the camcorder selected should have several automated features that help create good images for QMD. These features should also be manually adjustable. Most camcorders are designed to have low light sensitivity and autoexposure. The camcorder selected should be able to create clear video pictures with the low levels of illumination caused by poor lighting, camera position, or the small exposure times created by a shutter. Some camcorders have a special adjustment or setting for low-level lighting.

Other important image quality features include autofocus, auto white balance, and a backlighting adjustment. Camcorders automatically adjust the color sensitivity (white balance) to get uniform color in different lighting conditions. If the subject cannot be recorded with the sun behind the camera, the backlighting adjustment is needed to keep the light behind the subject from washing out the image.

Especially important is a manual override on the autofocus feature. Most camcorders use infrared beams directed at the center of the field of view or objects in the field of view to automatically focus the lens. Several conditions may not maintain a focused image, so the camcorder should have a manual override of the autofocus. A professional may want the subject not to be in the center of the video picture to keep something else in the field of view; otherwise, other objects behind or in front of the subject may interfere with the autofocus. To focus manually, zoom in on the subject and focus, then adjust the zoom lens to the desired field of view. Sometimes it may be necessary to let the camcorder autofocus and then deactivate the feature at the desired focus setting before recording the video. The camcorder should be as stable as possible to maximize image quality and maintain a stable perspective for viewing the replay. A tripod is highly desirable, but some camcorders have image-stabilizing features to help limit the effect of jerky motion by a hand-held operator on the field of view.

QMD Technology 8.1

Observe the high-speed (210 Hz) video clip of the golf drive in QMD Technology 8.1 in the web resource at www.HumanKinetics.com/QualitativeDiagnosisOfHuman Movement. Notice the extra detail of the movement observable in pause and slow-motion replay compared to the regular video in QMD Technology 8.2. Use the pause feature several times during playback on the swing and focus on the club head. What are the differences in the images of the club head throughout the swing? What video camera feature needs to be adjusted to obtain clear images of the club throughout the swing?

Format and Storage

One of the most important factors in choosing a video camera for QMD is format of the magnetic cassette tape or video file used for storing, reading, and transferring the video. The various sizes of tape (3/4 and 1/2 inch), cassettes (VHS, Beta, ED-Beta, Hi-8, and S-VHS), coding speeds, and media (chip, CD, DVD) all affect the video quality, equipment, and the amount of recording time. While many camcorders still use microcassettes to economically store larger-duration video sequences, many current cameras save video directly in digital files stored on chips, memory cards, or computer hard disks. Advances in low-cost memory cards and video compression–decompression routines (codecs) have made smaller camcorders possible and enabled those cameras to offer direct streaming of digital video to computer memory. The audio signals of video files are also often compressed to reduce file size. Large video files also can delay the transfer of these clips over wireless internet networks. The bitrate required to stream the video from over a wireless network must be within the bandwidth of the Internet service available for smooth real-speed playback.

Professionals need to be sure that the codecs used create images with sufficient quality for use in QMD and that the memory required for the clips they routinely use is sufficient. Unfortunately, the wide variety of video file formats (e.g., AVI, FLV, HDTV, MPG, MOV, VCD, WMV) and codecs (e.g., Cinepak, DivX, DVSD, Indeo, Sorenson, VP6) can create a confusing technical barrier for many kinesiology professionals. It is important to get professional advice and to use quality computer and software programs so that these electronic technology issues do not get in the way of video capture and storage of high-quality images for QMD.

> **»KEY POINT 8.4**
>
> Some of the most important camcorder features related to QMD are format, storage, shutter, low light sensitivity, and the zoom lens.

Lens and Other Features

Another important video camera feature for QMD is a **zoom lens**. Often a camera cannot be placed in a desirable position because of facility or competition limitations. A zoom lens, which can accommodate a variety of distances from the subject, solves that problem. Lens focal lengths are specified in millimeters, but the range of a zoom lens is expressed as a ratio. Video cameras often have a zoom lens with a ratio between 12:1 and 20:1 (12:1 means that an object will appear 12 times larger in full zoom than in full wide angle). Digital video cameras can augment this zoom capability by digitally enlarging the image, but the quality is inferior to optical. A zoom lens may allow the professional to create video images from locations distant from the performer or from a convenient electrical power source. Sometimes long imaging sessions, no spare batteries, or an unchanged battery may require reliance on an AC converter. Since electrical outlets are not always available, it's a good idea to include a long-life battery and a spare battery in your video equipment plans for QMD.

Some camcorders provide useful editing features. Date, titles, or time codes can be added to the image. Professionals should look up repair histories and reviews of these features in consumer publications (*Consumer Reports*) or trade magazines like *Videomaker* and *Video Review*. Editing features are discussed further in the following section on playback equipment.

When selecting a camcorder for QMD, think about how and where you will be making the videos. How much time do the events take? How will the video be viewed? How accessible is the equipment that will be needed to view, copy, or store the video? All these questions and others affect the choice of video format and camcorder features.

Playback Equipment

Once video is captured, there are numerous ways to play back the video. The recording media noted previously affect how replays are viewed, but the most common ways are viewing on built-in screens on the camera or camcorder, televisions, and importing to computers for presentation through software.

Some video cameras that code the video on magnetic tape have playback controls of VCRs. These features have similar playback limitations. The number of reading or recording heads affects the presentation of images in freeze-frame, pause, or still mode. If you plan to use the camcorder screen as the primary playback feature, try to get a model that has good freeze frame, frame advance, and a jog-shuttle dial. A **jog-shuttle dial** or feature can be turned forward or backward to change speeds from still to frame by frame, slow motion, normal speed, and search. This will facilitate the search for images of interest and slow-motion presentation movement for QMD.

Digital video cameras may not have some of the timing and replay problems that analog video cameras and VCRs have. Digital video can be quickly accessed on the magnetic media (hard drives, flash drives, and SD cards) they are stored on. The playback features on the digital camera, however, may not be designed to take advantage of these new media and provide the flexible playback desirable for QMD. When looking for a digital camera to be used with the camera screen for presentation or to input to a TV for presentation, be sure to check the playback features on the camera. Essential are good pause and slow-motion functions. The jog-shuttle feature with continuous forward and backward presentation of replay may be the most desirable feature for QMD.

> **» KEY POINT 8.5**
>
> The most important video playback features for QMD are freeze frame, frame advance, and variable-speed replay. A jog-shuttle dial feature is best because it facilitates searching video and replay at multiple speeds in both directions.

Often camcorders are hooked up to a television or a large computer monitor to provide a larger image for QMD. This is especially helpful in making the image larger and allowing easy viewing by multiple persons. Often the larger and brighter images from televisions help deal with bright light and glare that can be a problem with the use of video replay for QMD outdoors. Sometimes hoods and antiglare screens are helpful in using video replay outdoors. Remember that the replay features of the camera or VCR determine much of what can be presented in video replays on a TV.

A recent development that has been widely adopted for QMD is specialized video software that allows computers and tablets to play back video. Often these software programs allow for all the various timing of replays noted earlier, as well as drawing, enhancement, and presentation features not available on cameras. For example, if a time code was not imprinted by the camera on the video images, the software program can add this to the video file. A later section summarizes these programs commonly used in QMD.

MAKING VIDEO RECORDINGS FOR QMD

Recording patients and athletes for QMD has been popular for several decades. There are many different objectives for capturing human movement. Some people record fast skills like a golf swing to see details in the movement not visible in live observation. Others might record performance over time to track improvement. Some parents record their children's practice routines or competitions and send the recording to coaches to evaluate for possible athletic scholarships. These "daddy videos" are often of little value, however, because they may not be made from a vantage point of interest, or the conditions of the performances may not be adequate for QMD by the coaches.

Replay Options

Naturally the procedures for making a video to assess performance qualitatively depend on the objectives of the QMD. Video made to assess technique in the tennis service or volley may be very different from video made to evaluate court movement or service return ability. Video images for QMD can be made for real-time replay and feedback to the performer or for slow-motion replay to improve the observational power of the teacher, coach, or therapist.

Real-Time Replay

One approach to using video for QMD is to view or replay the recorded images at normal speed. This is real-time observation of video. Creating video for real-time replay for performers or QMD by a teacher or coach requires a certain approach. The field of view needs to be several times larger than the moving subject. This provides a stable background for the subject to move against. A good rule of thumb is to have the subject between one-third and one-half of the field height.

It is important to give the observers of your real-time video a field of view that accurately represents the performance environment. Some videos are made to diagnose how the performer responds to dynamic game conditions (tennis, basketball, volleyball), and QMD of the technique of the performer is not the primary interest. In this situation the field size is maximized to allow observation of the whole court. Most real-time replay of video for QMD is used in this fashion by coaches or motor skill instructors to assist in the observation of performance, although sometimes it is used as feedback or intervention to improve performance.

Research on the use of real-time video replay as intervention or feedback to performers has a long history in kinesiology. Numerous studies over the past 40 years show that video replay is no more effective than traditional teacher-augmented feedback (Rothstein and Arnold, 1976). Recent studies confirm these early results but also suggest that video replay may provide useful information as intervention to improve performance. On the basis of her meta-analysis of video feedback research, Rothstein (1980) proposed that seven factors are important for using video replay for augmented feedback. (1) It should be used with verbal cues that focus the performer's attention. (2) It should be used frequently (at least five replays). Video feedback is most effective for (3) advanced beginners and (4) intermediates, not complete novices or expert players. Some research has shown that unstructured video replay can negatively affect learning in beginners (Ross et al., 1985). (5) Provide practice immediately following the video feedback. (6) Possibly zoom in on the aspect of performance you want to provide feedback about. Finally, (7) when recording, vary the camera angle and capture a field of view that is consistent with the goals and nature of the activity. For example, a close-up shot of the tennis serve would be helpful in this closed motor skill, while a view of the whole court would be helpful in assessing how the ground stroke techniques selected relate to the environment (open motor skills).

Other articles have reviewed the use of video in the QMD or teaching of motor skills (Franks and Maile, 1991; Jambor and Weekes, 1995; Trinity and Annesi, 1996). One of the most important functions of video replay for performers has been its use for self-modeling (Dowrick, 1991b). A performer should be compared to a model of skilled performance who is similar to the performer in age, skill level, appearance, and the like. Some people exaggerate negative perceptions of themselves compared to models who are too different from themselves (Trower and Kiely, 1983). Recall that this is consistent with the research on observational learning summarized in chapter 7. Most motor skills instructors who have used video have experienced the keen interest people have in watching themselves on video.

Care must be taken to monitor subject motivation and the psychological impact of anyone else who may also be watching. Some rather

> **» KEY POINT 8.6**
>
> Video images created for real-time QMD should have a large field of view with background features for the subject to move against. This approach is designed to simulate live QMD.

QMD Demonstration 8.1

View the video clips in QMD Demonstration 8.1 in the web resource at www.HumanKinetics.com/Qualitative DiagnosisOfHumanMovement in normal speed and determine if there is a flight phase, meaning that the child is running rather than walking. Running has a flight phase (both feet off the ground). Use the regular speed and slow-motion versions of the clip to make this judgment, and then use the pause feature in the slow-motion clip to see if you were correct. Was your observation improved by the slowing or stopping of the video?

»KEY POINT 8.7

Video images created for slow-motion replay in QMD should zoom in on the subject to maximize the performer's size in the field of view.

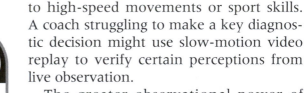

QMD Technology 8.2

Observe the video clip of the golf drive in QMD Technology 8.2 in the web resource at **www.Human Kinetics.com/Qualitative** DiagnosisOfHumanMovement. Are there aspects of this golfer's drive that are not visible in this regular video clip that were visible in the high-speed video clip in QMD Technology 8.1?

interesting comments and behaviors from an audience arise during video replay. The most desirable situation is one-on-one with the performer, eliminating extraneous effects of an audience. If other performers are watching the replay, be sure to focus group attention on good points of performance and try to limit their extraneous or negative comments.

Slow-Motion Replay

In general, video images generated for slow-motion replay should focus on higher-speed movements that are difficult to see with the naked eye. Videos made for this kind of replay should maximize the size of the performer in the field of view. Zooming in on the subject is important for two reasons. First, the resolution of the video picture will be improved in the pause and slow-motion modes. Second, perceptual problems with limited background information are reduced when the images are paused, slowed down, or repeated.

The ability to freeze the motion at key points in the movement and to slow down very fast movements greatly increases observational power. This is why slow-motion video replay should not be limited to high-speed movements or sport skills. A coach struggling to make a key diagnostic decision might use slow-motion video replay to verify certain perceptions from live observation.

The greater observational power of video replay, however, comes at a cost. Slow-motion video replay requires more time for the video to be positioned, played, and replayed. Time used in video capture, replay, and observation is often time lost in performer practice and the live QMD that could have been performed.

Camera Position

Camera location should be chosen carefully to make sure that subject motion is at a right angle to the camera, or as close to a right angle as possible. Ideally the video camera should be mounted on a tripod. Panning should generally be avoided, but limited panning may be needed in certain conditions. To avoid perspective or lens distortions, the video camera should be as far from the subject as possible, using the zoom lens to frame the subject. Extreme examples of such distortion are a peephole in a door and a video camera's view of a tennis court from above the fence. Objects in the foreground appear very large, objects far away appear much smaller than normal, and objects moving toward and away from the camera are distorted. Motion of objects in the distance is underestimated, while motion of objects near the camera is exaggerated.

Limit camera motion when capturing video images for any QMD. Observers of the video you generate are attempting to detect changes in the subject's position. Changes in camera angle or camera motion will distort the actual motion of the subject. It is important to understand how camera motion affects the images created. The most accurate video representation (2-D) of reality (3-D) would be a nonmoving camera positioned on a tripod, with the subject moving parallel to the camera lens. This situation will create images of the subject moving on a 2-D background that are representative of the actual motion.

It is not always possible to set up a stationary camera to record movements. A good real-world example of how camera motion can dramatically change the perception of motion is a NASCAR race or the Indianapolis 500. A stationary camera parallel to the straightaway would capture images (very few frames!) of the cars moving past. A camera mounted looking out the side of a car in the race would show a very different view of the speeds of the cars, like what you experience on the freeway.

Now imagine you are a race fan in the stands panning your camera as cars go by. Only at one instant, when the car is moving parallel to your seat, does your video picture accurately represent the movement of the car. The car moves toward you (and in the video) slowly, gets very fast near you, and slows again after it passes you. This situation is analogous to the gymnastics example presented earlier. The motion of the car is underestimated as the car approaches. In effect, it becomes the proportional speed of the car as the car nears the camera angle of 90 degrees. The apparent length of the car is another example of this distortion. The car looks short in the distance, appears to get longer as it approaches you, and then shrinks as it moves away.

Video Capture Protocol

Precise control of the setting and technique of video capture is needed to generate images suitable for the QMD of interest. Although modern video cameras do well in various lighting conditions, the most desirable condition is even lighting from behind the camera. The camera location should strive for this lighting and for a contrasting background with a horizontal or vertical reference. The background should not hide the motion of the subject and should provide a directional frame of reference on which to place the subject's motion. Figure 8.3 illustrates a field of view and background that hides the motion of the person in the foreground. The protocol should provide subject identification information and some way to identify trials. Verbal cueing to the recorded audio, time codes, or clocks in the field of view are all techniques that will help you identify key events or trials in later playback. Some people create numbers that can be put in the field of view to identify separate trials or movement conditions.

It is often important to make a written record of the subject, date, and notes on performance and keep it with the video. Any information about the movement or results of the movement that is not visible in the field of view should be written down. By keeping written records, the observer does not have to play the video to identify places to start. Time codes and other indexing information can be added to the written record to help access video images quickly. Some of these written records can be electronically entered with text tools and other features in computer-assisted replay systems that are discussed in the next section.

Figure 8.3 An example of a good versus bad background during video recording.

VIDEO AND COMPUTERS IN QMD

The observational power of observers can be improved by video replay, especially the slow-motion replay of human movement and computer-assisted viewing. Fortunately, advances in integrating video and computer or tablet technologies and increases in computing storage and speed have made many advanced video storage, presentation, and replay features available at lower costs. Computer software programs allow presentation (calculation, graphics, split screen, text) of video with consumer video camera hardware and video loading and editing software. A popular feature allows one to compare a performer's movement with a prototype performance by split screen, side-by-side displays, or overlays. These features may be motivational for a mover and help observers evaluate the performance; however, there has been little systematic research to see if these visual presentation options improve observational accuracy or motor skill performance. There are now several commercial computer-assisted video systems designed for qualitative and quantitative assessment that can place drawings or graphics on the video to illustrate key points in the movement (see figures 8.4 to 8.8). Advances in computer and video technology will eventually make many of these advanced video features affordable for most kinesiology professionals.

QMD Demonstration 8.2

View the video clip in QMD Demonstration 8.2 in the web resource at www.HumanKinetics.com/Qualitative DiagnosisOfHumanMovement and determine if the golf club stays above a horizontal orientation at the end of the backswing. Did the red horizontal line add to your ability to rate this critical feature?

»KEY POINT 8.8

Technological advances and the integration of video and computer technology may dramatically influence the impact and use of video for QMD.

The future of video imaging will be strongly related to developments in camera, computer, and Internet and communications technology. Technical advances in cameras to capture high-definition and high-speed video and in computers to store and enhance the presentation of video images could make the use of video more common for QMD. This section summarizes some of the features of common commercial and free (shareware) software programs used for QMD.

Several companies that specialize in high-speed and quantitative videography software for biomechanics faculty have also developed software for video replay that is useful for QMD of sport skills. The names of products and even the vendors themselves change rapidly. Thus, this section lists only a couple of examples of this technology. There are more and more vendors of this software, with a wide range of prices from free shareware to more expensive and complete systems.

The German biomechanics company SIMI (Unterschleissheim, Germany; www.simi.com/) has several software programs for capturing, editing, and presenting video for QMD and notational analysis. Notational analysis is used to create counts of particular events, success rates, and other variables related to strategy and game success (Franks and Goodman, 1986; Franks and Nagelkerke, 1988; Patrick and Lowdon, 1987).The SIMI Scout program focuses on this tactical use of video by coaches (figure 8.4). The SIMI MotionTwin program allows users to synchronize, zoom, split screen, and overlay video, as well as to make simple distance and angle calculations. Figure 8.5 shows tracking and drawing tools in the MotionTwin program.

One of the first commercial software programs for use of video with computers for QMD was Silicon COACH, developed by Joe Morrison in New Zealand in

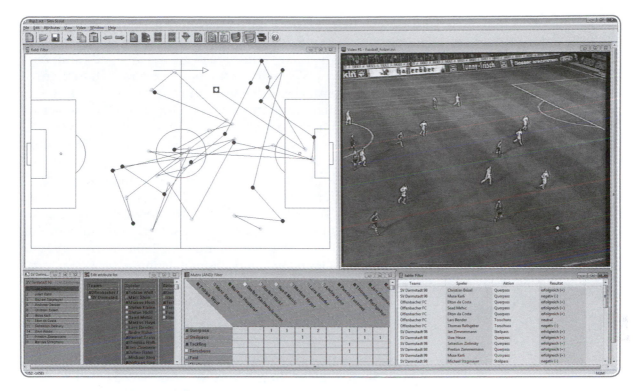

Figure 8.4 The Scout program by SIMI provides a platform for notational analysis as in this player motion in soccer, or the qualitative assessment of athlete performance in game situations.

Courtesy of SIMI, Unterschleissheim, Germany.

Figure 8.5 Tracking and drawing features in the SIMI MotionTwin program.

Courtesy of SIMI, Unterschleissheim, Germany.

1997. Renamed Siliconcoach, the software currently offers several programs that can capture, edit, and present video (Siliconcoach, Dunedin, New Zealand; www.siliconcoach.com). The Siliconcoach programs most relevant to QMD are TimeWarp, Siliconcoach Pro, and Siliconcoach LIVE. As with the other programs noted previously, program users can synchronize video views of a movement and use a variety of drawing, playback, and even quantification tools to highlight various aspects of performance. Siliconcoach has launched a web-based version of its software (Siliconcoach LIVE; http://live.siliconcoach.com/). Subscribers like coaches or instructors can create "zones" that are user groups where video can be shared, assessed, and discussed. The Basic Zone is free for people to try out and has numerous playback features useful for QMD (freeze frame, frame advance, slow-motion replay, drawing, measurement, angle, text tools). A $50 annual subscription gives a professional access to more advanced features like overlays and split-screen presentation. The new web-based Siliconcoach product provides substantial flexibility to the user and handles almost all video formats. Figure 8.6 shows how Siliconcoach LIVE can use video to create lessons on the property of inertia in human movement. Note how the circle illustration tool focuses the viewer's attention and the range of web-based tools below the video frame.

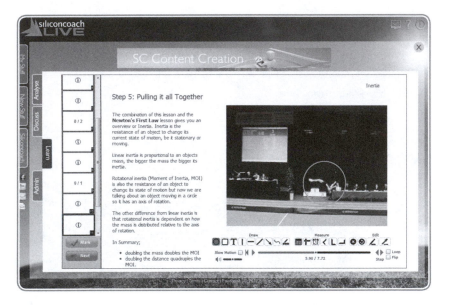

Figure 8.6   The Siliconcoach LIVE program provides a web-based platform for the presentation and QMD of video. This image shows one screen of a lesson designed for instruction about the property of inertia in human movement.

Courtesy of Siliconcoach, Dunedin, New Zealand.

The use of drawing tools in these computer software enhancements of video presentation allows the professional and athlete to see the motion relative to a line or point in space. This facilitates the comparison of motion relative to an important frame of reference. For example, one can qualitatively evaluate the stability of a golfer's swing plane by drawing a diagonal representation from the athlete at setup or impact.

One of the most popular qualitative video replay software systems in North American is Dartfish. A Swiss company working with sports television created SimulCam software in 1997. These innovations were adapted into the Dartfish program for QMD applications for sport coaching. As with Siliconcoach, use of Dartfish programs has expanded from individuals to sport governing bodies to instructional and business settings. Drawing and angle calculation within the analyzer feature are illustrated in figure 8.7.

Figure 8.7 Computer screen image of the Dartfish system for teaching QMD.

Photo courtesy of Duane Knudson.

The company has launched DartfishTV, an online environment where video can be used to create and share enhanced video (www.dartfish.com/en/videonetwork solutions/dartfishnet.htm). These videos can be used for QMD or instruction. This technology also allows professionals to capture video of movement, assess it, and publish the enhanced video for commercial instruction purposes.

Web-based software applications may be the future because of the hardware-independent nature of this new medium. These web-based QMD software programs, along with more powerful and small handheld computers, may allow for easier capture and presentation of enhanced video to athletes for QMD by professionals.

The rapid growth of tablet computers like the iPad has prompted the development of video programs for QMD that take advantage of the built-in cameras and large screens of these computers. The KinesioCapture program (Bethesda, MD; www.kinesiocapture.com/) can be installed on an iPad, iPhone, or PC tablet. Almost all of the presentation, drawing, and calculation features mentioned previously are available with KinesioCapture (figure 8.8).

Remember that any software program that allows for an instant calculation from a single video image can result in errors. Small digitizing errors in locating the two (angle of incline)- or three (joint angle)-segment endpoints for both horizontal and vertical coordinates can have a large influence on "instant" angle and velocity

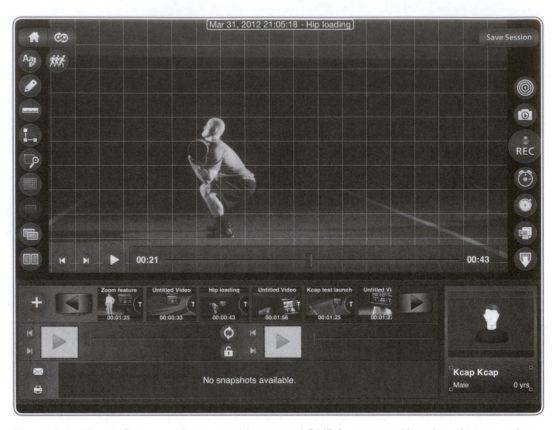

Figure 8.8 KinesioCapture software provides several QMD features and has the advantage of running on hand-held devices (iPhone, iPad, PC tablets).

Photo courtesy of Kinesio Capture LLC.

calculations. Since accurate calculations from the video require careful setup and have many technical limitations, especially in 3-D movements (Allard, Stokes, and Blanchi, 1995), special care must be taken to ensure that the angle or speed calculations are accurate. Consult the technical specifications of the software used or with biomechanics experts familiar with your system if you want to transition from QMD to quantifying movement with video.

Some biomechanists have developed their own video systems by combining inexpensive frame-grabber technology, computers, and their own custom software. These systems have the advantages of low cost and compatibility with common computer and video equipment. Some well-established systems are the Human Movement Analysis System (Hu-m-an) by Tom Duck of York University (HMA Technology, ON, Canada; www.hma-tech.com), the ATD (Análisis de la Técnica Deportiva) program by Francisco Garcia and Raúl Arellano (Granada, Spain), and the Kinematic Analysis (KA) system by Bob Schleihauf (San Francisco State University, http://userwww.sfsu. edu/~biomech/ForFaculty/kavideo.htm). These systems are designed for quantitative biomechanical analyses, so they have many features beyond what is needed for QMD.

Given the rapid advancement of video camera and computer technology, several programmers have developed low-cost or free versions of software to present video clips with computers. SkillCapture and SkillSpector are two free shareware programs created in Denmark (www.video4coach.com) for video capture and qualitative and quantitative video assessment. Other programs or apps suitable for QMD are Coach's Eye, CoachMyVideo, Kinovea, and UberSense.

Improvements in video technology, HDTV, and the increased integration of imaging and computers will make for an exciting future in the use of imaging for QMD. The video systems of the future designed for QMD will have many features attractive for teaching motor skills. Such systems will be portable, with extended battery power and wireless connectivity to the Internet. Screens will be large and flat, with hoods and antiglare monitors. The video camera may be run by remote control or programmed to follow the athlete. Currently the only barriers to this technology are the cost of the equipment and the expertise of the teacher, coach, or therapist. The cost of computer and video technology should continue to come down, and continued research will help us understand how to best use these tools to improve the QMD.

Professionals need to research these computer-enhanced video programs for QMD. Some programs are built to handle all aspects of video capture, storage, and input into the software for most off-the-shelf consumer video cameras. Others, however, require the user to know how to capture, edit, and upload video files to a computer before it is loaded into the software. Professionals need to explore what aspects of video capture, editing, and presentation are supported by the software program or system they plan to use for QMD.

FURTHER PRACTICE

This section lists some suggestions for QMD practice using video replay. Kinesiology professionals should develop their QMD skills through practice with video replay. Remember to carefully plan for the video and performance situation to get an accurate representation of performance.

- Project 1

  Video a friend who wants to improve performance in a closed motor skill such as bowling, golf, or archery. Video the performance from two perspectives: to get a close-up of the technique and to capture the whole performance (KP and KR). Which perspective was most helpful in your QMD?

- Project 2

  Video a friend who participates in an open-environment activity or team sport. Use the video to provide several suggestions to improve performance. Use the video replay to illustrate the points to your friend.

- Project 3

  Video a friend who is practicing a specific sport skill or technique. Use information from your friend to customize and capture video in several trials (e.g., intensity levels, targets). Use the video replay for QMD exploring the performance issues your friend is interested in and provide suggested improvements.

- Project 4

  Video a competitive sporting contest in a public place. Observe the video replay for coaching and strategic factors. Observe the movements of a specific player as would be done in performance analysis by a coach. Write down or type your strategic or tactical suggestions.

- Project 5

  Video yourself practicing a movement you enjoy. Use video replay and QMD to evaluate and diagnose your technique, and write a plan for improvement.

Implement the intervention during practice and video yourself again. Did the intervention created observable changes in the movement or performance outcomes?

SUMMARY

Video is an important tool in extending observational power for QMD. Although video replay has been used as feedback for performers, the research has not shown it to be superior to teacher or coach feedback. Slow-motion video replay, however, dramatically extends the observational ability of teachers, coaches, therapists, athletes, and clients. The ability to replay movements again and again also improves observation. Video and computer technology are converging and dramatically affect the kinesiology professional's ability to use video replay for qualitative and quantitative assessments. With advances in video, computer, and software technology, many advanced features for presenting and publishing video are currently available.

Discussion Questions

1. What video views would be desirable for a QMD of a pitcher's form versus the ball flight in baseball pitching?

2. What compromises must be made in setting up the field of view for video recording the approach to the long jump?

3. What kinds of human movements can be observed live? What kinds require normal and high-speed video replay?

4. A TV station would like to produce a special on the dramatic improvement of a local pole vaulter. What advice would you give the show's producer on how to capture some of the athlete's practice vaults? How do views that exaggerate the motion for TV viewers differ from views that are good for observation of performance?

5. If video clips are to be presented on a computer screen for QMD, what features are most important for the professional? The mover?

6. Does video replay improve the validity of the QMD? The reliability?

Tutorials in Qualitative Movement Diagnosis

Now it's your turn! This chapter leads you through selected tutorials, or practical examples, of the qualitative movement diagnosis of several human movements. These tutorials are of two kinds: QMD Practice and QMD Exploration. QMD Practices are guided practices. Photo sequences of the movements are included, and possible diagnoses and interventions are discussed. Videos of the same movements are provided in the web resource and in e-books that include video. QMD Explorations provide additional videos of other performances so you can try qualitative movement diagnosis on your own. If you have access to the web resource, you can view the video clips at www.HumanKinetics.com/Qualitative DiagnosisOfHumanMovement. If you are using a video-enhanced e-book, you'll find the videos embedded in the e-book. The QMD Practices illustrate several approaches to integrated qualitative movement diagnosis. Proposed explanations of the critical features, systematic observational strategies, evaluation and diagnosis, and intervention are presented for each movement.

Chapter Objectives

1. Illustrate an integrated approach to QMD of selected motor skills.
2. Discuss how the approaches to QMD of the presented skills differ.
3. Discuss how similar movement patterns may call for different QMD strategies.
4. Develop skill in performing QMD from video.
5. Generalize QMD skills to live movement conditions.

Research has conclusively shown that instruction in qualitative movement diagnosis (QMD) improves ability at QMD, but this ability tends to be sport or movement specific. Research has also shown that practice with many examples of good and bad performances is necessary for the development of skill in QMD. This chapter cannot provide enough experience to turn you into a skilled diagnostician of all kinds of human movement, but it does provide a progressive set of examples of interdisciplinary QMD from a variety of human movement activities. Examples of fundamental movement patterns (catching, kicking, and overarm throwing) and skills and exercise (Frisbee toss, squat) are presented in QMD Practices to illustrate the process of an integrated QMD. Additional opportunities for practice in QMD for several other movements are available in the QMD Explorations in the web resource and in video-enhanced e-books.

When critical features and sample teaching cues are identified in the tutorials in this chapter, remember that these are cues based on the author's assessment of the research on the given movement. They are not presented as the only right or perfect choices but simply as examples of an integrated QMD of human movement. As the performances get more complex, the discussion of the critical features and the diagnosis of the performances naturally become more extensive. As a professional knows more and more about the performer, the activity, and the context of the movement, the amount of knowledge to be integrated and the complexity of diagnosis will increase.

The size of subjects in the photos and video clips is normally maximized relative to the movement and goal to make them easier to see and because viewing time is essentially unlimited with images and video replay. In real-world QMD, the observer must be far enough away from the performer to be able to use background objects to judge the motion of the body and other outcomes of the movement. Where possible, an attempt has been made to note and standardize the time between images in the figures. Naturally, use of the real-time and slow-motion replay of the video is the most ecologically valid way to practice QMD.

For each movement sequence, systematically observe the changes in body position based on the critical features and observational strategies presented. Use as much information from the text, photos, videos, background, and figure captions as possible to inform your QMD. It is important for professionals interested in developing skill in QMD to practice various observational strategies and approaches to QMD using video and live movements. Professionals need to remember that QMD should be integrative; they should remain open to cues that help identify contextual factors and factors beyond biomechanics that might affect performance (table 9.1). Some of these are noted in the tutorials; such factors do not readily appear in images or video. Please take what you learn in this chapter, continue to practice, and be open to the integration of all information relevant to helping your clients.

Table 9.1 Examples of Nonbiomechanical Factors That Often Affect Performance

Psychological	Emotional
Fear	Excitement
Aggression	Boredom
Showing off	Happiness
Seeking attention	Sadness
Easy distractibility	Depression
Physiological	**Physical**
Strength	Vision
Flexibility	Maturity
Fatigue	Kinesthetic
Body composition	Haptic

QMD OF CATCHING

The fundamental pattern of catching is a good movement to use to illustrate several important points about an integrated QMD. QMD of this movement can be fine-tuned to apply to other specific catching skills. QMD of catching is somewhat simplified because the goal or mechanical purpose of catching is clear and easy to evaluate. This, combined with an understanding of the sequential nature of the movement, makes the diagnosis of performance within QMD easier than for other movements with complex objectives and contextual factors.

Critical Features

Several authors have reported models for the QMD of catching (Jones-Morton, 1991b; Kelly, Reuschlein, and Haubenstricker, 1989; Morrison and Harrison, 1985). The majority of the technique points mentioned by these authors can be summarized in terms of the four critical features of catching in table 9.2, listed roughly in order of their occurrence in catching.

Table 9.2 Critical Features and Cues for Catching

Critical features	Cues
Readiness, visual focus	Watch the ball
Intercept	Move to meet the ball; reach for the ball
Hand position	Thumbs in or out, fingers up or down
Ball momentum absorption	Give with the ball; retract hands and arms

Systematic Observational Strategy

A systematic observational strategy (SOS) for QMD of live or videotaped catching performances is based on the sequence or phases of the catching movement. Teachers should first observe the performer's state of readiness and attention to the object being caught. Next, observation focuses on the motion of the body and arms to intercept the object. The final two critical features to observe are the position of the hands and how the body, arms, and hands give to dissipate the motion energy of the object.

Subject 1

Figure 9.1 illustrates the performance of a child catching a foam ball. You may want to extend your observation by viewing the video clip of Catching, Subject 1 in QMD Practice 9.1 in the web resource at www.HumanKinetics.com/QualitativeDiagnosis OfHumanMovement. Use the critical features in table 9.2 to perform an integrated QMD of the movement illustrated before you read the following suggested QMD.

The subject has a typical immature catching pattern that is transitioning from trapping the ball against his body with some head turn, indicating apprehension, to using his hands proactively to catch. It looks as if his attention had been focused on the ball and he did not have to move to intercept the ball. His arms have limited

(continued) ➜

Figure 9.1 Sequence images of a child catching a foam ball. What would be the best intervention to help this child improve his catching ability? Time between pictures is 0.17 seconds.

Subject 2

Figure 9.2 illustrates an attempt to catch a football by a child playing catch. Perform an integrated QMD of this performance. (Assume that several trials showed similar strengths and weaknesses.) This child and his partner are passing the ball directly to each other from a moderate distance. You can extend your observation by viewing the video clip of Catching, Subject 2 in QMD Practice 9.2 in the web resource at www.HumanKinetics.com/QualitativeDiagnosisOfHumanMovement.

This boy exhibits several strengths and weaknesses in this attempt to catch a pass. He tries to give with his arms to absorb the energy of the ball, and he has good hand positioning for a reception point that is right between the approximate height of desirable "thumbs in" and a "thumbs-out" technique. Weaknesses have to do with attention (possible inattention), visual focus, and moving to intercept the ball. Notice that he flexes at the hips and reaches late to receive the ball, when a forward step and bending in all lower extremity joints would improve the ability to absorb

(continued) ➜

Figure 9.2 Sequence images of a child catching a football pass. What is the appropriate intervention in this case? Time between pictures is 0.17 seconds.

forward reach for intercepting the ball. Intervention could reinforce success to build confidence and then follow up with feedback to improve either hand positioning or reaching to intercept the ball earlier. Providing feedback about giving with the arms is inappropriate if the subject cannot first move the arms forward to intercept the ball in a position that allows him to "give with the ball." A good intervention if subsequent catching movements were similar might be "Good job! On the next try, watch the ball closely and try to move your hands forward to reach the ball earlier" or "Guess where the ball will be and move your hands there early." Is there any additional information that would improve the diagnosis and intervention? Do you think it would it be useful to observe additional trials or change the catching conditions? How would the technique be different if the ball was a softball rather than a foam ball?

the energy of the ball. If several trials showed similar technique and dropped balls, an intervention strategy might include task modification to see if visual acuity or attention contributes to poor interceptive skills and success. If the boy successfully catches easier passes, it is unlikely that visual acuity and readiness are weaknesses that require intervention.

It would be good feedback for this performer to praise his effort and attempts to give with the arms and absorb the motion energy of the ball. The professional could then use an indirect style of teaching and ask, "Does the ball sometimes bounce out of your grasp? What catching cues have we discussed that would help you?" Direct or indirect, the next intervention would likely be to encourage greater whole-body motion to intercept the ball. Encouraging proactive reaching to intercept the ball early to increase the time and space available to absorb the energy of the ball would be desirable. Remember, one can "sandwich" feedback like this with performer strengths to focus on a weakness and still reinforce strengths and motivation. How might you use this strategy in this case?

Subject 3

Figure 9.3 illustrates the performance of a child catching a softball. Perform an integrated QMD of this performance. You may want to extend your observation by viewing the video clip of Catching, Subject 3 in QMD Practice 9.3 in the web resource at www.HumanKinetics.com/QualitativeDiagnosisOfHumanMovement. Perform an integrated QMD of the movement illustrated before you read the following suggested QMD.

This girl is confident and exhibits strong catching technique. She proactively moves to intercept the ball, reaches, and uses the glove effectively to absorb the kinetic

(continued) ➔

Figure 9.3 Sequence images of a child catching a softball. What is the appropriate intervention in this case? Time between pictures is 0.17 seconds.

QMD OF THE SOCCER INSTEP KICK

Placekicking is an adapted striking pattern of the lower extremity in which the foot strikes the ball as it lies or rolls on the ground. A variety of sports require skill in this movement, referred to here as the *soccer instep kick*. How should an American football or soccer coach qualitatively diagnose placekicking technique? The examples in this section and the web resource include children kicking soccer balls.

Critical Features

High-speed kicking in a variety of sports has been extensively researched. The integrated QMD proposed is based on this large body of research and several models advanced by previous authors (Barfield, 1998; Jones-Morton, 1990a; Kellis and Katis, 2007; Kelly, Reuschlein, and Haubenstricker, 1990; Levanon and Dapena, 1998; Morrison and Reeve, 1986; Nunome et al., 2006; Tant, 1990). There are six critical features to be evaluated in a QMD of the soccer instep kick (table 9.3). These critical features are presented roughly in their order of occurrence in the movement. Does the movement sequence in kicking correspond to the potential influence of the critical features? If the objective of a placekick changes, do the critical features change in importance?

energy of the ball. She also uses her other hand as a backup in case there's an error in tracking the ball or absorbing its energy. If several catches showed the same technique, a coach might intervene by increasing the challenge, specifically increasing variability in ball speed and trajectory.

A good intervention strategy might be to compliment the player on any of the critical features she is doing well. One could say "Excellent forward motion to intercept the ball. You are doing so well that I want your partner to challenge you with more difficult catches. Because these will be faster and harder, I want you to focus on using 'soft' hands and giving with the motion of the ball with your glove hand."

Table 9.3   Critical Features and Cues for the Soccer Instep Kick

Critical features	Cues
Eye focus	Head down and watch the ball
Opposition	Turn your side to the target
Plant	Plant your foot next to the ball
Sequential coordination	Rotate your hip and leg
Solid impact	Kick through the center of the ball
Follow-through	Follow through toward the target

Systematic Observational Strategy

It is likely that several SOSs are effective for the QMD of human movement. It is, however, difficult to actually practice different SOSs with the figures in this chapter. Readers should try three SOSs when observing live or videotaped kicking performances.

The most common strategy is to observe according to the sequence or phases of the movement. A similar strategy is to observe from the origins of movement (from slower-moving to faster-moving segments). The third observational strategy is the gestalt approach, moving from general impressions to specific. Which observational strategy is easiest or most comfortable for you? Why do you prefer one over another? In the absence of evidence that a particular rationale for diagnosis is best, should novice soccer coaches observe in sequence and provide intervention according to the sequence of the movement?

The soccer kick is an excellent example of a motor skill involving sequential coordination for generating high speeds of a distal segment. Some soccer kick techniques have minor variations in these critical features to meet situational requirements. The kick is dramatically different if the purpose is to have the ball travel accurately 10 feet (3 meters) to a teammate than if the purpose is to get it as far from your own goal as possible. For example, the critical feature of solid impact would be less important in the short pass than in the long pass. Slow-speed kicks do not require sequential coordination and may use the side of the foot to push the ball to a nearby target. This leads to a common misconception that the side of the foot is the desirable striking point for high-speed kicks. In reality, the neutral position of the hip and foot and the flat surface of the instep (top of shoelaces) when the foot is pointed make the instep kick the best technique for high-speed placekicking. For the long pass, the professional would pay closer attention to the foot position at impact, the sound of the kick, and the trajectory of the ball to evaluate the quality of the impact.

Another common error is to minimize opposition by not approaching the ball at an angle from the anatomical sagittal plane. The kicker can make efficient use of hip rotation and the levers of the lower extremity by having an approach angle to the ball between 30 and 60 degrees (Tant, 1990). Still other common errors are not focusing attention on the ball, planting the foot behind or in front of the ball relative to the point of impact, and an exaggerated concern for accuracy that typically results in slow foot speed and a limited follow-through. This latter error is common in low developmental levels of kicking.

PRACTICAL APPLICATIONS
Observational Models for Kicking

Before we look at video clips of kicking, let's review some of the ways to organize critical features into an observational model. Look at the following examples and see if you prefer the gestalt approach (Dunham, 1994) or the spatial–temporal approach (Gangstead and Beveridge, 1984). Fill out the temporal–spatial model, based on the gestalt model, whether or not you plan to use the temporal–spatial model in QMD.

Gestalt Model (Dunham)

Body orientation:
Facing ball at 45-degree angle to line of kick

Preparation:
Feet: Weight on nonkicking foot, one foot (30 cm) from ball; kicking foot back away from ball

Knees: Slight bend of nonkicking, deep bend of kicking

Hips: Slight bend of nonkicking, slight extension of kicking, rotated back from ball

Trunk: Straight and slight extension, rotated back from ball

Shoulders: Nonkicking slightly forward, kicking back

Arms: Nonkicking forward, kicking back for balance

Hands: Relaxed

Head: Down, eyes on ball

Execution:
1. Rotate hips and trunk to ball.
2. Bend kicking hip and straighten leg forcefully.
3. Point toes; contact ball with laces.
4. Bring kicking leg through to target.
5. Balance with arms.
6. Follow through to target.

Temporal and Spatial Model (Gangstead and Beveridge)

Instep kick:

Body components	TEMPORAL PHASING		
	Preparation	Action	Follow-through
Path of hub			
Body weight			
Trunk action			
Head action			
Leg action			
Arm action			
Impact/release			

Subject 1

Perform an integrated QMD of the kick illustrated in figure 9.4. The child is kicking a stationary soccer ball toward a goal. What would be the best intervention in this situation? You may want to extend your observation by viewing the video clip of Soccer Instep Kick, Subject 1 in QMD Practice 9.4 in the web resource at www.Human Kinetics.com/QualitativeDiagnosisOfHumanMovement. For figure 9.4 and the rest of the movements illustrated in this chapter, assume that the performance strengths and weaknesses apparent in the figures and videos were consistent across several trials.

The child in figure 9.4 essentially achieves the goal of kicking the ball using technique typical of children at a low developmental level for kicking. Young children

(continued) ➜

Figure 9.4　Sequence images of a child kicking a stationary soccer ball. What correction would help this child improve the most? Time between pictures is 0.17 seconds.

Subject 2

Try another integrated QMD, this time of the instep kick illustrated in figure 9.5. You can view the video clip of Soccer Instep Kick, Subject 2 in QMD Practice 9.5 in the web resource at www.HumanKinetics.com/QualitativeDiagnosisOfHuman Movement. This subject is also kicking for speed and accuracy toward a soccer goal. Select appropriate cues for intervention. Remember that the strengths and weaknesses illustrated have been consistent across several kicking trials.

The player in figure 9.5 has the strengths of good focus and opposition given a rather straight approach. He also could be reinforced for a vigorous kick, striking the center of the ball, and a good follow-through. Weaknesses include a straight approach, a foot plant that is slightly behind the ball, and a side foot impact. Diagnosis of these weaknesses should focus on the likely interrelatedness of these factors. Taking advantage of greater opposition and solid, top-of-the-instep contact usually requires a more angled approach to the ball. It is difficult to increase range

(continued) ➜

Figure 9.5　Sequence images of a child kicking a soccer ball for a goal. What intervention would help this player improve the most? Time between pictures is 0.17 seconds.

typically use a stationary position next to the ball with a sagittal plane kick that is often focused around knee joint extension. Children at this age are often delighted to be out playing with a ball, focusing on the joy and wonder of sport and what happens rather than achieving goals or desirable technique.

Intervention in this case should, therefore, not overemphasize any of the six critical features in table 9.3. A soccer coach should reinforce the child for good effort and focus on the ball. If the child could focus on trying to improve his kicking, the coach could say, "Billy, let's try to kick faster like the big boys. Do you want to try? Let's try a 'step kick.' Step next to the ball and kick it hard!" Keeping the cues simple and using a demonstration will likely help this child improve by adding more body segments and opposition to his kicking technique.

of motion and get the top of the foot down and aligned with the center of the ball from a straight approach.

Given our assumption that this player prefers a straight approach, intervention can focus on either subtle adjustments (foot placement and lean) transitioning to angled approach kicking or a more direct change to an angled approach. A coach who chose the former could say "Nice, solid kick there, John! I bet you can get more speed on the ball with a few changes. Let's hit the ball with the top of your shoelaces. This will be easier if you get your plant foot even, but a little to the side of the ball, and really point your toe."

The coach can then follow up with cues to reinforce "planting the foot next to the ball" and making a "solid or hard" impact through the center of the ball. Subsequent intervention could lead the player to a more angled approach with cues like "turn your side to the target" or "approach the ball from the side." The coach also might say, "That was a great kick, John! I bet you could get more power. Try to approach the ball from the side." The professional always repeats the QMD to monitor how an intervention affected critical features and overall performance.

Subject 3

The child illustrated in figure 9.6 is kicking a rolling soccer ball for speed and accuracy toward a goal. Perform an integrated QMD of this kicking performance in order to provide appropriate intervention for this performer. You can view the video clip of Soccer Instep Kick, Subject 3 in QMD Practice 9.6 in the web resource at www.HumanKinetics.com/QualitativeDiagnosisOfHumanMovement.

The rolling ball makes this a more challenging environment for a fast and accurate kick. The critical features in this performance show only a few weaknesses. Note the strengths of small steps in the approach, anticipatory foot plant, opposition, and low kicking action to create a low ball trajectory. Professionals should remember to be careful in prescribing technique changes in more skilled performers and avoid

(continued) ➜

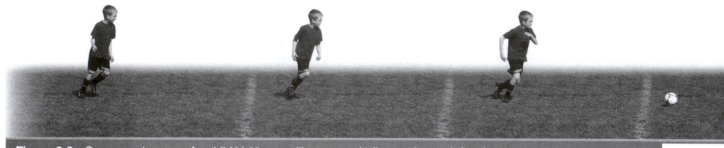

Figure 9.6 Sequence images of a child kicking a rolling soccer ball toward a goal. Are there any weaknesses in kicking technique? Time between pictures is 0.17 seconds.

QMD OF OVERARM THROWING

Overarm throwing is an important fundamental movement pattern in many sports. This section illustrates an integrated QMD of the overarm throw. As with the previous examples, professionals need to be sensitive to how performer and ecological factors affect the importance of critical features.

Critical Features

The high-speed overarm throw is another movement that has been extensively researched. Professionals interested in the overarm throw should read the classic reviews on this important movement (Atwater, 1979; Fleisig et al., 1996a). The extensive motor development research on overarm throwing is summarized in several sources (Haywood and Getchell, 2009; Roberton and Halverson, 1984; Wickstrom, 1983; Wild, 1938). Jones-Morton (1990b) proposed five critical elements for observing the overarm throw: (1) step with opposition, (2) open up, (3) rotate the trunk fully forward, (4) elbow leads with elbow extension, and (5) weight transfers from the back to the front foot. Kelly, Reuschlein, and Haubenstricker (1989) used five qualitative criteria for the forceful overhand throw of fourth graders: (1) side orientation, (2) nearly complete arm extension backward in the windup, (3) weight transfer to the opposite foot, (4) marked sequential hip and shoulder rotation, and (5) follow-through beyond ball release. Extensive biomechanical data are available on various levels of a specific technique of the overarm throw, baseball pitching (Fleisig et al., 1999; Nissen et al., 2007).

the tendency to overcorrect. Changes in technique should be prescribed only if they improve performance or prevent injury and when other interventions (conditioning, competition, practice, rest) will not produce similar results. Good intervention for this performer might be less frequent reinforcement and praise, followed by frequent modifications of the task. One weakness to consider working on is use of the instep; greater use of the instep of the foot could be encouraged to improve ball speed. The coach might say, "Nice low kick, Greg! It looks like you are ready for more difficult practice. On these kicks, try to use the top of your foot to strike the ball. This will give you a very hard impact point and more ball speed. Let's give it a try." More advanced players can understand more subtle feedback and instruction. The coach can begin to use cues during subsequent QMD and practice in these new practice tasks. The main challenges for the professional are maintaining the variety and intensity of practice.

Knudson and Morrison (1996) summarized the research and professional opinion on overarm throwing and presented an integrated model of QMD of throwing. Since forceful overarm throwing is a vital skill in many sports, this section reviews their six critical features of overarm throwing. These critical features and suggested cues are presented in table 9.4. Other major biomechanical studies have reported the effects of biomechanical feedback (Miyashita, Fukashiro, and Hirano, 1986) and another approach to QMD of baseball pitching technique (Nicholls et al., 1999).

Table 9.4 Critical Features and Cues for the High-Speed Overarm Throw

Critical features	Cues
Torso rotation, leg drive, and opposition	Turn your side to the target; step with the opposite foot
Sequential coordination	Uncoil the body
Strong throwing position	Align arm with shoulders
Inward rotation of arm	Roll the arm and wrist at release
Relaxation	Relax your upper body
Angle of release	Throw up an incline; throw over the cutoff's head

Adapted by permission from Knudson and Morrison 1996.

Overarm throwing should be qualitatively diagnosed relative to the goal of the throw. The conditions at release of the ball are most strongly associated with accomplishing the goal of the throw. A critical feature that the professional can easily observe in evaluating this aspect of performance is the angle of release created by the initial trajectory of the ball. As with the basketball jump shot, the outcome of the throw is largely determined by ball speed, spin, and, most importantly, angle of release. Biomechanical research has shown that air resistance plays a major role in the flight of most balls. In most sporting situations, optimal angles of projection for throwing balls for horizontal distance are between 35 and 42 degrees above the horizontal (Dowell, 1978). A common error in young baseball players is throwing the ball with a very high trajectory. This limits the throw's effectiveness for baseball in two ways: It limits the length of the throw and increases the time it takes to cover the horizontal distance thrown.

Leg drive and opposition is the critical feature that combines the thrower's stance and step, setting up the rotations of the body to provide most of the power of the throw. The athlete creates opposition by turning the nonthrowing side of the body to the target as well as pushing off the back leg to step toward the target. Mature high-speed throwing uses a forward step greater than half the person's height (Roberton and Halverson, 1984). Research has shown that body rotation from good opposition contributes 40% to 50% of the ball speed, while the step contributes about 10% to 20% in skilled throwers (Miller, 1980). A vigorous leg drive that is channeled into body rotation is an essential technique point in high-speed throwing. Common errors are to rely too heavily (overstride) on leg action and to not transfer the energy from the stride to hip and trunk rotation.

Sequential coordination is the precise timing of accelerations of a proximal segment that transfer energy to distal segments to increase their speed. The forward acceleration of a proximal segment eccentrically loads agonists, and the later negative acceleration of the proximal segment uses joint forces or segmental interactions to speed up the distal segment. This concept of the interaction of a linked system of body segments has come to be known as the *kinetic link* or the *kinetic chain* (Kreighbaum and Barthels, 1985; Steindler, 1955). The sequential action of the leg drive, hip rotation, spinal rotation, arm action, and forearm and hand action is required to generate high-speed throws. Roberton and Halverson (1984) have described the qualitative changes in various parts of the body in the development of sequential coordination in the overarm throw.

A strong throwing position maintains alignment of the humerus at approximately a right angle to the longitudinal axis of the spine (Plagenhoef, 1971). This position maximizes the speed transferred to the arm from the rotations of the hips and trunk. Atwater (1979) found that this alignment (90 ± 15 degrees) was a relatively invariant aspect of most throwing motions. In other words, there should be little difference in upper arm position between sidearm and overarm throws, just differences in the lean of the trunk. The throwing position and trunk lean at release are observable from a rear view of the thrower. The preparatory arm action in skilled overarm throwing typically involves a circular, downward backswing, with the upper arm aligned with the shoulders and the elbow maintaining a 90-degree angle. Some professionals might consider this important enough to be a critical feature, but these preparatory movements often develop naturally.

Inward rotation of the humerus and forearm is a major propulsive and injury-protective action in the overarm throw (Atwater, 1979). The critical feature of inward rotation of the arm should be understood as the combination of humeral inward rotation, radioulnar pronation, and wrist flexion that provides the final propulsion of the ball. While the fingers apply important forces to create speed and spin on the ball, actual finger flexion does not occur before release (Hore, Watts, and Martin, 1996). This and relaxation help create the final sequentially coordinated actions of the throw. A great deal of energy is transferred up the body, and good timing of the forearm pronation and wrist flexion can create that additional "pop," or great ball speed, of skilled throwers. Throwing with a dead wrist does not fully use the energy from the more proximal segments and fails to use the final joints in the kinetic chain of the throw. Recent biomechanical studies of the kinetics of baseball pitching have increased our understanding of the stresses that overarm throwing places on the body (Feltner and Dapena, 1986; Fleisig, 2001; Fleisig et al., 1995, 1996a, 1996b; Hong, Cheung, and Roberts, 2001).

Systematic Observational Strategy

Knudson and Morrison (1996) proposed a four-phase observational strategy for the QMD of the overarm throw. In the initial trials, look first for timing, rhythm, signs of tension, and the trajectory of the throws. This is difficult to do from illustrations alone but is feasible from replay of the video in the QMD Practice. Second, observe leg drive, opposition, and hip and trunk rotation in the next few trials. Third, observe the throwing position of the arm. Last, try to find evidence of sequential coordination in the fast actions of the arm by looking for lag in the trunk, humerus, and forearm. This kind of observational strategy, focusing on body segments and actions, is similar to what Gangstead and Beveridge (1984) suggest in their observational model.

Compare and contrast this observational strategy with the SOSs presented earlier using live or videotaped throwing performances. Compare differences in observational strategies for different critical features, especially the stages and components discussed in the motor development literature (Roberton and Halverson, 1984; Wickstrom, 1983). What aspects of other SOSs are incorporated into the Knudson and Morrison (1996) observational strategy?

Subject 1

Figure 9.7 illustrates the overarm throwing technique of a child. Perform an integrated QMD of this overarm throwing performance. You can view the video clip of Overarm Throwing, Subject 1 in QMD Practice 9.7 in the web resource at www.Human Kinetics.com/QualitativeDiagnosisOfHumanMovement. Use whatever observational strategy seems most natural for you.

The child in figure 9.7 shows several technique factors consistent with a low developmental level overarm throw. She appears to be throwing with good effort to a partner who is not too far away, and the ball does not seem to be too heavy for her. She does not use her lower extremities; however, she does use some trunk rotation and an intermediate-level arm action. Typical beginning throwers use a primarily sagittal plane, elbow extension–centered throwing motion. A good professional would be intrigued by this mix of beginning and intermediate technique and might begin by modifying the throwing conditions or making motivational comments to see if the girl's technique will change.

(continued) ➜

Figure 9.7 Sequence images of the overarm throw of a child throwing a softball to a partner. What intervention would help this child improve the most? Time between pictures is 0.17 seconds.

The best intervention may be to see if cues or modified practice can get this girl to use her lower extremity to throw the ball with more speed and accuracy. The instructor might praise her accuracy and arm action but then follow up with cues related to using her lower body and more opposition. A way to do this would be to say, "Nice throw, Sally. I would like you to try using your legs more to help you throw faster. Try another one and think about turning your side to the target and stepping with the opposite foot." The other critical features of a sequentially coordinated throw require mastery of opposition and energy transmitted from the lower extremities through the trunk to the throwing arm.

Good feedback for this performer would be to praise some aspect of her throw (arm alignment and angle of release are good) and provide cues such as "Turn your side to the target," or "Let's use that good leg drive more by first turning sideways to the target, stepping forward, and powerfully rotating your trunk into the throw." The age of the performer allows the professional to include more information with the feedback on this critical feature. Repeat the QMD of this girl's technique in QMD Exploration 9.16. This other performance was her first throw after she was asked if she ever turned her body sideways before throwing.

Subject 2

Figure 9.8 illustrates the overarm throwing performance of a young woman throwing a junior football. Throwing a football has the same critical features, but gripping constraints are added to allow the ball to be thrown with a spiral to streamline and stabilize it during flight. Perform an integrated QMD of this overarm throwing performance. You can view the video clip of Overarm Throwing, Subject 2 in QMD Practice 9.8 in the web resource at www.HumanKinetics.com/QualitativeDiagnosisOfHumanMovement.

This young woman has several strong critical features of overarm throwing. She has good opposition, keeps her arm aligned with the line of her shoulders, and has a good intermediate developmental level arm action. Improvements could be made in her leg drive, trunk rotation, sequential coordination, grip, and the angle of release. She uses little weight shift and has simultaneous (block) rotation of her hips and trunk. The ball is not full size, so she grips the ball across the middle. Diagnosis of these issues to optimize improvement is difficult and would certainly benefit from more knowledge based on observation of other throwing conditions. Having her throw harder or a longer distance might show that she knows how to use a longer, more rigorous leg drive.

(continued) →

Figure 9.8 Sequence images of the overarm throw of a young woman throwing a junior-size football to a partner. What intervention would help this person improve the most? Time between pictures is 0.17 seconds.

Assuming that task modification has not changed technique, the best intervention might be to focus the performer's attention on a more rigorous step and weight shift. More sequential coordination usually cannot develop unless energy is transferred from the lower extremities to the trunk. Focusing solely on more trunk rotation, therefore, would not likely be a good strategy for intervention. Focusing on weight shift and opposition might help develop greater trunk rotation and sequential hip and upper trunk rotation. The professional might say, "Ann, that was a very good pass. I believe you can throw it much faster if you use your legs more. Try five more passes thinking about pushing harder, followed by rotating your body into the throw." This might be followed by several cues like those in table 9.4.

Another rationale for diagnosis is to work on easy corrections to give clients small successes to build on. The grip on the ball is a minor weakness, but a professional could focus intervention on this technique point with the idea that a better spiral on the throw might build confidence. This person is older, so an explanation of the desired change might also be a good intervention strategy. "Ann, let's try moving your grip down to the end of the ball. I think you will be able to push more directly for greater speed and spin on the pass. Try five passes with this grip."

Subject 3

Figure 9.9 illustrates the overarm throwing performance of a young man throwing the same junior football. Perform an integrated QMD of this overarm throwing performance. You can view the video clip of Overarm Throwing, Subject 3 in QMD Practice 9.9 in the web resource at www.HumanKinetics.com/QualitativeDiag nosisOfHumanMovement. What is different about this video clip and the one in QMD Practice 9.8?

In many ways, this young man's overarm throwing technique is similar to that of the young woman in QMD Practice 9.8. He does not appear to be throwing with maximal effort, and it may be hard to diagnose whether minor weaknesses are real throwing technique limitations or symptomatic of an easy task. As in the previous example, he has a somewhat arm-centric throw with some opposition as well as good

(continued) ➜

Figure 9.9 Sequence images of the overarm throw of a young man throwing a junior football to a partner. What intervention would help this person improve the most? Time between pictures is 0.17 seconds.

QMD OF THE DISK (FRISBEE) THROW OR TOSS

Throwing a flying disk is a skill related to the sidearm throwing fundamental movement pattern. This skill is important in the sports of ultimate and disk golf. This section illustrates the QMD of the basic forehand disk throw or toss.

Critical Features

The basic forehand disk toss, similar to the overarm skill of football passing, requires skilled coordination to keep the disk aligned as speed and spin are imparted to it. Very little biomechanical research on disk throwing has been published, so the critical features are generally based on popular literature on disk and ultimate golf (Baccarini and Booth, 2008; Gregory, 2003). The critical features and cues for disk throwing are presented in table 9.5.

upper arm alignment and arm throwing motion. If asking him to throw harder or for a longer distance didn't change technique, then diagnosis would indicate that the best intervention might be to work on the lower extremity and trunk rotation contributions to the pass. The professional might say, "Your passes are looking good, George. They will even be more effective and less likely to be intercepted if we increase the speed of the pass. I want you to use more leg drive and body rotation for your pass. In the next five passes, think about a longer, faster leg drive and trunk rotation to start your pass." Following up with several cues and focusing the player's attention on the effect of these changes might help motivate him to practice and adopt the changes.

Hopefully you noticed that the vantage point or perspective for observation in this clip was more oblique than in the video in QMD Practice 9.8. This is less of a side view, making it a little more difficult to judge forward (from the performer's perspective) movement.

Table 9.5   Critical Features and Cues for Sidearm Disk Throwing

Critical features	Cues
Grip	Pinch the edge of the disk
Opposition	Turn your side to the target
Preparation	Tuck the disk under your opposite arm
Arm extension	Extend your arm toward the target
Follow-through	Extend your wrist and point to the target

Adapted by permission from Knudson, Luedke, and Fairbault 1994.

Systematic Observational Strategy

As with overarm throwing, an SOS for a sidearm disk or Frisbee throw typically proceeds sequentially or by phases of the movement. Readers are encouraged to use whatever SOS they feel most comfortable with when observing video of performances in QMD Practices 9.10 through 9.12.

Subject 1

Figure 9.10 illustrates the disk throwing technique of a young boy throwing to a partner. Perform an integrated QMD of this disk throwing performance. You can view the video clip of Disk Throw, Subject 1 in QMD Practice 9.10 in the web resource at www.HumanKinetics.com/QualitativeDiagnosisOfHumanMovement. Use whatever observational strategy seems most natural for you.

This young boy is having a good time trying to throw the disk as fast as he can. Like many young children, his enthusiasm translates into considerable body rotation to drive the throw, with little forward body motion, which creates a corresponding loss of control over the direction of the disk. He has a good grip on the disk, opposition, and a long follow-through to drive the throw. Weaknesses show up in the exaggerated preparatory actions and trunk rotations that drive the disk in an indi-

(continued) ➜

Figure 9.10   Sequence images of a young boy throwing a disk (Frisbee) to a partner. What intervention would help this child improve the most? Time between pictures is 0.17 seconds.

Subject 2

Figure 9.11 illustrates the disk throwing technique of a girl throwing to a partner. Perform an integrated QMD of this disk throwing performance. You can view the video clip of Disk Throw, Subject 2 in QMD Practice 9.11 in the web resource at www.HumanKinetics.com/QualitativeDiagnosisOfHumanMovement. Use whatever observational strategy seems most natural for you.

This girl has several strengths in the critical features of a disk throw. Notice that she also has good opposition and a small step and weight transfer that subject 1 did not. In addition, she has a good grip on the disk and a relatively straight motion in accelerating the disk. She has a small follow-through toward the target, but this could also be a minor weakness. Weaknesses to work on are limited wrist flexion in

(continued) ➜

Figure 9.11   Sequence images of a girl throwing a disk (Frisbee) to a partner. What intervention would help this child improve the most? Time between pictures is 0.17 seconds.

rect and high path. This limits not only the accuracy of the throw but also the spin on the disk that is necessary to stabilize flight. The rotational follow-through is not elongated toward the target, increasing the difficulty in timing release of the disk to fly toward the target.

Good intervention for a young beginner should focus on one critical feature and should be expressed in a positive fashion to maintain the child's enthusiasm. Relating actions to previous actions in diagnosis, it is possible that focusing the child's attention on extension of the arm to the target would increase disk spin, but it might also have secondary effects in taming exaggerated body rotation and change in direction of the disk as it is accelerated. The professional might say, "That was a super effort and a nice high throw, Josh. Remember when we talked about extending your arm right to your partner or target? Let's throw a few more and think about extending toward your partner. Try to point to your target after you throw."

the preparation phase and a rather abrupt termination of the follow-through. Diagnosis of the weaknesses indicates that these minor improvements would be equally important. An abbreviated follow-though affects accuracy, but it is not likely an injury issue given the low speed and the mass of the disk. The disk had little rotation, and improving on either of these weaknesses should help improve disk rotation at release. It would seem that intervention focused on either of these would be appropriate.

An older child might be able to handle feedback on two technique points, but here the best intervention would focus on only one and then let QMD on the subsequent trials determine additional intervention. The professional might say, "Nice opposition and accuracy on that one, Jane. See if you can focus on tucking the disk under your opposite arm. Remember to bend your wrist so when you do that good throw of yours there is both speed and spin on the disk. Any questions? Give it a try."

Subject 3

Figure 9.12 illustrates the disk throwing technique of an adult throwing to a partner. Perform an integrated QMD of this disk throwing performance. You can view the video clip of Disk Throw, Subject 3 in QMD Practice 9.12 in the web resource at www.HumanKinetics.com/QualitativeDiagnosisOfHumanMovement. Use whatever observational strategy seems most natural for you.

This young woman has good performance on all the critical features of disk throwing. She shows a strong grip, body opposition, disk preparation, arm extension, and follow-through toward the target. She appears to be throwing with moderate effort

(continued) ➜

Figure 9.12 Sequence images of an adult throwing a disk (Frisbee) to a partner. What intervention would help this woman improve the most? Time between pictures is 0.17 seconds.

QMD OF THE SQUAT EXERCISE

A particularly important area of QMD for kinesiology professionals involves exercise technique, especially exercises with free weights. Positioning and technique strongly influence the loads the body experiences; and with additional weight, errors in technique can push stresses on the body beyond training levels to injurious levels. This section focuses on the QMD of the basic squat exercise.

Critical Features

The squat is one of the most popular whole-body exercises, and clinicians have developed rating scales for the movement to be used as part of a functional movement screening assessment (Butler et al., 2010). Conditioning textbooks (e.g., Chandler and Brown, 2013) and extensive research provide detailed information about the biomechanics of the squat and its variations (Abelbeck, 2002; Escamilla et al., 2001; Fry, Smith, and Shilling, 2003; Russell and Phillips, 1989; Wretenberg, Feng, and Arborelius, 1996). Table 9.6 lists typical critical features and cues for the squat exercise.

Almost all the critical features are also exercise safety issues and are important to teach and provide intervention on after QMD. People performing squats are usually instructed to keep their head up and eyes focused on a fixed point to stabilize any bar on the shoulders and assist with balance. In squat technique it is important to maintain a neutral spine, controlled breathing, and slow simultaneous coordination as well as to limit range of motion. Limiting range of knee flexion to a global reference of thighs parallel to the ground helps keep knee compression and shear forces toward a partner. A minor weakness may be alignment of the disk with arm and wrist extension. Since no major weaknesses are apparent, the plan for intervention might be general encouragement and praise and a plan for modifying practice. Strong performance on a task means that the person is likely ready for increase in the challenge or difficulty of practice tasks. The professional could say something like "Excellent technique and accuracy, Amy. Let's have both partners back up and try throwing to a jogging partner. Think about a target ahead of your partner's jog." A faster or more difficult throw will bring out potential weaknesses in arm extension and disk alignment.

Table 9.6 Critical Features and Cues for the Squat Exercise

Critical features	Cues
Bar position	Head up and shoulders back
Neutral spine	Keep back straight and stomach tight
Simultaneous coordination	Equalize trunk and knee rotation
Balance control	Keep eyes up and speed slow
Range of motion control	Thigh parallel or higher for safety
Breathing	Synchronize inhalation and exhalation

low. Trunk lean is another important segment angle to focus on, as the greater the lean, the greater the gravitational torque and stresses on the lower back.

Systematic Observational Strategy

Several SOSs would likely be desirable for QMD of the squat exercise. Since good balance control is so important, especially when the performer is using weight, professionals might elect to focus observation on balance and base of support. Others might choose to observe based on the importance of critical features. For example, a professional working with an athlete using heavy weight might choose to emphasize attention to safety-related factors like neutral spine, knee range of motion, and trunk lean. For QMD Practices 9.13 through 9.15, use whatever SOS you think is relevant to the situation or feels comfortable to you.

Subject 1

Figure 9.13 illustrates an adult doing a warm-up squat. Perform an integrated QMD of this squat performance. Notice that the slower movement of the squat results in a substantially longer period of time between images (0.5 seconds) compared to the previous QMD Practice clips (0.17 seconds). You can view the video clip of Squat, Subject 1 in QMD Practice 9.13 in the web resource at www.HumanKinetics.com/QualitativeDiagnosisOfHumanMovement. Use whatever observational strategy seems most natural for you.

The performer of this squat has fairly good technique, and the bar weight does not represent a heavy load. He performs most of the critical features of the squat well. He has a neutral spine, complete range of motion, approximately equal joint flexion in the eccentric phase and extension in the concentric phase, and a good range of motion (thigh parallel to the floor at reversal position). There are two minor weaknesses. One weakness is related to bar position, in that he should keep his head and his eye focus higher. The second potential weakness is not visible in the figure but

(continued) ➜

Figure 9.13 Sequence images of the squat as an adult warm-up exercise with an Olympic bar. What intervention would be most appropriate for this person? Time between pictures is 0.5 seconds.

can be perceived in the video replay. Some lack of balance control or smoothness of knee motion (wobble) seems apparent. Additional observation in the frontal plane would be useful in evaluation and diagnosis of this weakness. Both of these weaknesses relate to skills that are often absent in beginners starting complex free weight training exercises like the squat.

The best intervention in this case would depend on all the knowledge the professional had about this client (e.g., strength, experience) and the situation. Most likely, the professional would make a mental note to check on the apparent unsteadiness and provide feedback about the head and eye position. It is possible that poor visual focus decreases balance and contributes to the lack of smoothness and control of the legs in the exercise. The professional might reinforce the good range of motion and straight back, following up with a cue on the head and eyes: "Those are nice reps, John. Good range of motion and a perfect straight back. The next set I want you to remember the cue 'head up.' Keeping your head up and eyes focused on a spot on the wall will help you balance the bar and your body." Adult clients can handle and often appreciate a little more explanation on the selected interventions than children do.

Subject 2

Figure 9.14 illustrates the single-leg squat technique of a young adult. Perform an integrated QMD of this squat. You can view the video clip of Squat, Subject 2 in QMD Practice 9.14 in the web resource at www.HumanKinetics.com/QualitativeDiagnosisOf HumanMovement. Use whatever observational strategy seems most natural for you.

The single-leg squat exercise is quite difficult because it places great demands on strength and balance over a tiny base of support. Sometimes the exercise is performed with the support foot on an elevated surface to allow for greater lowering without placing great demands on controlling the opposite leg. The person performing certainly appears to be challenged by this body-weight exercise (the wooden bar does not provide substantial load). Her strengths are a straight back and slow, controlled movement. Weaknesses include a limited range of motion and limited hip joint flexion and extension and, possibly, small weaknesses in balance control of knee motion

(continued) ➜

Figure 9.14 Sequence images of the single-leg squat of a young adult. What intervention would be most appropriate for this person? What elements of the critical features are difficult to evaluate from a frontal plane view? Time between pictures is 0.33 seconds.

Subject 3

Figure 9.15 illustrates an adult doing a squat exercise with an Olympic bar. Perform an integrated QMD of this squat. You can view the video clip of Squat, Subject 3 in QMD Practice 9.15 in the web resource at www.HumanKinetics.com/Qualitative DiagnosisOfHumanMovement. Use whatever observational strategy seems most natural for you.

The subject performing this squat has good form and, given his apparent level of conditioning, this too is likely a warm-up set and not a training load. He performs all the critical features of the squat well. He has complete range of motion as well as smooth, equal joint flexion in the eccentric phase and extension in the concentric

(continued) ➜

Figure 9.15 Sequence images of a person doing a squat exercise. What intervention would be most appropriate for this person? Time between pictures is 0.5 seconds.

and bar position (head and eye focus). Frontal plane observation would normally be indicated for single-leg squats.

As in the first example, diagnosis and intervention of this performance would usually be aided by additional knowledge that the professional has about the client and the situation. The professional would most likely consider changing the exercise task for intervention. If this client has difficulty with balance and completing the exercise, easier squat variations might provide challenge and still allow good technique. A professional who believed that the performer was skilled and strong enough to do this exercise regularly might select this intervention: "That was a nice set of a difficult exercise, Amy. I would like you to try another set using an elevated surface. This will give you some space to try and get more range of motion. I want you to think about keeping your head up and eyes focused on the wall to help you balance." An adult might be able to handle two items of feedback (range of motion and eye focus) in subsequent repetitions.

phase. Very minor technique adjustments that are not necessarily about weaknesses at this point could be to keep the head and eye focus up and not let the thigh get below parallel to the ground. The best intervention in this case could certainly be silence or general encouragement. The professional might make a mental note to check on the technique points once this subject starts using training weights. Other professionals might choose to give reinforcement or a reminder about safety issues and encourage the performer to "keep head up and eyes focused on a spot on the wall," or ask "Bob, did those reps feel like your thighs were parallel to the floor or beyond parallel?" Questioning is a good intervention strategy with advanced performers, who are likely to be fairly good at monitoring the intrinsic feedback from their kinesthetic senses during movement.

QMD EXPLORATIONS

The rest of this chapter presents open-ended opportunities for QMD with several movements. These QMD Explorations allow you to practice QMD with human movements and situations other than those discussed in the chapter. Each QMD Exploration includes a video clip of an activity in real time and in slow motion.

For each QMD Exploration, consider the following questions related to each task of QMD:

1. **Preparation:** What are the critical features of the movement?
2. **Observation:** What is a good systematic observational strategy for live and slow-motion observation of the movement?
3. **Evaluation and diagnosis:** What are the most important strengths and weaknesses that you see in this performance?
4. **Intervention:** What would you recommend to help this person improve performance and why?

If you have access to the web resource, you can download a form with these same questions and fill in your judgments about the best approach to diagnosing and improving performance. The web resource is at www.HumanKinetics.com/QualitativeDiagnosisOfHumanMovement.

The introductory information provided for QMD Explorations is purposefully limited. A few sources are recommended to assist you in the preparation task for QMD. As with the theory-into-practice examples in chapter 10, however, you are now in the driver's seat with regard to gathering relevant knowledge about the movement, mover, and environment. You get to design the observational strategy, evaluate the performance, and diagnose strengths and weaknesses. The intervention you select should integrate all the knowledge you have or can access from all the subdisciplines of kinesiology.

Since there are no suggested answers for the QMD Explorations as there are for the QMD Practices, readers are encouraged to discuss these examples with other kinesiology professionals. Comparing assessments, evaluations, diagnoses, and proposed interventions can be quite enlightening. After you work with the QMD Explorations, it would be a good idea also to practice QMD with live performances. You may be able to observe unobtrusively at a neighborhood park, gym, or athletic field. Another good way to practice is to review your own video recorded performances.

QMD of Basketball Shooting

Shooting in the game of basketball takes a variety of forms. The most common shot, however, is the jump shot. Teachers and coaches of basketball need to be aware of the open nature of basketball. The uncertainty imposed by the defense, teammates, and the clock can dramatically affect how a player shoots the ball. Factors like shot distance and release height also affect the optimal release conditions for a shot. Professionals should note that there are numerous studies of the interaction of biomechanical factors, skill, and age on basketball shooting (Arias, 2012; Hamilton and Reinschmidt, 1997; Huston and Grau, 2003; Miller and Bartlett, 1993, 1996; Liu and Burton, 1999; Rojas et al., 2000). A model for QMD of the jump shot was proposed by Knudson (1993).

Basketball Shot, Subject 1

View the video clips of Basketball Shot, Subject 1 in QMD Exploration 9.1 in the web resource at www.HumanKinetics.com/QualitativeDiagnosisOfHuman Movement. Perform an integrated QMD of this performance. What intervention will improve the performance the most?

Basketball Shot, Subject 2

View the video clips of Basketball Shot, Subject 2 in QMD Exploration 9.2 in the web resource at www.HumanKinetics.com/QualitativeDiagnosisOfHuman Movement. Perform an integrated QMD of this performance. What intervention will improve the performance the most?

Basketball Shot, Subject 3

View the video clips of Basketball Shot, Subject 3 in QMD Exploration 9.3 in the web resource at www.HumanKinetics.com/QualitativeDiagnosisOfHuman Movement. Perform an integrated QMD of this performance. What intervention will improve the performance the most?

PRACTICAL APPLICATIONS
Similar Skills

Have a partner perform overarm throws and tennis serves. How are the overarm throw and tennis serve similar and how are they different? Do similar critical features have the same importance in these two skills? Do differences in the serve and throw remain consistent across different performers? Does the ability to perform QMD of these two skills transfer to similar motor skills?

Perform your own mini-experiment by practicing QMD of live or videotaped tennis serve and overarm throwing performances. Both activities are difficult to observe because of their speed. What outcome variables (ball accuracy, speed, spin) may be useful for gaining information about performance elements that are difficult to see? Do you think skill in QMD of the overarm throw will transfer to QMD of the tennis serve? Do you think skill in observation of the overarm throw helps in the QMD of the tennis serve?

QMD of the Tennis Serve

The QMD of the tennis serve provides an interesting contrast to the overarm throw. The overarm throw is the fundamental movement pattern associated with the tennis serve, and several authors have attempted to document the similarities between these two movements in order to examine the potential transfer of learning (Adrian and Enberg, 1971; Anderson, 1979; Miyashita et al., 1979; Rose and Heath, 1990; Wilkinson, 1996; Zebas and Johnson, 1989). Much of the tennis service action is different because of the use of a racket and also rules restrictions on how the serve can be delivered. The ball must be tossed, must be hit in the air without the body's changing position on the court, and must travel over the net and land in the correct service court.

Classic tennis instruction calls for a specific pattern of good form in the tennis serve, with small variations for different ball spins and match tactics. There are two proposed models for QMD of the tennis serve (Knudson, Luedtke, and Faribault, 1994; Rose, Health, and Meagle, 1990). Professionals should also review an extensive body of biomechanics research on the tennis serve (Bahamonde, 2000; Chow et al., 1999; Elliott, Marshall, and Noffal, 1995; Knudson, 2006; Sheet et al., 2011).

Tennis Serve, Subject 1

View the video clips of Tennis Serve, Subject 1 in QMD Exploration 9.4 in the web resource at www.HumanKinetics.com/QualitativeDiagnosisOfHumanMovement. Perform an integrated QMD of this performance. What intervention will improve the performance the most?

Tennis Serve, Subject 2

View the video clips of Tennis Serve, Subject 2 in QMD Exploration 9.5 in the web resource at www.HumanKinetics.com/QualitativeDiagnosisOfHumanMovement. Perform an integrated QMD of this performance. What intervention will improve the performance the most?

Tennis Serve, Subject 3

View the video clips of Tennis Serve, Subject 3 in QMD Exploration 9.6 in the web resource at www.HumanKinetics.com/QualitativeDiagnosisOfHumanMovement. Perform an integrated QMD of this performance. What intervention will improve the performance the most?

QMD of Batting

Baseball and softball batting are skills of the striking fundamental movement pattern. While you can find coaching opinions about hitting technique for baseball and softball (Robson, 2003), not much research exists on the biomechanics of batting (Miyashita et al., 1986; Inkster et al., 2011; Mcintyre and Pfautsch, 1982; McLean and Reeder, 2000; Welch et al., 1995) considering the popularity of the sports. As in basketball shooting, the psychology of competition can affect batting technique. For young players, fear of being hit by the pitched ball sometimes has a major influence on their technique.

Batting, Subject 1

View the video clips of Batting, Subject 1 in QMD Exploration 9.7 in the web resource at www.HumanKinetics.com/QualitativeDiagnosisOfHumanMovement. Perform an integrated QMD of this performance. What intervention will improve the performance the most?

Batting, Subject 2

View the video clips of Batting, Subject 2 in QMD Exploration 9.8 in the web resource at www.HumanKinetics.com/QualitativeDiagnosisOfHumanMovement. Perform an integrated QMD of this performance. What intervention will improve the performance the most?

Batting, Subject 3

View the video clips of Batting, Subject 3 in QMD Exploration 9.9 in the web resource at www.HumanKinetics.com/QualitativeDiagnosisOfHumanMovement. Perform an integrated QMD of this performance. What intervention will improve the performance the most?

QMD of In-Line Skating

In-line skating is a skill that presents an interesting adaptation of the fundamental motor skill of locomotion. Adults often do not appreciate the high level of skill required for walking and running (highly efficient push-off, support, and balance over a very small base of support) until they are injured or are interacting with a toddler. In-line skating requires adapted skills of push-off and balance over a very small base of support in dramatically different surface friction conditions. There is very little research on in-line skating, but several books describe techniques, safety, and progressions (Millar, 1996; Miller, 1998; Powell and Svensson, 1998). Critical features for forward skating would typically focus on stance, arm position for balance and safety, push-off, and recovery and gliding.

In-Line Skating, Subject 1

View the video clips of In-Line Skating, Subject 1 in QMD Exploration 9.10 in the web resource at www.HumanKinetics.com/QualitativeDiagnosisOfHuman Movement. Perform an integrated QMD of this performance. What intervention will improve the performance the most?

In-Line Skating, Subject 2

View the video clips of In-Line Skating, Subject 2 in QMD Exploration 9.11 in the web resource at www.HumanKinetics.com/QualitativeDiagnosisOfHuman Movement. Perform an integrated QMD of this performance. What intervention will improve the performance the most?

In-Line Skating, Subject 3

View the video clips of In-Line Skating, Subject 3 in QMD Exploration 9.12 in the web resource at www.HumanKinetics.com/QualitativeDiagnosisOfHuman Movement. Perform an integrated QMD of this performance. What intervention will improve the performance the most?

QMD of the Lunge Exercise

The lunge is a common conditioning and rehabilitation exercise used by kinesiology professionals. The basic technique is outlined in most strength and conditioning texts (e.g., Chandler and Brown, 2013), and some biomechanics studies have documented the movement and its variations in greater detail (Cronin, McNair, and Marshall, 2003; Jonhagen, Halvorsen, and Benoit, 2009).

Lunge, Subject 1

View the video clips of Lunge, Subject 1 in QMD Exploration 9.13 in the web resource at www.HumanKinetics.com/QualitativeDiagnosisOfHumanMovement. Perform an integrated QMD of this performance. What intervention will improve the performance the most?

Lunge, Subject 2

View the video clips of Lunge, Subject 2 in QMD Exploration 9.14 in the web resource at www.HumanKinetics.com/QualitativeDiagnosisOfHumanMovement. Perform an integrated QMD of this performance. What intervention will improve the performance the most?

Lunge, Subject 3

View the video clips of Lunge, Subject 3 in QMD Exploration 9.15 in the web resource at www.HumanKinetics.com/QualitativeDiagnosisOfHumanMovement. Perform an integrated QMD of this performance. What intervention will improve the performance the most?

Additional QMD Explorations

The following QMD Explorations provide additional practice opportunities for QMD of familiar and some less familiar movement skills.

Young Girl Throwing

View the video clips of throwing in QMD Exploration 9.16 in the web resource at www.HumanKinetics.com/QualitativeDiagnosisOfHumanMovement. Perform an integrated QMD of this performance. What intervention will improve the performance the most?

Young Boy Catching

View the video clips of catching in QMD Exploration 9.17 in the web resource at www.HumanKinetics.com/QualitativeDiagnosisOfHumanMovement. Perform an integrated QMD of this performance. What intervention will improve the performance the most?

Juggling

View the video clips of juggling in QMD Exploration 9.18 in the web resource at www.HumanKinetics.com/QualitativeDiagnosisOfHumanMovement.Perform an integrated QMD of this performance. What intervention will improve the performance the most?

Lacrosse Stick Handling

View the video clips of lacrosse stick handling in QMD Exploration 9.19 in the web resource at www.HumanKinetics.com/QualitativeDiagnosisOfHumanMovement. Perform an integrated QMD of this performance. What intervention will improve the performance the most?

SUMMARY

Training has been shown to improve QMD ability. This chapter presents sequence images and videos of several human movements to illustrate application of the integrated model of QMD. Readers should compare their diagnoses of the examples with diagnoses of peers and the diagnoses presented in the text and critically examine differences. Discussion of the QMD Practice and QMD Explorations with professional peers is another way to practice this important skill.

Discussion Questions

1. What SOSs are difficult to employ with the sequential images presented in this chapter?

2. Discuss the advantages of diagnosing and providing intervention based on critical features that have been prioritized according to their importance to performance.

3. Identify possible problems of diagnosing and providing intervention based on critical features that have been prioritized according to their importance.

4. What information is missing from the sequence images of the movements that can be perceived from video replay? Do the video clips and your ability to replay them have limitations? How would your evaluation and diagnosis be different if additional information were available?

chapter 10

Theory-Into-Practice Situations

An adult at your health club is becoming very interested in weight training and has asked for your help in doing lunges and lunge variations. He has done lunges before but wants you to evaluate his technique and give him a lesson on the finer points. Precise body positioning and technique are essential in weight training. They ensure that the muscle groups targeted by the exercises are in fact the muscle groups that are primarily involved. Use the science of biomechanics to decide how you might qualitatively diagnose a lunge. Imagine that on the first lunge the client has minimal forward trunk lean, but as he does several more repetitions he begins to have a more forward trunk lean. What differences in muscle group involvement would you expect in the later lunges? Consider changes he would experience in his arm movement and the horizontal distance from the load to the joints. What back position (amount of lumbar arch) is associated with even or safe spinal loading?

Chapter Objectives

1 Develop skill in applying integrated QMD to a variety of real-world human movement problems.

2 Identify subdisciplines of kinesiology that provide relevant information for making decisions within QMD of human movement.

Many situations arise in which a professional must integrate information from several subdisciplines of kinesiology. This chapter presents some scenarios from various kinesiology professions in which an integrated qualitative movement diagnosis (QMD) can be applied to help a performer. Think through each of the tasks of an integrated QMD. These situations are useful for class exercises, assignments, and for generating discussions with other professionals. The first theory-to-practice situation is developed as an example. Following this example you will be presented with other QMD challenges. Each theory-to-practice situation provides a few specific questions to consider. The suggested format for responding to the situations is naturally organized around the comprehensive model proposed in the text: preparation, observation, evaluation and diagnosis, and intervention. Review chapters 4 through 7 on the main tasks of QMD. The answer sheets here and on the web resource provide only a reminder of major subtasks within each task of QMD.

"CLEARED TO PLAY"

A sophomore coming off an anterior cruciate ligament (ACL) knee injury has asked to return to your club volleyball team. She has been cleared to play by her physician. As head coach, you have to decide if she is ready for practice and match play because there is no athletic trainer to monitor her rehabilitation. What aspects of volleyball skills should QMD focus on? What physical tests can you give this athlete to help you make this decision? Do the athlete's personality traits have any bearing on this decision?

Preparation

In the preparation task of QMD, you should review any preinjury information you may have on this athlete. Your research on her injury, rehabilitation, and previous history should suggest which critical features of volleyball movements you should observe to evaluate her ability to play. Since the ACL is an important knee ligament limiting forward motion of the tibia on the femur, you should plan to observe the control of knee flexion and extension. Important volleyball movements that may stress the ACL are jumping, landing, and making vigorous changes of direction. You would likely want to do some of the common drills at lower intensity and gradually increase the intensity and challenge. The athlete would perform these separately from actual team practice so that you can pay proper attention to her and decrease the possibility of reinjury. You should also plan to focus on the player's facial expressions and body language for signs of knee pain.

Beyond observing for signs of injury, you would want to evaluate how much previous ability has been recovered. If quantitative information (jump height, speed trials, agility measures) is not available, use your memory to compare the athlete's volleyball movements to those of other players. Frontal plane observation of landing movement allows the evaluation of hip abduction strength (controlling hip adduction), and sagittal plane observation allows you to evaluate initial knee angle at foot contact and flexion. You might also plan to perform your typical QMD of selected volleyball skills to look for changes in technique.

Observation

The plan would be to carefully observe the athlete's control of knee flexion in landing, jumps, and changes of direction in submaximal and maximal efforts. The observational strategy should also focus on any signs of discomfort. A good plan for observation might begin with the usual team warm-up routine so that you'll know what movements are coming and can be sure of a gradual, safe increase in movement intensity. Volleyball movements you should plan to observe are jumps and landings in spiking and blocking. Intensity should be gradually increased. Since the ACL checks forward motion of the tibia, you might ask the athlete to hop forward and backward as far as possible on each foot six times. Landing in backward hopping creates a forward force on the lower leg that tends to stress the ACL. You would plan to observe landing from sagittal and frontal planes. Major differences in distance jumped or control of the knee between the injured and uninjured legs are important points for observation. Finally, plan on visual and tactile monitoring of the affected knee for swelling after practice or physical testing.

Evaluation and Diagnosis

The focus in evaluation would be on any signs of weakness in the athlete's control of the knee during the high-energy phases of jumping, landing, and cutting movements. Evaluation in this situation may be simpler than in many QMDs because as a coach you are primarily interested in detecting injury-related weaknesses that would prohibit participation. It may be that merely the buckling or giving of the knee or unusual knee motions need to be detected. Evaluating the athlete's recovery of ability relative to other athletes, or evaluating her technique in the various volleyball skills, is more difficult. Judging the quality of a spike is more difficult than merely detecting differences in distances or heights of jumps. This is not to say that strengths of lower extremity performance are not of interest, but these are secondary in this case in which you want to make sure that the athlete is ready for full-speed practice and competition.

The diagnosis of the weaknesses identified in evaluation would depend on your philosophy regarding the return of injured players. Since tissues that have not fully healed are more easily reinjured, some coaches might keep diagnosis simple by not allowing a player to return if she shows any signs of weakness or loss of knee control. This diagnosis of performance could also be tricky because the athlete might experience some discomfort associated with higher-intensity activity rather than with overuse of tissue that is not fully healed. Here you must apply your knowledge of sports medicine and biomechanics to make sure that critical features identified (control of hip adduction and knee flexion) in evaluation are not related to reinjury or to limiting volleyball performance. Ethically, the injury risk to the athlete is the first priority of diagnosis; level of performance is the second priority.

Intervention

As the coach, you have many options in dealing with this athlete. If your diagnosis indicates that she is ready to return, you should provide positive feedback on her effort in rehabilitating the injury. It would also be wise to express confidence in the player's recovery and ability. Remind her that her timing and teamwork will take time to recover. Express confidence in her, but ask her to be careful and caution her that most athletes tend to rush the recovery process.

If your diagnosis indicates that the athlete should delay her return to the team, there are also several approaches to intervention. It is very important to be sensitive when you tell the athlete that she needs to continue rehab before she can rejoin the team. You might say, "You've made a lot of progress, but I would like you to come back in two weeks for another tryout." Another option would be to allow limited practice with the team. You could also recommend that the athlete consider a knee brace during this practice and offer to reevaluate her situation soon to help her stay motivated.

THEORY-INTO-PRACTICE SITUATIONS FOR QMD

Read the following situations and then carefully decide what would be most appropriate within each task of a comprehensive QMD. Given the limited information provided, some scenarios may lead you to focus on one task of the QMD over the others, but even in these situations, try to consider the other tasks of QMD as well.

"But I Can Score!"

A high school basketball player is having great success driving to the left and scoring during a preseason practice. But he uses his right hand to execute layups and is successful only because of his speed and superior jumping ability. What feedback or intervention would be appropriate? What will happen if taller or faster opponents guard this player or if weak-side defense rotates to defend the layup?

1. **Preparation:** What knowledge do you need to gather about the movement, the performer, and the situation in preparing for QMD of this case? Why?

2. **Observation:** What would be an effective systematic observational strategy (SOS) for gathering information for QMD? Is there a need for extended observation? Why?

3. **Evaluation and diagnosis:** What evaluation method (e.g., sequential or mechanical) and rationale for diagnosing performance would be most effective in this case? Why?

4. **Intervention:** What mode of intervention would likely be best? Why?

From D.V. Knudson, 2013, *Qualitative diagnosis of human movement,* 3rd ed. (Champaign, IL: Human Kinetics).

The Slump

A college softball player is struggling with her hitting. At midseason she was leading the team, but now she has been hitless in several games. What would be a good approach to the QMD of her hitting? Assume there is video of her hitting from earlier in the season. Should intervention focus on technique or the hitter's confidence?

1. **Preparation:** What knowledge do you need to gather about the movement, the performer, and the situation in preparing for QMD of this case? Why?

2. **Observation:** What would be an effective systematic observational strategy (SOS) for gathering information for QMD? Is there a need for extended observation? Why?

3. **Evaluation and diagnosis:** What evaluation method (e.g., sequential or mechanical) and rationale for diagnosing performance would be most effective in this case? Why?

4. **Intervention:** What mode of intervention would likely be best? Why?

From D.V. Knudson, 2013, *Qualitative diagnosis of human movement,* 3rd ed. (Champaign, IL: Human Kinetics).

"But the Champ Does It This Way"

As a tennis coach, you notice that a promising junior tennis player is using more open-stance forehand drives during practice and matches. You think he is emulating many tennis professionals who are using open-stance strokes rather than the traditional square stance. What intervention do you provide to this 12-year-old player?

1. **Preparation:** What knowledge do you need to gather about the movement, the performer, and the situation in preparing for QMD of this case? Why?

2. **Observation:** What would be an effective systematic observational strategy (SOS) for gathering information for QMD? Is there a need for extended observation? Why?

3. **Evaluation and diagnosis:** What evaluation method (e.g., sequential or mechanical) and rationale for diagnosing performance would be most effective in this case? Why?

4. **Intervention:** What mode of intervention would likely be best? Why?

From D.V. Knudson, 2013, *Qualitative diagnosis of human movement,* 3rd ed. (Champaign, IL: Human Kinetics).

"But I Can Lift More!"

An athlete is struggling with a plateau in chest strength and asks for your help with training. You notice that her hips and low back lift off the bench during her bench press. What intervention is appropriate for an athlete of any age with this technique? The athlete senses that she can lift more with this technique, but what parts of the body does it put at risk of injury? Are there bench-press variations that would be safer? Are there alternative exercises isolating specific muscle groups that can be used together to replace the bench press? What intervention strategy will you use to motivate the use of a safer lifting technique?

1. **Preparation:** What knowledge do you need to gather about the movement, the performer, and the situation in preparing for QMD of this case? Why?

2. **Observation:** What would be an effective systematic observational strategy (SOS) for gathering information for QMD? Is there a need for extended observation? Why?

3. **Evaluation and diagnosis:** What evaluation method (e.g., sequential or mechanical) and rationale for diagnosing performance would be most effective in this case? Why?

4. **Intervention:** What mode of intervention would likely be best? Why?

From D.V. Knudson, 2013, *Qualitative diagnosis of human movement,* 3rd ed. (Champaign, IL: Human Kinetics).

The Parent Distraction

An overzealous parent keeps yelling at one of your players, "Box out!" Defensive rebounding has not been a problem in this game, and you have coached the team to emphasize the fast break against this opponent. How can QMD be used to identify which player the parent is yelling at and evaluate whether the players are being affected? What intervention would be appropriate?

1. **Preparation:** What knowledge do you need to gather about the movement, the performer, and the situation in preparing for QMD of this case? Why?

2. **Observation:** What would be an effective systematic observational strategy (SOS) for gathering information for QMD? Is there a need for extended observation? Why?

3. **Evaluation and diagnosis:** What evaluation method (e.g., sequential or mechanical) and rationale for diagnosing performance would be most effective in this case? Why?

4. **Intervention:** What mode of intervention would likely be best? Why?

From D.V. Knudson, 2013, *Qualitative diagnosis of human movement,* 3rd ed. (Champaign, IL: Human Kinetics).

The Final Authority

You are asked to be a judge for a diving demonstration given by a local swim club. The divers will be children ages 10 to 16 who are first- and second-year students preparing for future competition. How can you increase the accuracy and consistency of your QMD of the dives? How can you avoid bias related to dive difficulty, age of the performer, and the order within the demonstration?

1. **Preparation:** What knowledge do you need to gather about the movement, the performer, and the situation in preparing for QMD of this case? Why?

2. **Observation:** What would be an effective systematic observational strategy (SOS) for gathering information for QMD? Is there a need for extended observation? Why?

3. **Evaluation and diagnosis:** What evaluation method (e.g., sequential or mechanical) and rationale for diagnosing performance would be most effective in this case? Why?

4. **Intervention:** What mode of intervention would likely be best? Why?

From D.V. Knudson, 2013, *Qualitative diagnosis of human movement,* 3rd ed. (Champaign, IL: Human Kinetics).

Machine Versus Free Weights

You are asked to be part of a panel discussion on weight training at the state convention. The panel has chosen to discuss the differences between weight training with free weights and with a machine. How can you use QMD to compare free-weight arm curls with machine arm curls? What subdisciplines of kinesiology are most relevant in comparing each exercise?

1. **Preparation:** What knowledge do you need to gather about the movement, the performer, and the situation in preparing for QMD of this case? Why?

2. **Observation:** What would be an effective systematic observational strategy (SOS) for gathering information for QMD? Is there a need for extended observation? Why?

3. **Evaluation and diagnosis:** What evaluation method (e.g., sequential or mechanical) and rationale for diagnosing performance would be most effective in this case? Why?

4. **Intervention:** What mode of intervention would likely be best? Why?

From D.V. Knudson, 2013, *Qualitative diagnosis of human movement,* 3rd ed. (Champaign, IL: Human Kinetics).

Finishing Strong

You are a basketball coach whose best shooter has been shooting the lights out the entire game, but suddenly near the end of the game his shots are falling short. You may have only one or two looks at his form before you can call a time-out in the final minutes of the game. What physical factors are involved and how can they be evaluated? How can you evaluate whether game pressure is affecting the athlete? What critical factors in shooting and other basketball skills should be observed? What interventions are best for different causes of this problem? How does this athlete's psychological makeup affect your intervention?

1. **Preparation:** What knowledge do you need to gather about the movement, the performer, and the situation in preparing for QMD of this case? Why?

2. **Observation:** What would be an effective systematic observational strategy (SOS) for gathering information for QMD? Is there a need for extended observation? Why?

3. **Evaluation and diagnosis:** What evaluation method (e.g., sequential or mechanical) and rationale for diagnosing performance would be most effective in this case? Why?

4. **Intervention:** What mode of intervention would likely be best? Why?

Choke?

Two of the best tennis players in your junior development program have signed up for group lessons through the summer. One player excels in practice, where there are few observers and little crowd noise. This player has had difficulty in tournaments because of the crowd noise and pressure. The other player is just the opposite. She seems unmotivated in practice but thrives on pressure and has pulled off some big wins in tournaments. What can be done in practice and competition to help both performers? How can QMD be used to address the needs of each player?

1. **Preparation:** What knowledge do you need to gather about the movement, the performer, and the situation in preparing for QMD of this case? Why?

2. **Observation:** What would be an effective systematic observational strategy (SOS) for gathering information for QMD? Is there a need for extended observation? Why?

3. **Evaluation and diagnosis:** What evaluation method (e.g., sequential or mechanical) and rationale for diagnosing performance would be most effective in this case? Why?

4. **Intervention:** What mode of intervention would likely be best? Why?

From D.V. Knudson, 2013, *Qualitative diagnosis of human movement,* 3rd ed. (Champaign, IL: Human Kinetics).

Push and Glide

You are teaching inline skating at a summer camp. One student is having difficulty with turns to the left. Your initial intervention focused on the push-off and recovery of his right leg. Like a good professional, you immediately return to observation from intervention and find that the student is still having difficulty. Could your initial evaluation and diagnosis have been incorrect? Should you plan to evaluate the right leg again, or focus on the left leg? What aspects of leg action and balance might provide clues to the skater's problem? Could changing the task provide other clues for your evaluation?

1. **Preparation:** What knowledge do you need to gather about the movement, the performer, and the situation in preparing for QMD of this case? Why?

2. **Observation:** What would be an effective systematic observational strategy (SOS) for gathering information for QMD? Is there a need for extended observation? Why?

3. **Evaluation and diagnosis:** What evaluation method (e.g., sequential or mechanical) and rationale for diagnosing performance would be most effective in this case? Why?

4. **Intervention:** What mode of intervention would likely be best? Why?

From D.V. Knudson, 2013, *Qualitative diagnosis of human movement,* 3rd ed. (Champaign, IL: Human Kinetics).

Head Over Heels, Please

One of your newer gymnastics students is having difficulty learning the front handspring. Your observation of the first few lessons makes you think this young person is afraid of the more difficult skills. As you plan your next lesson, you set up your approach for the four tasks of QMD. What psychological cues about the gymnast's attitude do you plan to look for? What intervention will be most effective for this gymnast, motivational or technique? Could modifying the task build skills and confidence in this gymnast?

1. **Preparation:** What knowledge do you need to gather about the movement, the performer, and the situation in preparing for QMD of this case? Why?

2. **Observation:** What would be an effective systematic observational strategy (SOS) for gathering information for QMD? Is there a need for extended observation? Why?

3. **Evaluation and diagnosis:** What evaluation method (e.g., sequential or mechanical) and rationale for diagnosing performance would be most effective in this case? Why?

4. **Intervention:** What mode of intervention would likely be best? Why?

From D.V. Knudson, 2013, *Qualitative diagnosis of human movement,* 3rd ed. (Champaign, IL: Human Kinetics).

Coaching Clinic

You are a football coach worried about a player who is struggling to participate after several lower extremity injuries in earlier years. This athlete really wants to play and is the kind of person who will ignore the pain in order to play and help the team. You plan to secretly watch this player during the school day and in physical education class to evaluate movement for signs of injury. What cues in walking gait or other locomotion might hint of an ankle or knee injury? If your initial QMD suggests that the athlete is hurt, how could you plan a more formal QMD of several movements during practice to determine if you should rest this player?

1. **Preparation:** What knowledge do you need to gather about the movement, the performer, and the situation in preparing for QMD of this case? Why?

2. **Observation:** What would be an effective systematic observational strategy (SOS) for gathering information for QMD? Is there a need for extended observation? Why?

3. **Evaluation and diagnosis:** What evaluation method (e.g., sequential or mechanical) and rationale for diagnosing performance would be most effective in this case? Why?

4. **Intervention:** What mode of intervention would likely be best? Why?

From D.V. Knudson, 2013, *Qualitative diagnosis of human movement,* 3rd ed. (Champaign, IL: Human Kinetics).

Glide

You have a class of beginning swimmers. You notice during freestyle practice that one of your swimmers plunges her hand into the water directly in front of her head. She does not roll her body onto the entry and glide before she begins to pull. What instructional cues can you use to help her get a feel for proper technique? Are there any drills you could prescribe to help her roll and glide more?

1. **Preparation:** What knowledge do you need to gather about the movement, the performer, and the situation in preparing for QMD of this case? Why?

2. **Observation:** What would be an effective systematic observational strategy (SOS) for gathering information for QMD? Is there a need for extended observation? Why?

3. **Evaluation and diagnosis:** What evaluation method (e.g., sequential or mechanical) and rationale for diagnosing performance would be most effective in this case? Why?

4. **Intervention:** What mode of intervention would likely be best? Why?

From D.V. Knudson, 2013, *Qualitative diagnosis of human movement,* 3rd ed. (Champaign, IL: Human Kinetics).

Check It Out

You are the coach of the local high school hockey team. You have a defenseman who has played on your team for three years. This player understands the game well, especially the spatial relationship of players on the ice. This understanding extends to his position in relation to an oncoming forward carrying the puck across the blue line toward the goal. He is usually in good position to force the forward to the boards close to the blue line, but he rarely completes the play successfully with a solid, clean check into the boards. Since this player understands correct positioning, how can you help him optimize his checking? In your preparation phase, consider all the important technical information related to skating and checking. Also consider motivation and psychological factors. Plan how you will observe skating and checking technique. How will you diagnose and evaluate checking technique? What types of intervention could you use?

1. **Preparation:** What knowledge do you need to gather about the movement, the performer, and the situation in preparing for QMD of this case? Why?

2. **Observation:** What would be an effective systematic observational strategy (SOS) for gathering information for QMD? Is there a need for extended observation? Why?

3. **Evaluation and diagnosis:** What evaluation method (e.g., sequential or mechanical) and rationale for diagnosing performance would be most effective in this case? Why?

4. **Intervention:** What mode of intervention would likely be best? Why?

From D.V. Knudson, 2013, *Qualitative diagnosis of human movement,* 3rd ed. (Champaign, IL: Human Kinetics).

Surf's Up

You are at your favorite surfing spot when you notice a young surfer trying to catch a wave. She appears to be located at the correct takeoff point and gets good speed paddling to get on the wave. She also seems to be able to choose good waves. Once the wave starts moving her board, she appears to be able to make a quick and accurate transition from paddling to standing. Regardless of wave size or shape, however, she seems to fall off the board quickly, rarely gaining her balance or establishing herself in the correct standing position. How could you set yourself up to be able to observe her more closely? What will be your approach to observation? What things do you think she should know about surfing and balancing technique that would help her?

1. **Preparation:** What knowledge do you need to gather about the movement, the performer, and the situation in preparing for QMD of this case? Why?

2. **Observation:** What would be an effective systematic observational strategy (SOS) for gathering information for QMD? Is there a need for extended observation? Why?

3. **Evaluation and diagnosis:** What evaluation method (e.g., sequential or mechanical) and rationale for diagnosing performance would be most effective in this case? Why?

4. **Intervention:** What mode of intervention would likely be best? Why?

From D.V. Knudson, 2013, *Qualitative diagnosis of human movement,* 3rd ed. (Champaign, IL: Human Kinetics).

Hurdle

You are a track coach with several young athletes. You have a number of good sprinters and wish to start developing some of them into hurdlers. One of your better sprinters does not seem to be as excited about the prospects of running the hurdles as some of your less talented sprinters. Instead of approaching the hurdle with a great deal of speed, he seems to chop his steps and hesitate. This athlete seems to have all the tools of a good hurdler (height, speed, long legs, and so on). What would you do to ascertain what is limiting his ability to hurdle? How could you prepare to gather information upon which to base a decision? Could you plan an observation strategy to gather relevant information? What would you evaluate and diagnose? Could you suggest some forms of intervention based on your judgment of the problem?

1. **Preparation:** What knowledge do you need to gather about the movement, the performer, and the situation in preparing for QMD of this case? Why?

2. **Observation:** What would be an effective systematic observational strategy (SOS) for gathering information for QMD? Is there a need for extended observation? Why?

3. **Evaluation and diagnosis:** What evaluation method (e.g., sequential or mechanical) and rationale for diagnosing performance would be most effective in this case? Why?

4. **Intervention:** What mode of intervention would likely be best? Why?

From D.V. Knudson, 2013, *Qualitative diagnosis of human movement,* 3rd ed. (Champaign, IL: Human Kinetics).

Preactivity Screening

You are a junior high school girls' soccer coach planning off-season practice and tryouts. You are familiar with the research on the higher risk of ACL injuries in maturing young women compared to men. Should some of your warm-up, conditioning, or even direct testing of these athletes be specifically targeted to evaluating potential risk for this injury? To what extent do positive indicators for risk factor into decisions on making the team? Future practice and conditioning?

1. **Preparation:** What knowledge do you need to gather about the movement, the performer, and the situation in preparing for QMD of this case? Why?

2. **Observation:** What would be an effective systematic observational strategy (SOS) for gathering information for QMD? Is there a need for extended observation? Why?

3. **Evaluation and diagnosis:** What evaluation method (e.g., sequential or mechanical) and rationale for diagnosing performance would be most effective in this case? Why?

4. **Intervention:** What mode of intervention would likely be best? Why?

From D.V. Knudson, 2013, *Qualitative diagnosis of human movement,* 3rd ed. (Champaign, IL: Human Kinetics).

Short- or Long-Term?

You are a teacher, coach, or therapist working with a young client who wants to improve his technique in a sport skill (select one of personal interest). The client is an intermediate-level athlete who is a little impatient for improvement. Under what conditions would you generally focus on short-term improvement through focused instruction and training? Under what conditions would you generally focus the athlete on more open instruction and practice? How would you handle the disappointment or the resistance to slow improvement to reach more permanent long-term performance?

1. **Preparation:** What knowledge do you need to gather about the movement, the performer, and the situation in preparing for QMD of this case? Why?

2. **Observation:** What would be an effective systematic observational strategy (SOS) for gathering information for QMD? Is there a need for extended observation? Why?

3. **Evaluation and diagnosis:** What evaluation method (e.g., sequential or mechanical) and rationale for diagnosing performance would be most effective in this case? Why?

4. **Intervention:** What mode of intervention would likely be best? Why?

From D.V. Knudson, 2013, *Qualitative diagnosis of human movement,* 3rd ed. (Champaign, IL: Human Kinetics).

FURTHER PRACTICE

Many readers will have assisted others in learning some motor skill. Think back to an interesting experience you have had in teaching someone to move. Translate that experience into a "theory into practice" QMD scenario. Share this scenario with your instructor or another student.

SUMMARY

This chapter presents some real-life situations in which an integrated QMD would be appropriate. Simultaneous consideration of information from several subdisciplines of kinesiology is needed in QMD of each situation. The uniqueness of each situation demands thoughtful consideration in integrating information from many subdisciplines. The QMD of a squat, for example, relies heavily on biomechanics, but information from exercise physiology and psychology is also vital in shaping the appropriate intervention. These situations are useful in stimulating discussion among students and professionals about QMD.

Discussion Questions

1. What subdisciplines of kinesiology were most relevant to you in each of the scenarios presented in this chapter? Why?

2. Which scenarios were most interdisciplinary? What subdisciplines of kinesiology had to be integrated?

3. What other kinds of information would you want to observe in each of the scenarios presented in this chapter?

4. Propose another "theory into practice" QMD scenario and determine if your solution is consistent with the diagnoses of your peers.

Glossary

augmented feedback—Feedback that goes beyond the information normally available to a performer.

closed motor skills—A classification of motor skills that that are not strongly influenced by their environment because it has few temporal or environmental restrictions.

common errors—Errors that are typically observed in people learning a motor skill.

criterion-referenced validity—Accuracy of an assessment that is established by comparison to a criterion measure or "gold" standard.

critical features—The key features of a movement that are necessary for optimal performance.

cues—Short, descriptive words or phrases used to communicate ideas about movement to learners.

deterministic model—A model linking mechanical variables with the goal of the movement, used in the Hay and Reid (1988) approach to qualitative movement diagnosis.

developmental sequence—The classification of typical phases or stages that people exhibit in the development of a movement.

diagnosis—Critical scrutiny of the strengths and weaknesses identified in evaluation to prioritize possible intervention in qualitative movement diagnosis.

dynamic visual acuity (DVA)—The accuracy of visual discrimination when relative movement is occurring between the observer and an object.

error detection—An outdated model of qualitative assessment in which observers merely identify errors in performance by comparing perceived motion to a mental model of technique.

evaluation—The judgment of the quality of human movement to identify strengths and weaknesses of performance within qualitative movement diagnosis.

exaggeration—An intervention technique in which the performer is encouraged to over-correct or overcompensate for a persistent movement error and this feedback may bring about the small, desired change in technique.

feedback—Information about human movement.

field—Half of a video picture, composed of the even- or odd-numbered horizontal lines of a video frame (but not both).

field dependence—A way of processing information that depends on the background information for accurate interpretation. An analyst might use this perceptual style since it relies heavily on the frame of reference.

field independence—A way of processing information without reference to any source of information except the object or movement being observed. An observer using this perceptual style does not need background information to make sense of the movement.

fixation—The focusing of the eyes on an object in the visual field.

frame—A single video picture that is composed of two halves called "fields." There are 25 or 30 frames for each second of normal video.

freeze frame—A mode of video replay that stops and holds a video image so that it can be shown on a monitor. *Freeze frame* can be a misnomer because the pause or still functions of a VCR usually show one field of video in freeze-frame mode.

fundamental movement pattern—A general category of human movements (for example, lift, run, jump, throw).

fusion—The putting together of the 2-D visual information from the eyes to create the 3-D perception of vision.

gestalt—A way of processing visual–spatial information to get an overall impression. In a gestalt approach, the whole is greater than the sum of its parts. This impression of the event uses features such as region, proximity, continuation, and closure. Gestalt is a field of study in psychology.

high-speed video—Special video technology used to create more video pictures (frames) per second than standard TV video (U.S.: 30 frames, U.K.: 25 frames).

information processing—The cognitive process of organizing and making sense of sensory information, and the decision making based on that sensory information.

interdisciplinary—Referring to a process involving the integration of several kinds of disciplinary knowledge. This is analogous to synergy of factors where their combination can be greater than the sum of individual factors.

intervention—The fourth task of the integrated model of qualitative movement diagnosis. It involves the administration of feedback, corrections, or other change in the environment by the analyst to improve performance.

intrinsic feedback—Information about movement that is readily available to the performer (for example, sensory, kinesthetic, and proprioceptive information).

jog-shuttle dial—A video playback feature that uses a dial or controls to allow pause and multiple-speed playback, either forward or backward.

knowledge—The systematic, theoretical, data-supported ideas that make the best current explanation of reality for a particular academic discipline.

knowledge of performance (KP)—Information about the movement of the body.

knowledge of results (KR)—Information about the outcome or results of a movement.

logical validity—Accuracy of an assessment as established by logic or expert opinion.

manual guidance—A mode of intervention within qualitative movement diagnosis, in which the analyst physically positions or assists the performer in making the desired movement.

mechanical guidance—A mode of intervention within qualitative movement diagnosis that uses an aid or mechanical device to help the performer make or feel the desired movement.

motor program—The essential cognitive information needed to perform a movement.

observation—The second task of an integrated qualitative movement diagnosis, in which sensory information is gathered about performance with a systematic observational strategy.

observational learning—The use of visual models (pictures or demonstrations) to provide information about a movement.

open motor skills—A classification of motor skills that are strongly influenced by their environment because it is unpredictable or variable.

perception—The organization and interpretation of stimuli from our environment, mediated by our senses.

performance—The quality of a movement in achieving a goal, or the short-term and long-term effectiveness of a person's movement in achieving a goal.

pixel—Short for "picture element"; the small dots of light that form a video image.

preparation—The first task of an integrated qualitative movement diagnosis, in which prerequisite knowledge is gathered about the movement, client, and situation.

qualitative movement diagnosis—The systematic observation and introspective judgment of the quality of human movement for the purpose of providing the most appropriate intervention to improve performance.

reinforcement—Feedback that supports what was done so that the behavior will be repeated and eventually learned.

reliability—The consistency of a measurement or a qualitative assessment.

resolution—The number of lines of pixels that make up a video image.

saccade—The quick motion of the eyes from one fixation to another.

sampling rate—The temporal resolution of a measurement of a continuous event. The sampling rate of normal video in the United States is 30 frames (pictures) per second.

scanning strategy—A plan of focus for visual attention in observation of movement. This is the opposite of an unstructured, gestalt approach to observation.

shutter—A photo or video feature (electronic or mechanical) that limits the exposure time when an image is captured. This limits the chances of significant motion of the object during image capture, which would create a blurry image.

skill—An adapted fundamental movement pattern for a specific activity or goal. A baseball pitch is a skill related to overarm throwing.

slow motion—The replay of video at a speed relative to the frame rate of capture that is slower than the live motion.

smooth pursuit—The simultaneous rotation of both eyes to track a slow-moving object.

spatial ability—The ability to deal with spatial relationships and to use this information in different contexts.

static visual acuity (SVA)—The accuracy of visual discrimination in static, high-contrast conditions.

style—Aspects of movement that are personal differences, idiosyncrasies, or actions related to a specific performer.

systematic observational strategy (SOS)—A plan to gather all the relevant information about a human movement within qualitative movement diagnosis.

task modification—An intervention strategy that improves performance by changing practice to a task more appropriate for the performer.

technique—A kind of motor skill that has a specific purpose. A curveball is a technique related to the skill of baseball pitching.

validity—The extent to which a measurement or a qualitative assessment approaches the true amount of some variable.

vergence—Eye movements medially and laterally to adjust for movements of objects toward and away from the observer.

zoom lens—A key feature in video camcorders that allows the field of view to be adjusted for objects at different distances.

Bibliography

Abelbeck, K.G. 2002. Biomechanical model and evaluation of a linear motion squat type exercise. *Journal of Strength and Conditioning Research* 16: 516-524.

Abendroth-Smith, J., and J. Kras. 1999. More B-BOAT: The volleyball spike. *Journal of Physical Education, Recreation and Dance* 70 (3): 56-59.

Abendroth-Smith, J., J. Kras, and B. Strand. 1996. Get aboard the B-BOAT: Biomechanically based observation and analysis for teachers. *Journal of Physical Education, Recreation and Dance* 67 (8): 20-23.

Abernethy, B. 1988. Visual search in sport and ergonomics: Its relationship to selective attention and performer expertise. *Human Performance* 1: 205-235.

Abernethy, B. 1989. Expert-novice differences in perception: How expert does the expert have to be? *Canadian Journal of Sport Sciences* 14: 27-30.

Abernethy, B. 1993. Searching for the minimal essential information for skilled perception and action. *Psychological Research* 55: 131-138.

Abernethy, B., and D.G. Russell. 1987. The relationship between expertise and visual search strategy in a racquet sport. *Human Movement Science* 6: 283-319.

Abernethy, B., and K. Zawi. 2007. Pick-up of essential kinematics underpins expert perception. *Journal of Motor Behavior* 39: 353-368.

Abernethy, B., and R.J. Neal. 1999. Visual characteristics of clay target shooters. *Journal of Science and Medicine in Sport* 2: 1-19.

Abernethy, B., J.M. Wood, and S. Parks. 1999. Can the anticipatory skills of experts be learned by novices? *Research Quarterly for Exercise and Sport* 70: 313-318.

Ackermann, B.J., and R. Adams. 2004. Interobserver reliability of general practice physiotherapists in rating aspects of movement patterns of skilled violinists. *Medical Problems of Performing Artists* 19: 3-11.

Adrian, M., and G. House. 1987a. Sporting miscues: Part one. *Strategies* 1 (1): 11-14.

Adrian, M., and G. House. 1987b. Sporting miscues: Part two. *Strategies* 1 (2): 13-15.

Adrian, M.J., and J.M. Cooper. 1989. *Biomechanics of human movement.* Indianapolis: Benchmark Press.

Adrian, M.J., and J.M. Cooper. 1995. *Biomechanics of human movement.* 2nd ed. Madison, WI: Brown & Benchmark.

Adrian, M.J., and M.L. Endberg. 1971. Sequential timing of three overhead patterns. In *Kinesiology review,* edited by C. Widule. Washington, DC: AAHPER, 1-9.

Ae, M., Y. Muraki, H. Koyama, and N. Fujii. 2007. A biomechanical method to establish an averaged motion and identify critical motion by motion variability: With examples of high jump and sprint running. Bulletin of the Institute of Health and Sport Science, University of Tsukuba 30: 5-12. (In Japanese)

Aleshinsky, S.Y. 1986. An energy "sources" and "fractions" approach to the mechanical energy expenditure problem—I. Basic concepts, description of the model, analysis of a one-link system movement. *Journal of Biomechanics* 19: 287-293.

Alfano, P.L., and G.F. Michel. 1990. Restricting field of view: Perceptual and performance effects. *Perceptual and Motor Skills* 70: 35-45.

Allard, P., I. Stokes, and J. Blanchi, eds. 1995. *Three-dimensional analysis of human movement.* Champaign, IL: Human Kinetics.

Allison, P.C. 1985a. The development of the skill of observing during field experiences of pre-service physical education teachers. Paper presented at the meeting of the Association Internationale des Ecoles Superieures d'Education Physique, August, Garden City, NY.

Allison, P.C. 1985b. Observing for competence. *Journal of Physical Education, Recreation and Dance* 56 (6): 50-51, 54.

Allison, P.C. 1986. Laban's movement framework as a descriptor of change in students' movement response observations. Paper presented at the national convention of the American Alliance for Health, Physical Education, Recreation and Dance, April, Cincinnati.

Allison, P.C. 1987a. The impact of varying amounts of lesson responsibility on pre-service physical education teachers' ability to observe. Paper presented at the national convention of the American Alliance for Health, Physical Education, Recreation and Dance, April, Las Vegas, NV.

Allison, P.C. 1987b. What and how pre-service physical education teachers observe during an early field experience. *Research Quarterly for Exercise and Sport* 58: 242-249.

Allison, P.C. 1988. Strategies for observing during field experiences. *Journal of Physical Education, Recreation and Dance* 59 (2): 28-30.

Allison, P.C. 1990. Classroom teachers' observations of physical education lessons. *Journal of Teaching in Physical Education* 4: 272-283.

Ammons, R.B. 1956. Effect of knowledge of performance: A survey and tentative theoretical formulation. *Journal of General Psychology* 54: 279-299.

Amonette, W.E., K.L. English, and K.J. Ottenbacher. 2010. Nullius in verba: A call for the incorporation of evidence-based practice into the discipline of exercise science. Sports Medicine 40: 449-457.

Anderson, J.R. 1990. *Cognitive psychology and its implications.* New York: Freeman.

Anderson, M.B. 1979. Comparison of muscle patterning in the overarm throw and tennis serve. *Research Quarterly* 50: 541-553.

Annett, J. 1993. The learning of motor skills: Sports science and ergonomics perspectives. *Ergonomics* 37: 5-16.

Arend, S., and J.R. Higgins. 1976. A strategy for the classification, subjective analysis and observation of human movement. Journal of Human Movement Studies 2: 36-52.

Arias, J.L. 2012. Performance as a function of shooting style in basketball players under 11 years of age. Perceptual and Motor Skills 114: 446-456.

Armstrong, C.W. 1977. Skill analysis and kinesthetic experience. In *Research and practice in physical education*, edited by R.E. Stadulis. Champaign, IL: Human Kinetics, 13-18.

Armstrong, C.W. 1986. Research on movement analysis: Implications for the development of pedagogical competence. In *The 1984 Olympic scientific congress proceedings VI, sports pedagogy*, edited by M. Pieron and G. Graham. Champaign, IL: Human Kinetics, 27-32.

Armstrong, C.W., and S.J. Hoffman. 1979. Effects of teaching experience, knowledge of performer competence, and knowledge of performance outcome on performance error identification. *Research Quarterly* 50: 318-327.

Arnell, P., and P. Bauker. 1991. The accuracy of visual gait assessment. In *Proceedings of the 11th international congress of the World Confederation for Physical Therapy*. London: World Confederation for Physical Therapy, 431.

Arnold, P.J. 1993. Kinesiology and the professional preparation of the movement teacher. *Journal of Human Movement Studies* 25: 203-231.

Arnold, R.K. 1978. Optimizing skill learning: Moving to match the environment. *Journal of Physical Education, Recreation and Dance* 49 (9): 84-86.

Arrighi, M.A. 1974. The nature of game strategy observation in field hockey with respect to selected variables. Ph.D. diss., University of North Carolina Greensboro. Abstract in *Dissertation Abstracts International* 35: 2030A.

Ashford, D., K. Davids, and S.J. Bennett. 2007. Developmental effects influencing observational modeling: A meta-analysis. Journal of Sports Sciences 25: 547-559.

Ashford, D., S.J. Bennett, and K. Davids. 2006. Observational modeling effects for movement dynamics and movement outcome measures across differing task constraints: A meta-analysis. Journal of Motor Behavior 38: 185-205.

Attinger, D., S. Luethi, and E. Stuessi. 1987. Comparison of subjective gait observation with measured gait asymmetry. In *Biomechanics: Basic and applied research*, edited by G. Bergmann et al. Boston: M. Nijhoff, 563-568.

Atwater, A.E. 1979. Biomechanics of overarm throwing movements and of throwing injuries. *Exercise and Sport Sciences Reviews* 7: 43-85.

Austin, S., and L. Miller. 1992. An empirical study of the sybervision golf videotape. *Perceptual and Motor Skills* 74: 875-881.

Avila, F., and F.J. Moreno. 2003. Visual search strategies elaborated by tennis coaches during execution error detection processes. *Journal of Human Movement Studies* 44: 209-224.

Baccarini, M., and T. Booth. 2008. *Essential ultimate: Teaching, coaching, playing*. Champaign, IL: Human Kinetics.

Bahamonde, R. 2000. Angular momentum changes during the tennis serve. *Journal of Sports Sciences* 18: 579-592.

Bahill, A.T., and T. LaRitz. 1984. Why can't batters keep their eyes on the ball? *American Scientist* 72: 249-253.

Balan, C.M., and W.E. Davis. 1993. Ecological task analysis: An approach to teaching physical education. *Journal of Physical Education, Recreation and Dance* 64 (9): 54-61.

Baluyut, R., A.M. Genaidy, L.S. Davis, R.L. Sehll, and R.J. Simmons. 1995. Use of visual perception in estimating static postural stresses: Magnitudes and sources of error. *Ergonomics* 38: 1841-1850.

Bampton, S. 1979. *A guide to the visual examination of pathological gait*. Philadelphia: Temple University Rehabilitation and Research Training Center No. 8, Moss Rehabilitation Hospital.

Bao, S., N. Howard, P. Spielholz, and B. Silverman. 2010. Inter-observer reliability of forceful exertion analysis based on video-recordings. *Ergonomics* 53: 1129-1139.

Bard, C., and M. Fleury. 1976. Analysis of visual search activity during sport problem situations. *Journal of Human Movement Studies* 3: 214-222.

Bard, C., M. Fleury, L. Carriere, and M. Halle. 1980. Analysis of gymnastics judges' visual search. *Research Quarterly for Exercise and Sport* 51: 267-273.

Barfield, W.R. 1998. The biomechanics of kicking in soccer. *Clinics in Sports Medicine* 17: 711-728.

Barnett, L., E. van Beurden, P. Morgan, D. Lincoln, A. Zask, and J. Beard. 2009. Interrater objectivity for field-based fundamental motor skill assessment. *Research Quarterly for Exercise and Sport* 80: 363-368.

Barrett, K.R. 1977. We see so much but perceive so little: Why? In *Proceedings of the NAPECW/NCPEAM national conference*, edited by L.I. Gedvilas and M.E. Kneer. Chicago: University of Illinois Chicago Circle.

Barrett, K.R. 1979a. Observation for teaching and coaching. *Journal of Physical Education and Recreation* 50 (1): 23-25.

Barrett, K.R. 1979b. Observation of movement: An assumed teaching/coaching behavior. Unpublished manuscript, University of North Carolina Greensboro.

Barrett, K.R. 1979c. Observation of movement for teachers: A synthesis and implications. *Motor Skills: Theory into Practice* 3: 67-76.

Barrett, K.R. 1980. A system for describing and observing a spatial relationship movement task. Paper presented at the pre-convention symposium on physical education for children, southern district American Alliance for Health, Physical Education and Recreation, February, Nashville.

Barrett, K.R. 1981. Observation as a teaching behavior—research to practice. Paper presented at the national meeting of the Canadian Association for Physical Education and Health Education, June, Victoria, BC.

Barrett, K.R. 1982. The content of observing: A model for curriculum development. Paper presented at the national convention of the American Alliance for Health, Physical Education, Recreation and Dance, April, Houston, TX.

Barrett, K.R. 1983. A hypothetical model of observing as a teaching skill. *Journal of Teaching in Physical Education* 3: 22-31.

Barrett, K.R., P.C. Allison, and R. Bell. 1987. What preservice physical education teachers see in an unguided field experience: A follow-up study. *Journal of Teaching in Physical Education* 7: 12-21.

Barthels, K. 1989. Applying mechanics to swimming performance analysis. *Strategies* 3 (1): 17-19.

Barthels, K., and E. Kreighbaum. 1988. A Western system of analysis and observation. Paper presented to the national convention of the American Alliance for Health, Physical Education, Recreation and Dance, April, Kansas City, MO.

Bartlett, R. 2007. Introduction to sports biomechanics: Analyzing human movement patterns. 2nd ed. New York: Routledge.

Bayless, M.A. 1980. The effect of style of teaching on ability to detect errors in performance. *Wyoming Journal for Health, Physical Education, Recreation and Dance* 3: 2-4, 15.

Bayless, M.A. 1981. Effect of exposure to prototypic skill and experience in identification of performance error. *Perceptual and Motor Skills* 52: 667-670.

Behar, I., and W. Bevan. 1961. The perceived duration of auditory and visual intervals: Cross-model comparison and interaction. *American Journal of Psychology* 74: 17-26.

Behets, D. 1996. Comparison of visual information processing between preservice students and experienced physical education teachers. *Journal of Teaching in Physical Education* 16: 79-87.

Belka, D.E. 1988. What preservice physical educators observe about lessons in progressive field experiences. *Journal of Teaching in Physical Education* 7: 311-326.

Bell, F.I. 1987. The effects of two training programs on the ability of preservice physical education majors to observe the developmental steps in the overarm throw for force. Ph.D. diss., University of North Carolina Greensboro. Abstract in *Dissertation Abstracts International* 48: 1144A.

Bell, R., K.R. Barrett, and P.E. Allison. 1985. What preservice physical education teachers see in an unguided, early field experience. *Journal of Teaching in Physical Education* 4: 81-90.

Berg, K. 1975. Functional approach to undergraduate kinesiology. *Journal of Physical Education and Recreation* 46 (7): 43-44.

Berger, C.G. 1999. The camcorder as a teaching tool in the weight room. *Strength and Conditioning Journal* 21 (6): 70-72.

Berger, J. 1987. *Ways of seeing.* London: British Broadcasting Corp. and Penguin Books.

Bernhardt, J., P.J. Bate, and T.A. Matyas. 1998. Accuracy of observational kinematic assessment of upper-limb movements. *Physical Therapy* 78: 259-270.

Bernhardt, J., T. Matyas, and P. Bate. 2002. Does experience predict observational kinematic assessment accuracy? *Physiotherapy Theory and Practice* 18: 141-149.

Best, J.B. 1986. *Cognitive psychology.* St. Paul: West.

Beveridge, S.K., and S.K. Gangstead. 1984. A comparative analysis of the effects of instruction on the analytical proficiency of physical education teachers and undergraduates. Annual convention of the American Alliance for Health, Physical Education, Recreation and Dance, Anaheim, CA. ERIC Document Reproduction Service, ED 244-939.

Beveridge, S.K., and S.K. Gangstead. 1988. Teaching experience and training in the sports skill analysis process. *Journal of Teaching in Physical Education* 7: 103-114.

Bian, W., and P.G. Schempp. 2004. Examination of expert and novice coaches' diagnostic ability. *Research Quarterly for Exercise and Sport* 75: A60.

Bideau, B., R. Kulpa, N. Vignais, S. Brault, F. Multon, and C. Craig. 2010. Using virtual reality to analyze sports performance. *Computer Graphics and Applications* 30 (2): 14-21.

Bilodeau, I.M. 1969. Information feedback. In *Principles of skill acquisition,* edited by E.A. Bilodeau. London: Academic Press.

Biomechanics Academy. 2003. *Guidelines for undergraduate biomechanics.* Reston, VA: NASPE. www.aahperd.org/naspe/pdf_files/Guidelines_Biomechanics_Approved%20April%202003.pdf

Bird, M., and J. Hudson. 1990. Biomechanical observation: Visually accessible variables. In *Proceedings of the VIIIth international symposium of the Society of Biomechanics in Sports,* edited by M. Nosek, D. Sojka, W.E. Morrison, and P. Susanka. Prague: Conex, 321-326.

Biscan, D.U., and S.J. Hoffman. 1976. Movement analysis as a generic ability of physical education teachers and students. *Research Quarterly* 47: 161-163.

Blundell, N.L. 1985. The contribution of vision to the learning and performance of sports skills: Part 1. The role of selected visual parameters. *Australian Journal of Science and Medicine in Sport* 17: 3-11.

Boehm, A.E., and R.A. Weinberg. 1987. *The classroom observer: Developing observation skills in early childhood settings.* New York: Teachers College Press.

Bolt, B.R. 2000. Using computers for qualitative analysis of movement. *Journal of Physical Education, Recreation and Dance* 71 (3): 15-18.

Bowers, L., and S. Klesius. 1991. Use of interactive videodisk in the analysis of skills for teaching. Paper presented at the national convention of the American Alliance for Health, Physical Education, Recreation and Dance, April, San Francisco.

Boyce, W.F., C. Gowland, P. Rosenbaum, M. Lane, N. Plews, C. Goldsmith, D. Russell, V. Wright, S. Poter, and D. Harding. 1995. The gross motor performance measure: Validity and responsiveness of a measure of quality of movement. *Physical Therapy* 75: 603-613.

Boyer, E.L. 1990. *Scholarship reconsidered: Priorities of the professoriate*. Princeton, NJ: Carnegie Foundation for the Advancement of Teaching.

Braden, V. 1983. Vic Braden's startling revelations about line calls. *Tennis* (May): 37-39.

Branta, C., J. Haubenstricker, and V. Seefeldt. 1984. Age changes in motor skills during childhood and adolescence. *Exercise and Sport Sciences Reviews* 12: 467-520.

Bressan, E.S., and M.R. Weiss. 1982. A theory of instruction for developing competence, self-confidence and persistence in physical education. *Journal of Teaching in Physical Education* 2 (1): 38-47.

Brisson, T.A., and C. Alain. 1996. Should common optimal movement patterns be identified as the criterion to be achieved? *Journal of Motor Behavior* 28: 211-223.

Broadbent, D. 1958. *Perception and communication*. Oxford: Pergamon.

Broer, M.R. 1960. *Efficiency of human movement*. Philadelphia: Saunders.

Broker, J.P., R.J. Gregor, and R.A. Schmidt. 1989. Extrinsic feedback and the learning of cycling kinetic patterns. Abstract of XII Congress ISB. *Journal of Biomechanics* 22: 991.

Brophy, J., and T. Good. 1986. Teacher behavior and student achievement. In *Handbook of research on teaching*, edited by M. Wittrock. 3rd ed. New York: Macmillan, 328-375.

Brosvic, G.M., and S. Finizio. 1995. Inaccurate feedback and performance on the Muller-Lyer illusion. *Perceptual and Motor Skills* 80: 896-898.

Brown, E.W. 1982. Visual evaluation techniques for skill analysis. *Journal of Physical Education, Recreation and Dance* 53 (1): 21-26, 29.

Brown, E.W. 1984. Kinesiological analysis of motor skills via visual evaluation techniques. In *Proceedings: Second national symposium on teaching kinesiology and biomechanics in sports*, edited by R. Shapiro and J.R. Marett. Colorado Springs, CO: NASPE, 95-96.

Brown, S. 1995. The effects of limited and repeated demonstrations of the development of fielding and throwing in children. Ph.D. diss., University of South Carolina, 1994. Abstract in *Dissertation Abstracts International* 55: 1868A.

Bruton, A., B. Ellis, and J. Goddard. 1999. Comparison of visual estimation and goniometry for assessment of metacarpophalangeal joint angle. *Physiotherapy* 85: 201-208.

Buckolz, E., H. Prapavesis, and J. Fairs. 1988. Advance cues and their use in predicting tennis passing shots. *Canadian Journal of Sport Sciences* 13: 20-30.

Buizza, A., and R. Schmidt. 1986. Velocity characteristics of smooth pursuit eye movements to different patterns of target motion. *Experimental Brain Research* 63: 395-401.

Buizza, A., and R. Schmidt. 1989. The influence of smooth pursuit dynamics on eye tracking: A mathematical approach. *Medical and Biological Engineering and Computing* 27: 617-622.

Bunn, J.W. 1955. *Scientific principles of coaching*. Englewood Cliffs, NJ: Prentice-Hall.

Burdorf, A., J. Derksen, B. Naaktgenboren, and M. van Riel. 1992. Measurement of trunk bending during work by direct observation and continuous measurement. *Applied Ergonomics* 23: 263-267.

Burg, A. 1966. Visual acuity as measured by static and dynamic tests: A comparative evaluation. *Journal of Applied Psychology* 50: 460-466.

Butler, R.J., P.J. Plisky, C. Southers, C. Scoma, and K.B. Kiesel. Biomechanical analysis of the different classifications of the Functional Movement Screen deep squat test. *Sports Biomechanics* 9: 270-279.

Camcorders: Time to go digital? (Camcorders test). 1999. *Consumer Reports* (October): 34-37.

Campbell, F.W., and R.H. Wurtz. 1978. Saccadic omission: Why we do not see a gray-out during a saccadic movement. *Vision Research* 18: 1297-1303.

Cappozzo, A., M. Marchetti, and V. Tosi, eds. 1992. *Biolocomotion: A century of research using moving pictures*. Rome: Promograph.

Carpenter, R.H.S. 1988. *Movements of the eyes*. 2nd ed. London: Pion.

Catalano, J. 1995. The eyes have it. *Training & Conditioning* (June): 6-14.

Causer, J., P.S. Holmes, and A.M. Williams. 2011. Quiet eye training in a visuomotor control task. *Medicine and Science in Sports and Exercise* 43: 1042-1049.

Cavanagh, P.R. 1990. Biomechanics: A bridge builder among the sport sciences. *Medicine and Science in Sports and Exercise* 22: 546-557.

Cavanagh, P.R., and R. Kram. 1985. The efficiency of human movement: A statement of the problem. *Medicine and Science in Sports and Exercise* 17: 304-308.

Cayer, L. 1992. The skill of analysis and correction. In *Proceedings: Third national tennis seminar*. Melbourne, Australia: Tennis Australia, 1-15.

Chabris, C., and D. Simons. 2010. *The invisible gorilla: And other ways our intuitions deceive us*. New York: Crown.

Chandler, T.J., and L.E. Brown. eds. 2013. *Conditioning for strength and human performance* 2nd ed. Philadelphia: Wolters Kluwer.

Chang, C., R.W. McGorry, L. Jia-Hua, X. Xu, and S.M. Hsiang. 2010. Prediction accuracy in estimating joint angle trajectories using a video posture coding method for sagittal lifting tasks. *Ergonomics* 53: 1039-1047.

Chavez, E.J. 2008. Flow in sport: A study of college athletes. *Imagination, Cognition and Personality* 28: 69-91.

Chen, D.D. 2001. Trends in augmented feedback research and tips for the practitioner. *Journal of Physical Education, Recreation and Dance* 72 (1): 32-36.

Chen, L. 1982. Topological structure in visual perception. *Science* 128 (12): 699-700.

Chow, J., L. Carlton, W. Chae, Y. Lim, and J. Shim. 1999. Pre- and post-impact ball and racquet characteristics during tennis serves performed by elite male and female players. In *Scientific proceedings of the XVIII international symposium on biomechanics in sports.* Perth, Western Australia: Edith Cowan University, 45-48.

Christina, R.W., and D.M. Corcos. 1988. *Coaches guide to teaching sport skills.* Champaign, IL: Human Kinetics.

Chung, T.W. 1993. The effectiveness of computer-based interactive video instruction on psychomotor skill analysis competency of preservice physical education teachers in tennis teaching. Ph.D. diss., University of Northern Colorado, 1992. Abstract in *Dissertation Abstracts International* 53: 2290A.

Ciapponi, T. 1999. Skill analysis. *Strategies* 12 (5): 13-16.

Clark, J.E., C.L. Stamm, and M.F. Urquia. 1979. Developmental variability: The issue of reliability. In *Psychology of motor behavior and sport—1978,* edited by G.C. Roberts and K.M. Newell. Champaign, IL: Human Kinetics, 253-257.

Cloes, M., A. Deneve, and M. Pieron. 1995. Interindividual variability of teacher's feedback: Study in simulated teaching conditions. *European Physical Education Review* 1: 83-93.

Cloes, M., J. Premuzak, and M. Pieron. 1995. Effectiveness of a video training program used to improve error identification and feedback processes by physical education student teachers. *International Journal of Physical Education* 32 (3): 4-10.

Cohn, T.E., and D.D. Chaplik. 1991. Visual training in soccer. *Perceptual and Motor Skills* 72: 1238.

Coker, C. 2005. Faulty mechanics: To fix or not to fix? *Strategies* 19 (1): 29-31.

Coker, C.A. 1998. Observation strategies for skill analysis. *Strategies* 11 (4): 17-19.

Cole, J.L. 1981. Teacher-augmented feedback: Shortening the "fairway" between theory and instruction. *Motor Skills: Theory into Practice* 5: 81-87.

Collins, H., and R. Evans. 2008. You cannot be serious! Public understanding of technology with special reference to "Hawk-Eye." *Public Understanding of Science* 17: 283-308.

Cook, D.A. 1990. Using the basic skills to organize your movement analysis. *Professional Skier* (winter): 50-52.

Cook, E.G., L. Burton, and B. Hogenboom. 2006a. The use of fundamental movements as an assessment of function—Part 1. *North American Journal of Sports Physical Therapy* 1: 62-72.

Cook, E.G., L. Burton, and B. Hogenboom. 2006b. The use of fundamental movements as an assessment of function—Part 2. *North American Journal of Sports Physical Therapy* 1: 132-139.

Cools, W., C. De Martelaer, C. Samnaey, and C. Andries. 2009. Movement skill assessment of typically developing preschool children: A review of seven movement skill assessment tools. *Journal of Sports Science and Medicine* 8: 154-168.

Cooper, J.M., and R.B. Glassow. 1963. *Kinesiology.* St. Louis: Mosby.

Cooper, L.K., and A.L. Rothstein. 1981. Videotape replay and the learning of skills in open and closed environments. *Research Quarterly for Exercise and Sport* 52: 191-199.

Corlett, H. 1973. Movement analysis. In *Human movement—a field of study,* edited by J.D. Brooke and H.T.A. Whiting. Lafayette, IN: Balt, 238-250.

Cox, R.H. 1987. An exploratory investigation of a signal discrimination problem in tennis. *Journal of Human Movement Studies* 13: 197-210.

Cozzallio, E.R. 1986. The development of assessment instruments for screening selected gross motor skills in kindergarten children. Ph.D. diss., Florida State University, 1985. Abstract in *Dissertation Abstracts International* 47: 118A.

Craft, A.H. 1977. The teaching of skills for the observation of movement: Inquiry into a model. Ph.D. diss., University of North Carolina Greensboro. Abstract in *Dissertation Abstracts International* 38: 1975A.

Craik, R.L., and C.A. Oatis. 1995. *Gait analysis: Theory and application.* St. Louis: Mosby.

Cronin, J., P.J. McNair, and R.N. Marshall. 2003. Lunge performance and its determinants. *Journal of Sports Sciences* 21: 49-57.

Cutting, J.E., and D.R. Proffitt. 1981. Gait perception as an example of how we may perceive events. In *Intersensory perception and sensory integration,* edited by R.D. Walk and H.L. Pick. New York: Plenum Press, 249-273.

Dahle, L.K., M. Mueller, A. Delitto, and J.E. Diamond. 1991. Visual assessment of foot type and relationship of foot type to lower extremity injury. *Journal of Orthopedic and Sports Physical Therapy* 14: 70-74.

Dale, E. 1984. *The educator's quotebook.* Bloomington, IN: Phi Delta Kappa Educational Foundation.

Daniels, D.B. 1984. Basic movements and modeling: An approach to teaching skill analysis in the undergraduate biomechanics course. In *Proceedings: Second national symposium on teaching kinesiology and biomechanics in sports,* edited by R. Shapiro and J.R. Marett. Colorado Springs, CO: NASPE, 243-246.

Daniels, D.B. 1987. Qualitative analysis: The coach's most important tool. In *World identification systems for gymnastic talent,* edited by B. Petiot et al. Montreal: Sport Psyche Editions, 163-172.

Darden, G., and J. Shimon. 2000. Revisit an "old" technology: Videotape feedback for motor skill learning and performance. *Strategies* 13 (4): 17-21.Davids, K., C. Button, and S. Bennett. 2008. *Dynamics of skill acquisition—a constraints-led approach.* Champaign, IL: Human Kinetics.

Darden, G.F. 1999. Videotape feedback for student learning and performance: A learning-stages approach. *Journal of Physical Education, Recreation and Dance* 70 (9): 40-45, 62.

Davids, K.W., D.R. DePalmer, and G.J.P. Savelsbergh. 1989. Skill level, peripheral vision and tennis volleying performance. *Journal of Human Movement Studies* 16: 191-202.

Davis, J. 1980. Learning to see: Training in observation of movement. *Journal of Physical Education and Recreation* 51 (1): 89-90.

Davis, W.E. 1984. Motor ability assessment of populations with handicapping conditions: Challenging basic assumptions. *Adapted Physical Activity Quarterly* 1: 125-140.

Davis, W.E., and A.W. Burton. 1991. Ecological task analysis: Translating movement behavior theory into practice. *Adaptive Physical Education Quarterly* 8: 154-177.

Day, M.C. 1975. Developmental trends in visual scanning. In *Advances in child development and behavior,* edited by H.W. Reese. New York: Academic Press, 154-193.

DeBruin, H., D.J. Russell, J.E. Latter, and J.T.S. Sadler. 1982. Angle-angle diagrams in monitoring and quantification of gait patterns for children with cerebral palsy. *American Journal of Physical Medicine* 61: 176-192.

Dedeyn, K. 1991. Error identification comparison between three modes of viewing a skill. In *Teaching kinesiology and biomechanics in sports,* edited by J. Wilkerson, E. Kreighbaum, and C. Tant. Ames, IA: NASPE Kinesiology Academy, 21-25.

DeLooze, M.P., H.M. Toussaint, J. Ensink, C. Mangnus, and A.J. Van der Beek. 1994. The validity of visual observation to assess posture in a laboratory-simulated manual material handling task. *Ergonomics* 37: 1335-1343.

Denis, D., M. Lortie, and M. Bruxelles. 2002. Impact of observers' experience and training on reliability of observations for a manual handling task. *Ergonomics* 45: 441-454.

DePauw, K., and G. Goc Karp. 1989. Systematic infusion of knowledge into the undergraduate curriculum. Paper presented at the National Association for Physical Education in Higher Education conference, January, San Antonio.

DeRenne, C., and D.J. Szymanski. 2009. Effects of baseball weighted implement training: A brief review. *Journal of Strength and Conditioning Research* 31: 30-37.

DeRenne, C., and T. House. 1987. The four absolutes of pitching mechanics. *Scholastic Coach* (March): 79-83.

DeRenne, C., K. Ho, and A. Blitzblau. 1990. Effects of weighted implement training on throwing velocity. *Journal of Applied Sport Science Research* 4: 16-19.

DeRenne, C., K. Ho, and J. Murphy. 2001. Effects of general, special, and specific resistance training on overarm throwing velocity in baseball: A brief review. *Journal of Strength and Conditioning Research* 15: 148-156.

DiCicco, G.L. 1990. The effect of tennis playing and teaching experience on ability to perform a diagnostic task. Ph.D. diss., University of Pittsburgh. Abstract in *Dissertation Abstracts International* 51: 1581A.

Donkelaar, P., and R.G. Lee. 1994. The role of vision and eye motion during reaching to intercept moving targets. *Human Movement Science* 13: 765-783.

Douwes, M., and J. Dul. 1991. Validity and reliability of estimating body angles by direct and indirect observations. In *Designing for everyone: Proceedings of the Inter-national Ergonomics Association,* edited by Y. Queinnec and F. Daniellou. London: Taylor & Francis, 885-887.

Dowell, L.J. 1978. Throwing for distance: Air resistance, angle of projection, and ball size and weight. *Motor Skills: Theory into Practice* 3 (1): 11-14.

Downing, J.H., and J.E. Lander. 2002. Performance errors in weight training and their correction. *Journal of Physical Education, Recreation and Dance* 73 (9): 44-52.

Dowrick, P.W. 1991. Feedback and self-confrontation. In *Practical guide to using video in the behavioral sciences.* Edited by P.W. Dowrick. New York: Wiley, 92-108.

Draper, J. 1986. Analyzing skill to improve performance. *Sports Coach* 9 (4): 33-35.

Drummond, J.L., and S.K. Gangstead. 1996. The effects of experience on analytical proficiency and observational strategy among female collegiate athletes and coaches. *Applied Research in Coaching and Athletics Annual* 11: 75-94.

Duck, T. 1986. Applied sport biomechanics for advanced coaching. *Sports Science Periodical on Research and Technology in Sport* (June): 1-6.

Dunham, P. 1986. Evaluation for excellence. *Journal of Physical Education, Recreation and Dance* 57 (6): 34-36, 60.

Dunham, P. 1994. *Evaluation for physical education.* Englewood, CO: Morton.

Dunham, P., E.J. Reeve, and C.S. Morrison. 1989. *DISPE: A total instructional system for physical education.* Edina, MN: Alpha Editions.

Eastlack, M.E., J. Arvidson, L. Snyder-Mackler, J.V. Canoff, and C.L. McGarvey. 1991. Interrater reliability of videotaped observational gait analysis assessments. *Physical Therapy* 71: 465-472.

Eastman Kodak Company. 1979. *High-speed photography standard book.* No. 0-87985-165-1.

Eckrich, J., C.J. Widule, R.A. Shrader, and J. Maver. 1994. The effects of video observational training on video and live observational proficiency. *Journal of Teaching in Physical Education* 13: 216-227.

Eckrich, J.R. 1991. The effects of video observational training on video and live observational proficiency. Ph.D. diss., Purdue University, 1990. Abstract in *Dissertation Abstracts International* 52: 465A.

Edwards, I., M. Jones, J. Carr, A. Braunack-Mayer, and G.M. Jensen. 2004. Clinical reasoning strategies in physical therapy. *Physical Therapy* 84: 312-335.

Ehrlenspiel, F. 2001. Paralysis by analysis? A functional framework for the effects of attentional focus on the control of motor skills. *European Journal of Sport Science* 1 (5): 1-11.

Elliott, B. 2004. Analyzing serve and groundstroke technique on court. *ITF Coaching & Sports Science Review* 32: 2-4.

Elliott, B.C., R.N. Marshall, and G.J. Noffal. 1995. Contributions of upper limb segment rotations during the power serve in tennis. *Journal of Applied Biomechanics* 11: 433-442.

Emmen, H.H., L.G. Wesseling, R.J. Bootsma, H.T.A. Whiting, and P.C.W. van Wieringen. 1985. The effect of video-modelling and video-feedback on the learning of the tennis service by novices. *Journal of Sports Sciences* 3: 127-138.

Ericson, M., A. Kilbom, C. Wiktorin, and J. Winkel. 1991. Validity and reliability in the estimation of trunk, arm and neck inclination by observation. In *Proceedings of the International Ergonomics Association conference.* Paris: International Ergonomics Association, 245-247.

Eriksen, C.W., and J.M. Webb. 1989. Shifting of attentional focus within and about a visual display. *Perception and Psychophysics* 45: 175-183.

Escamilla, R.F., G.S. Fleisig, T.M. Lowry, S.W. Barrentina, and J.R. Andrews. 2001. A three-dimensional biomechanical analysis of the squat during varying stance widths. *Medicine and Science in Sports and Exercise* 33: 984-998.

Faigenbaum, A.D., W.J. Kraemer, C.J.R. Blimkie, I. Jeffreys, L.J. Micheloi, M. Nitka, and T.W. Rowland. 2009. Youth resistance training: Updated position statement paper from the National Strength and Conditioning Association. *Journal of Strength and Conditioning Research* 23: S60-S79.

Farah, M.J., K.M. Hammond, D.N. Levine, and R. Calvanio. 1988. Visual and spatial mental imagery: Dissociable systems of representation. *Cognitive Psychology* 20: 439-462.

Faulkner, G., A. Taylor, R. Ferrence, S. Munro, and P. Selby. 2006. Exercise science and the development of evidence-based practice: A "better practices" framework. *European Journal of Sport Science* 6: 117-126.Feldman, L. 1988. Sony VCR ED-V9000. *Video Review* (April): 66-67.

Feltner, M.E. 1989. Three-dimensional interactions in a two-segment kinetic chain. Part II: Application to the throwing arms in baseball pitching. *International Journal of Sport Biomechanics* 5: 420-450.

Feltner, M.E., and J. Dapena. 1986. Dynamics of the shoulder and elbow joints of the throwing arm during the baseball pitch. *International Journal of Sport Biomechanics* 2: 235-259.

Fife, S.E., L.A. Roxborough, R.W. Armstrong, S.R. Harris, J.L. Gregson, and D. Field. 1991. Development of a clinical measure of postural control for assessment of adaptive seating in children with neuromotor disabilities. *Physical Therapy* 71: 981-993.

Fisher, G.H. 1981. Human information processing and a taxonomy of sporting skills. In *Vision and sport,* edited by I.M. Cockerill and W.W. MacGillivary. Cheltenham, England: Stanley Thornes.

Fisk, S.F. 1993. Seeing is believing. *Tennis* (August): 33.

Fitts, P.M. 1965. Factors in complex skill training. In *Training research and education,* edited by R. Glasser. New York: Wiley.

Fitts, P.M., and M.I. Posner. 1967. *Human performance.* Belmont, CA: Brooks/Cole.

Fleisig, G. 2001. The biomechanics of throwing. In *Proceedings of oral sessions: XIX international symposium on biomechanics in sports,* edited by J. Blackwell. San Francisco: University of San Francisco, 91-94.

Fleisig, G.S., J.R. Andrews, C.J. Dillman, and R.F. Escamilla. 1995. Kinetics of baseball pitching with implications about injury mechanisms. *American Journal of Sports Medicine* 23: 233-239.

Fleisig, G.S., R.F. Escamilla, J.R. Andrews, T. Matsuo, Y. Satterwhite, and S.W. Barrentine. 1996b. Kinematic and kinetic comparison between baseball pitching and football passing. *Journal of Applied Biomechanics* 12: 207-224.

Fleisig, G.S., S.W. Barrentine, N. Zheng, R.F. Escamilla, and J.R. Andrews. 1999. Kinematic and kinetic comparison of baseball pitching among various levels of development. *Journal of Biomechanics* 32:1371-1375.

Fleisig, G.S., S.W. Barrentine, R.F. Escamilla, and J.R. Andrews. 1996a. Biomechanics of overhand throwing with implications for injuries. *Sports Medicine* 21: 421-437.

Fleming, L.K. 1980. Identifying performance errors and teaching cues in tennis: The effectiveness of an instruction program. Master's thesis, Brigham Young University, Provo, UT.

Foster, S.L., and J.D. Cone. 1986. Design and use of direct observation procedures. In *Handbook of behavioral assessment,* edited by A.R. Ciminero, K.S. Calhounm, and H.E. Adams. New York: Wiley, 253-324.

Franck, F. 1979. *The awakened eye.* New York: Random House.

Franks, I.M. 1993. The effect of experience on the detection and location of performance differences in a gymnastic technique. *Research Quarterly for Exercise and Sport* 64: 227-231.

Franks, I.M., and D. Goodman. 1986. A systematic approach to analyzing sports performance. *Journal of Sports Sciences* 4: 49-59.

Franks, I.M., and G. Miller. 1991. Training coaches to observe and remember. *Journal of Sports Sciences* 9: 285-297.

Franks, I.M., and L.J. Maile. 1991. The use of video in sport skill acquisition. In *Practical guide to using video in the behavioral sciences,* edited by P.W. Dowrick. New York: Wiley, 231-243.

Franks, I.M., and P. Nagelkerke. 1988. The use of computer interactive video in sport analysis. *Ergonomics* 31: 1593-1603.

Frederick, A.B. 1977. Using checklists and templates for analyzing gymnastic skills. In *Research and practice in physical education,* edited by R.E. Stadulis. Champaign, IL: Human Kinetics, 28-31.

Frederick, A.B., and M.U. Wilson. 1973. Web graphics and the qualitative analysis of movement. In *Kinesiology III.* Washington, DC: AAHPER, 1-11.

French, C.A., and J.J. Plack. 1982. Effective communication: A rationale for skill instruction techniques. *Motor Skills: Theory into Practice* 6: 59-66.

Fronske, H. 2012. *Teaching cues for sports skills for secondary school students.* 5th ed. San Francisco: Benjamin Cummings.

Fronske, H., and S.E. Dunn. 1992. Cue your students in on good swimming. *Strategies* 5 (5): 25-27.

Fronske, H., J. Abendroth-Smith, and C. Blakmore. 1995. The effect of critical cues on throwing efficiency of elementary school children. *Research Quarterly for Exercise and Sport* 66 (suppl.): A53.Fronske, H.R., and R. Wilson. 2002. *Teaching cues for basic sports skills for elementary and middle school students.* San Francisco: Benjamin Cummings.

Fronske, H., R. Wilson, and S.E. Dunn. 1992. Visual teaching cues for tennis instruction. *Journal of Physical Education, Recreation and Dance* 63 (5): 13-14.

Fry, A.C., J.C. Smith, and B.K. Schilling. 2003. Effect of knee position on hip and knee torques during the barbell squat. *Journal of Strength and Conditioning Research* 17: 629-633.

Gangstead, S.K. 1984. A comparison of three methodological approaches to skill specific analytical training. Annual convention of the Northern Rocky Mountain Research Association, Jackson, WY. ERIC Document Reproduction Service, ED 255-471.

Gangstead, S.K. 1987. Toward a pedagogical kinesiology: A training paradigm. Paper presented at the meeting of the International Congress of Health, Physical Education, and Recreation, June, Vancouver, BC.

Gangstead, S.K. 1995. Development of observational and diagnostic competence: A training paradigm. *International Journal of Physical Education and Sport Science* 7: 31-41.

Gangstead, S.K., and S.K. Beveridge. 1984. The implementation and evaluation of a methodical approach to qualitative sports skill analysis instruction. *Journal of Teaching in Physical Education* 3 (winter): 60-70.

Gangstead, S.K., C. Cashel, and S.K. Beveridge. 1987. Perceptual style, visual retention and visual discrimination in qualitative sports skill analysis. Paper presented to the annual convention of the American Alliance for Health, Physical Education, Recreation and Dance, April, Kansas City, MO.

Garceau, L., D. Knudson, and W. Ebben. 2011. Fourth North American survey of undergraduate biomechanics instruction in kinesiology/exercise science. In *Biomechanics in sports 29,* edited by J.P. Villas-Boras, L. Machado, W. Kim, and A.P. Veloso. *Portuguese Journal of Sport Science* 11 (suppl. 2): 951-954.

Gassner, G.J. 1999. Using metaphors for high-performance teaching and coaching. *Journal of Physical Education, Recreation and Dance* 70 (7): 33-35.

Gavriyski, V. 1969. The colours and colour vision in sport. *Journal of Sports Medicine* 4: 49-53.

Genaidy, A.M., R.J. Simmons, L. Guo, and J.A. Hidalgo. 1993. Can visual perception be used to estimate body part angles? *Ergonomics* 36: 323-329.

Gentile, A.M. 1972. A working model of skill acquisition with application to teaching. *Quest* 17: 3-23.

Ghasemi, A., M. Momeni, M. Rezaee, and A. Gholami. 2009. The difference in visual skills between expert versus novice soccer referees. *Journal of Human Kinetics* 22: 15-20.

Gibson, E.J. 1969. *Principles of perceptual development.* New York: Appleton-Century-Crofts.

Girardin, Y., and D. Hanson. 1967. Relationship between ability to perform tumbling skills and ability to diagnose performance errors. *Research Quarterly* 38: 556-561.

Glazier, P.S., and K. Davids. 2009. Constraints on the complete optimization of human motion. *Sports Medicine* 39: 15-28.

Gleeson, N.P., F. Parfitt, J. Doyle, and D. Rees. 2005. Reproducibility and efficacy of the performance profile technique. *Journal of Exercise Science and Fitness* 3: 66-73.

Gluck, M., and L. Kerr. 1982. *Mechanics for gymnastics coaching: Tools for skill analysis.* Springfield, IL: Charles C Thomas.

Godwin, S. 1975. Training powers of observation. In *Human movement behavior: Conference report,* edited by G.F. Curl. West Midlands, England: Association of Principals of Women's Colleges of Physical Education.

Goodkin, R., and L. Diller. 1973. Reliability among physical therapists in diagnosis and treatment of gait deviations in hemiplegics. *Perceptual and Motor Skills* 37: 727-734.

Gopher, D., and E. Donchin. 1986. Workload: An examination of the concept. In *Handbook of perception and human performance,* vol. 2, edited by K.R. Boff, L. Kaufman, and J.P. Thomas. New York: Wiley.

Gould, D., and G. Roberts. 1982. Modeling and motor skill acquisition. *Quest* 33: 214-230.

Goulet, C., M. Fleury, C. Bard, M. Yerles, D. Michaud, and L. Lemire. 1988. Analyse des indices visuels preleves en reception de service au tennis. *Canadian Journal of Sport Sciences* 13: 79-87.

Gowland, C., W.F. Boyce, V. Wright, D.J. Russell, C.H. Goldsmith, and P.L. Rosenbaum. 1995. Reliability of the gross motor performance measure. *Physical Therapy* 75: 597-602.

Graham, K.C. 1988. A qualitative analysis of an effective teacher's movement task presentations during a unit of instruction. *Physical Educator* 11: 187-195.

Graham, K.C., K. Hussey, K. Taylor, and P. Werner. 1993. A study of verbal presentations of three effective teachers. *Research Quarterly for Exercise and Sport* 64 (suppl.): 87A. (abstract)

Gregg, J.R. 1987. *Vision and sports: An introduction.* Stoneham, MA: Butterworth.

Gregory, M.S. 2003. *Disk golf.* Duluth: Trellis.

Griffin, M.R. 1985. The utilization of product and process measures to compare the throwing, striking, and kicking proficiency of third and fifth grade students. Ph.D. diss., Florida State University, 1984. Abstract in *Dissertation Abstracts International* 45: 2797A.

Groot, C., F. Ortega, and F.S. Beltran. 1994. Thumb rule of visual angle: A new confirmation. *Perceptual and Motor Skills* 78: 232-234.

Groves, R., and D.N. Camaione. 1983. *Concepts in kinesiology.* 2nd ed. Philadelphia: Saunders.

Grunwald, H.A., ed. 1986. *Artificial intelligence: Understanding computers.* Alexandria, VA: Time-Life Books.

Guadagnoli, M., W. Holcomb, and M. Davis. 2002. The efficacy of video feedback for learning the golf swing. *Journal of Sports Sciences* 20: 615-622.

Haight, H.J., D.L. Dahm, J. Smith, and D.A. Kraus. 2005. Measuring standing hindfoot alignment: Reliability of goniometric and visual measurements. *Archives of Physical Medicine and Rehabilitation* 86: 571-575.

Hall, J. 1993. Toward outcome-based movement analysis. *Professional Skier* (spring): 10-12.

Hall, S.J. 2007. *Basic biomechanics.* 5th ed. Boston: McGraw-Hill.

Halverson, L.E. 1983. Observing children's motor development in action. Paper presented at the national convention of the American Alliance for Health, Physical Education, Recreation and Dance, April, Minneapolis.

Halverson, L.E., M.A. Roberton, and C.J. Harper. 1979. Learning to observe children's motor development. Paper presented to the national convention of the American Alliance for Health, Physical Education and Recreation, April, New Orleans.

Halverson, P.D. 1988. The effects of peer tutoring on sport skill analytic ability. Ph.D. diss., Ohio State University, 1987. Abstract in *Dissertation Abstracts International* 48: 2274A.

Hamburg, J. 1995. Coaching athletes using Laban movement analysis. *Journal of Physical Education, Recreation and Dance* 66 (2): 34-37.

Hamilton, G.R., and C. Reinschmidt. 1997. Optimal trajectory for the basketball free throw. *Journal of Sports Sciences* 15: 491-504.

Handford, C., K. Davids, S. Bennett, and C. Button. 1997. Skill acquisition in sports: Some applications of an evolving practice ecology. *Journal of Sports Sciences* 15: 621-640.

Harano, Y. 1986. Biomechanical analysis of baseball hitting. In *Proceedings of the 4th International Symposium on Biomechanics in Sports,* edited by J. Terauds, B.A. Gowitzke, and L.E. Holt. Halifax.: Dalhousie University, 21-28.

Harari, I., and D. Siedentop. 1990. Relationships among knowledge, experience and skill analysis ability. In *Integration or diversification of physical education and sport studies,* edited by D. Eldar and U. Simri. Netanya, Israel: Emmanuel Gill Publishing House, 197-204.

Harper, R.C. 1995. Effects of a videodisk instructional program on physical education majors' ability to observe the quality of performances of selected motor activities of pre-adolescents. Ph.D. diss., University of Alabama, 1994. Abstract in *Dissertation Abstracts International* 55: 3446A.

Harris, J.C. 1993. Using kinesiology: A comparison of applied veins in the subdisciplines. *Quest* 45: 389-412.

Harrison, J.M. 1973. A comparison of a videotape program and a teacher directed program of instruction in teaching the identification of archery errors. Ph.D. diss., Brigham Young University, Provo, UT.

Harrison, J.M. 1999. Are you a super teacher? A review of 10 years of research on teaching and learning in physical education. Paper presented at the southwest district American Alliance for Health, Physical Education, Recreation and Dance convention, February, Tucson, AZ.

Hartman, B.O., and G.E. Secrist. 1991. *Situational awareness is more than exceptional vision.* Alexandria, VA: Aerospace Medical Association.

Hatfield, F.C. 1972. Effects of prior experience, access to information, and level of performance on individual and group performance rating. *Perceptual and Motor Skills* 35: 19-26.

Hatze, H. 1976. Biomechanical aspects of a successful motion optimization. In *Biomechanics V-B,* edited by P.V. Komi. Baltimore: University Park Press, 5-12.

Haubenstricker, J.L., C.F. Branta, and V.D. Seefeldt. 1983. *Standards of performance for throwing and catching.* East Lansing, MI: American Society for the Psychology of Sport and Physical Activity.

Hay, J. 1983. A system for the qualitative analysis of a motor skill. In *Collected papers on sports biomechanics,* edited by G.A. Wood. Perth, Western Australia: University of Western Australia Press, 97-116.

Hay, J.G. 1984. The development of deterministic models for qualitative analysis. In *Proceedings: Second national symposium on teaching kinesiology and biomechanics in sports,* edited by R. Shapiro and J.R. Marett. Colorado Springs, CO: NASPE, 71-83.

Hay, J.G. 1993. *The biomechanics of sports techniques.* 4th ed. Englewood Cliffs, NJ: Prentice-Hall.

Hay, J.G., and J.G. Reid. 1982. *The anatomical and mechanical bases of human motion.* Englewood Cliffs, NJ: Prentice-Hall.

Hay, J.G., and J.G. Reid. 1988. *Anatomy, mechanics, and human motion.* 2nd ed. Englewood Cliffs, NJ: Prentice-Hall.

Hayes, S.J., N.J. Hodges, M.A. Scott, R.R. Horn, and A.M. Williams. 2006. Scaling a motor skill through observation and practice. *Journal of Motor Behavior* 38: 357-366.

Haywood, K.M. 1984. Use of image-retina and eye-head movement visual systems during coincidence-anticipation performance. *Journal of Sports Sciences* 2: 139-144.

Haywood, K.M., and K. Williams. 1995. Age, gender, and flexibility differences in tennis serving among experienced older adults. *Journal of Aging and Physical Activity* 3: 54-66.

Haywood, K.M., and N. Getchell. 2009. *Life span motor development.* 5th ed. Champaign, IL: Human Kinetics.

Haywood, K.M., K. Williams, and A. Van Sant. 1991. Qualitative assessment of the backswing in older adult throwing. *Research Quarterly for Exercise and Sport* 62: 340-343.

Henderson, D. 1971. The relationship among time, distance, and intensity as determinants of motion discrimination. *Perception and Psychophysics* 10: 313-320.

Hendriks, E., J. Brandsma, Y. Heerkens, R. Oostendorp, and R. Nelson. 1997. Intraobserver and interobserver reliability of assessments of impairments and disabilities. *Physical Therapy* 77: 1097-1106.

Hensley, L.D. 1983. Biomechanical analysis. *Journal of Physical Education, Recreation and Dance* 54 (8): 21-23.

Hensley, L.D., J.R. Morrow, and W.B. East. 1990. Practical measurement to solve practical problems. *Journal of Physical Education, Recreation and Dance* 61 (3): 42-44.

Heptulla-Chatterjee, S.H., J.J. Freyd, and M. Shiffrar. 1996. Configural processing in the perception of apparent biological motion. *Journal of Experimental Psychology: Human Perception and Performance* 22: 916-929.

Herbert, R., S. Moore, A. Moseley, K. Schurr, and A. Wales. 1993. Making inferences about muscle forces from clinical observations. *Australian Journal of Physiotherapy* 39: 195-202.

Hermassi, S., M.S. Chelly, M. Fathloun, and R.J. Shepard. 2010. The effect of heavy- versus moderate-load training on the development of strength, power, and throwing ball velocity in male handball players. *Journal of Strength and Conditioning Research* 24: 2408-2418.

Hernandez, M., F. Javier, R. Raul, and L. del Campo. 2006. Visual behavior of tennis coaches in a court and video-based conditions. *Revista Internacional del Ciencias del Deporte* 5 (2): 28-41.

Higgins, J., and S. Higgins. 1988. An Eastern system of analysis and observation. Paper presented at the national convention of the American Alliance for Health, Physical Education, Recreation and Dance, April, Kansas City, MO.

Higgins, J.R. 1977. *Human movement: An integrated approach.* St. Louis: Mosby.

Higgins, J.R., and R.K. Spaeth. 1972. Relationship between consistency of movement and environmental condition. *Quest* 17: 61-69.

Higgs, J., and M. Jones, eds. 2000. *Clinical reasoning in the health professions.* 2nd ed. New York: Butterworth-Heinemann.

Hirashima, M., K. Yamane, Y. Natamura, and T. Ohtsuki. 2008. Kinetic chain of overarm throwing in terms of joint rotations revealed by induced acceleration analysis. *Journal of Biomechanics* 41: 2874-2883.

Hirth, C.J. 2007. Clinical movement analysis to identify muscle imbalances and guide exercise. *Athletic Therapy Today* 12 (4): 10-14.

Hoare, D. 1992. Screening for gross motor co-ordination problems in primary school children. *Sports Coach* 15 (2): 13-15.

Hodges, N.J., R. Chua, and I.M. Franks. 2003. The role of video in facilitating perception and action of a novel coordination movement. *Journal of Motor Behavior* 35: 247-260.

Hoffman, S.J. 1974. Toward taking the fun out of skill analysis. *Journal of Health, Physical Education and Recreation* 45 (9): 74-76.

Hoffman, S.J. 1977a. Competency based training in skill analysis. In *Research and practice in physical education,* edited by R.E. Stadulis. Champaign, IL: Human Kinetics, 3-12.

Hoffman, S.J. 1977b. Observing and reporting on learner responses: The teacher as a reliable feedback agent. In *Proceedings of the NAPECW/NCPEAM national conference,* edited by L.I. Gedvilas and M.E. Kneer. Chicago: University of Illinois Chicago Circle, 153-160.

Hoffman, S.J. 1977c. Toward a pedagogical kinesiology. *Quest* 28: 38-48.

Hoffman, S.J. 1983. Clinical diagnosis as a pedagogical skill. In *Teaching in physical education,* edited by T.J. Templin and J.K. Olson. Champaign, IL: Human Kinetics, 35-45.

Hoffman, S.J. 1984. The contributions of biomechanics to clinical competence: A view from the gymnasium. In *Proceedings: Second national symposium on teaching kinesiology and biomechanics in sports,* edited by R. Shapiro and J.R. Marett. Colorado Springs, CO: NASPE, 67-70.

Hoffman, S.J., and C.W. Armstrong. 1975. Effects of pretraining on performance error identification. In *Mouvement, Actes du 7e symposium en apprentissage psycho-motor et psychologie du sport.* Quebec City, 207-214.

Hoffman, S.J., and J.C. Harris, eds. 2000. *Introduction to kinesiology.* Champaign, IL: Human Kinetics.

Hoffman, S.J., and J.L. Sembiante. 1975. Experience and imagery in movement analysis. In *British proceedings of sports psychology,* edited by G.J.K. Alderson and D.A. Tyldesley. Salford, England: British Society of Sports Psychology, 288-293.

Hoffman, S.J., C.H. Imwold, and J.A. Kohler. 1983. Accuracy and prediction in throwing: A taxonomic analysis of children's performance. *Research Quarterly for Exercise and Sport* 54: 33-40.

Holekamp, M.J. 1987. The effect of training on physical therapy student raters evaluating videotaped motor skill performances. Ph.D. diss., University of Missouri-Columbia, 1986. Abstract in *Dissertation Abstracts International* 48: 1145A.

Hong, D., T.K. Cheung, and E.M. Roberts. 2001. A three-dimensional, six-segment chain analysis of forceful overarm throwing. *Journal of Electromyography and Kinesiology* 11: 95-112.

Hore, J., S. Watts, and J. Martin. 1996. Finger flexion does not contribute to ball speed in overarm throws. *Journal of Sports Sciences* 14: 335-342.

Hoshizaki, T.B. 1984. The application of models to understanding the biomechanical aspects of performance. In *Proceedings: Second national symposium on teaching kinesiol-*

ogy and biomechanics in sports, edited by R. Shapiro and J.R. Marett. Colorado Springs, CO: NASPE, 85-93.

Housner, L.D., and D.C. Griffey. 1985. Teacher cognition: Differences in planning and interactive decision making between experienced and inexperienced teachers. *Research Quarterly for Exercise and Sport* 56: 45-53.

Housner, L.D., and D.C. Griffey. 1994. Wax on, wax off: Pedagogical content knowledge in motor skill acquisition. *Journal of Physical Education, Recreation and Dance* 65 (2): 63-68.

Howell, M.L. 1956. Use of force-time graphs for performance analysis in facilitating motor learning. *Research Quarterly* 27: 12-22.

Hsiang, S.M., G.E. Brogmus, S.E. Martin, and I.B. Bezverkhny. 1998. Video based lifting technique coding system. *Ergonomics* 41: 239-256.

Hubbard, A., and C.N. Seng. 1954. Visual movements of batters. *Research Quarterly* 25: 42-57.

Hudson, J. 1985. POSSUM: Purpose/observation system for studying and understanding movement. Paper presented at the American Alliance for Health, Physical Education, Recreation and Dance national convention, April, Atlanta.

Hudson, J. 1987. What goes up. . . . Paper presented at the American Alliance for Health, Physical Education, Recreation and Dance national convention, April, Las Vegas, NV.

Hudson, J.L. 1990a. Biomechanical observation: Visually accessible variables. In *Proceedings of the VIIIth international symposium of the Society of Biomechanics in Sports,* edited by M. Nosek, D. Sojka, W.E. Morrison, and P. Susanka. Prague: Conex, 321-326.

Hudson, J.L. 1990b. Drop, stop, pop: Keys to vertical jumping. *Strategies* 3 (6): 11-14.

Hudson, J.L. 1990c. The value of visual variables in biomechanical analysis. In *Proceedings of the 6th International Symposium on Biomechanics in Sports,* edited by E. Kreighbaum and A. McNeill. Bozeman, MT: Color World Printers, 499-509.

Hudson, J.L. 1995. Core concepts of kinesiology. *Journal of Physical Education, Recreation and Dance* 66 (5): 54-55, 59-60.

Hudson, J.L. 2006. Applied biomechanics in an instructional setting. *Journal of Physical Education, Recreation and Dance* 77 (8): 25-27.

Huelster, L.J. 1939. Learning to analyze performance. *Journal of Health and Physical Education* 10 (2): 84, 120-121.

Hume, D. 2003. Using videotape to analyze the tennis serve. *Strategies* 16 (5): 15-16.

Huston, R.L., and C.A. Grau. 2003. Basketball shooting strategies—the free throw, direct shot and layup. *Sports Engineering* 6: 49-64.

Ignico, A.A. 1995. A comparison of videotape and teacher-directed instruction on knowledge, performance, and assessment of fundamental motor skills. *Journal of Educational Technology Systems* 23: 363-368.

Ignico, A.A. 1997. The effects of interactive videotape instruction on knowledge, performance, and assessment of sport skills. *Physical Educator* 54: 58-63.

Imwold, C.H., and S.J. Hoffman. 1983. Visual recognition of a gymnastic skill by experienced and inexperienced instructors. *Research Quarterly for Exercise and Sport* 54: 149-155.

Inkster, B., A. Murphy, R. Bower, and M. Watsford. 2011. Differences in the kinematics of the baseball swing between hitters of varying skill. *Medicine and Science in Sports and Exercise* 43: 1050-1054.

Ishigaki, H., and M. Miyao. 1993. Differences in dynamic visual acuity between athletes and nonathletes. *Perceptual and Motor Skills* 77: 835-839.

Ishigaki, H., and M. Miyao. 1994. Implications for dynamic visual acuity with changes in age and sex. *Perceptual and Motor Skills* 78: 363-369.

Ishikura, T., and K. Inomata. 1995. Effects of angle of model demonstration on learning of motor skill. *Perceptual and Motor Skills* 80: 651-658.

Jackson, S.A. 2007. Factors influencing the occurrence of flow state in elite athletes. In *Essential readings in sport and exercise psychology,* edited by D. Smith and M. Bar-Eli. Champaign, IL: Human Kinetics, 144-154.

Jambor, E.A., and E.M. Weekes. 1995. Videotape feedback: Make it more effective. *Journal of Physical Education, Recreation and Dance* 66 (2): 48-50.

James, C.P.A., K.L. Harburn, and J.F. Kramer. 1997. Cumulative trauma disorders in the upper extremities: Reliability of the postural and repetitive risk-factors index. *Archives of Physical Medicine and Rehabilitation* 78: 860-866.

James, G. 2002. Diagnosis in physical therapy: Insights from medicine and cognitive science. *Physical Therapy Reviews* 7: 17-31.

James, R., and J.S. Dufek. 1993. Movement observation: What to watch . . . and why. *Strategies* 6 (2): 17-19.

Janda, D.H., and P. Loubert. 1991. A preventative program focusing on the glenohumeral joint. *Clinics in Sports Medicine* 10: 955-971.

Janelle, C.M., D.A. Barba, S.G. Frehlich, L.K. Tennant, and J.H. Cauraugh. 1997. Maximizing performance feedback effectiveness through videotape replay and a self-controlled learning environment. *Research Quarterly for Exercise and Sport* 68: 269-279.

Janelle, C.M., J. Kim, and R.N. Singer. 1995. Subject-controlled performance feedback and learning of a closed motor skill. *Perceptual and Motor Skills* 81: 627-634.

Johansson, G. 1973. Visual perception of biological motion and a model for its analysis. *Perception and Psychophysics* 14: 202-211.

Johansson, G. 1975. Visual motion perception. *Scientific American* 232 (6): 76-88.

Johnson, R. 1990. Effects of performance principle training upon skill analysis competency. Ph.D. diss., Ohio State University.

Jones, P. 1987. The overarm baseball pitch: A kinesiological analysis and related strength-conditioning programming. *NSCA Journal* 9 (1): 5-13, 78.

Jones-Morton, P. 1990a. Skills analysis series: Part 1. Analysis of the place kick. *Strategies* 3 (5): 10-11.

Jones-Morton, P. 1990b. Skills analysis series: Part 2. Analysis of the overarm throw. *Strategies* 3 (6): 22-23.

Jones-Morton, P. 1990c. Skills analysis series: Part 3. Analysis of running. *Strategies* 4 (1): 22-24.

Jones-Morton, P. 1990d. Skills analysis series: Part 4. The standing long jump. *Strategies* 4 (2): 26-27.

Jones-Morton, P. 1991a. Skills analysis series: Part 5. Striking. *Strategies* 4 (3): 28-29.

Jones-Morton, P. 1991b. Skills analysis series: Part 6. Catching. *Strategies* 4 (4): 23-24.

Jonhagen, S., K. Halvorsen, and D.L. Benoit. 2009. Muscle activation and length changes during two lunge exercises: Implications for rehabilitation. *Scandinavian Journal of Medicine and Science in Sports* 19: 561-568.

Juul-Kristensen, B., G. Hanson, N. Fallentin, J.H. Andersen, and C. Eckdahl. 2001. Assessment of work postures and movements using a video-based observation method and direct technical measurements. *Applied Ergonomics* 32: 517-524.

Kamieneski, C.D. 1980. The effectiveness of an instructional unit in the analysis and correction of basketball skills. Ph.D. diss., Brigham Young University. Abstract in *Dissertation Abstracts International* 40: 6191A.

Karn, K.S., and M.M. Hayhoe. 2000. Memory representations guide targeting eye movements in a natural task. *Visual Cognition* 7 (6): 673-703.

Kay, H. 1970. Analyzing motor skill performance. In *Mechanisms of motor skill development*, edited by K.J. Connaly. London: Academic Press, 139-159.

Kazmierczak, K., S.E. Mathiassen, P. Neumann, and J. Winkel. 2006. Observer reliability of industrial activity analysis based on video recordings. *International Journal of Industrial Ergonomics* 36: 275-282.

Keenan, A.M., and T.M. Bach. 1996. Video assessment of rear-foot movements during walking: A reliability study. *Archives of Physical Medicine and Rehabilitation* 77: 651-655.

Kellis, E., and A. Katis. 2007. Biomechanical characteristics and determinants of instep soccer kicking. *Journal of Sports Science and Medicine* 6: 154-165.

Kelly, L.E. 1990. The effectiveness of computer managed interactive videodisk training on the development of sport skill competencies in teachers. Paper presented at the American Alliance for Health, Physical Education, Recreation and Dance national convention, March, New Orleans.

Kelly, L.E., J. Dagger, and J. Walkley. 1989. The effects of an assessment-based physical education program on motion skill development in preschool children. *Education and Treatment of Children* 12 (2): 152-164.

Kelly, L.E., J. Walkley, and M.R. Tarrant. 1988. Developing an interactive videodisk application. *Journal of Physical Education, Recreation and Dance* 59 (4): 22-26.

Kelly, L.E., P. Reuschlein, and J. Haubenstricker. 1989. Qualitative analysis of overhand throwing and catching motor skills: Implications for assessing and teaching. *Journal of the International Council for Health, Physical Education and Recreation* 25: 14-18.

Kelly, L.E., P. Reuschlein, and J.L. Haubenstricker. 1990. Qualitative analysis of bouncing, kicking, and striking motor skills: Implications for assessing and teaching. *Journal of the International Council for Health, Physical Education, and Recreation* 26 (2): 28-32.

Kenney, W.L., J.H. Wilmore, and D.L. Costill. 2012. *Physiology of sport and exercise*. 5th ed. Champaign, IL: Human Kinetics.

Kerner, J.F., and J. Alexander. 1981. Activities of daily living: Reliability and validity of gross vs. specific ratings. *Archives of Physical Medicine and Rehabilitation* 62: 161-166.

Kernodle, M.W., and E.T. Turner. 1998. The effective use of guidance techniques in teaching racquet sports. *Journal of Physical Education, Recreation and Dance* 69 (5): 49-54.

Kerrigan, D.C., P.O. Riley, J.L. Lelas, and U.D. Croce. 2001. Quantification of pelvic rotation as a determinant of gait. *Archives of Physical Medicine and Rehabilitation* 82: 217-220.

Keyserling, W.M. 1986. Postural analysis of the trunk and shoulders in simulated real time. *Ergonomics* 29: 569-583.

Kilani, H., D. Too, and M.J. Adrian. 1989. Visual perception of biomechanical characteristics of walking, jumping, and landing. In *Biomechanics in sports V*, edited by J. Tsarouchas, J. Terauds, B.A. Gowitzke, and L.E. Holt. Athens, Greece: Hellenic Sports Research Institute, 380-392.

Kindig, L.E., and E.J. Windell. 1984. Analysis of sport skills. In *Proceedings: Second national symposium on teaching kinesiology and biomechanics in sports*, edited by R. Shapiro and J.R. Marett. Colorado Springs, CO: NASPE, 231-232.

Kinesiology Academy. 1980. Guidelines and standards for undergraduate kinesiology. *Journal of Physical Education, Recreation and Dance* 51 (2): 19-21.

Kinesiology Academy. 1992. Guidelines and standards for undergraduate biomechanics/kinesiology. *Kinesiology Academy Newsletter* (spring): 3-6.

Kirtley, C. 2006. *Clinical gait analysis: Theory and practice*. London: Elsevier.

Klatt, L.A. 1992. Biomechanics: Analyzing skills and performance. In *Science of coaching baseball*, edited by J. Kindall. Champaign, IL: Leisure Press, 49-83.

Klavora, P., P. Gaskovski, and R.D. Forsyth. 1994. Test-retest reliability of the dynavision apparatus. *Perceptual and Motor Skills* 79: 448-450.

Klavora, P., P. Gaskovski, and R.D. Forsyth. 1995. Test-retest reliability of three dynavision tasks. *Perceptual and Motor Skills* 80: 607-610.

Klesius, S., and L. Bowers. 1990. The "I'm special" interactive videodisc for training teachers of handicapped children. *Florida Educational Computing Quarterly* 2 (4): 73-76.

Kluger, A.N., and A. DeNisi. 1996. The effects of feedback interventions on performance: A historical review, a meta-analysis, and a preliminary feedback intervention theory. *Psychological Bulletin* 119: 254-284.

Kluka, D.A. 1987. Visual skill enhancement. *Strategies* 1 (1): 20-24.

Kluka, D.A. 1991. Visual skills: Considerations in learning motor skills for sport. *ASAHPERD Journal* 14 (1): 41-43.

Kluka, D.A. 1994. Visual skills related to sport performance. *Research Consortium Newsletter* (winter): 3.

Kniffin, K.M. 1985. The effects of individualized videotape instruction on the ability of undergraduate physical education majors to analyze select sport skills. Ph.D. diss., Ohio State University. Abstract in *Dissertation Abstracts International* 47: 119A.

Knudson, D. 1991. The tennis topspin forehand drive: Technique changes and critical elements. *Strategies* 5 (1): 19-22.

Knudson, D. 1993. Biomechanics of the basketball jump shot: Six key teaching points. *Journal of Physical Education, Recreation and Dance* 64 (2): 67-73.

Knudson, D. 1999a. Using sport science to observe and correct tennis strokes. In *Applied proceedings of the XVII international symposium on biomechanics in sports: Tennis*, edited by B. Elliott, B. Gibson, and D. Knudson. Perth, Western Australia: Edith Cowan University, 7-16.

Knudson, D. 1999b. Validity and reliability of visual ratings of the vertical jump. *Perceptual and Motor Skills*: 89: 642-648.

Knudson, D. 2000. What can professionals qualitatively analyze? *Journal of Physical Education, Recreation and Dance* 71 (2): 19-23.

Knudson, D. 2003. An integrated approach to the introductory biomechanics course. *Physical Educator* 60: 122-133.

Knudson, D. 2005. Evidence-based practice in kinesiology: The theory to practice gap revisited. *Physical Educator* 62: 212-221.

Knudson, D. 2006. *Biomechanical principles of tennis technique*. Vista, CA: Racquet Tech Publishing.

Knudson, D. 2007a. *Fundamentals of biomechanics*. 2nd ed. New York: Springer Science.

Knudson, D. 2007b. Qualitative biomechanical principles for coaches. *Sports Biomechanics* 6: 108-117.

Knudson, D. 2008. Warm-up and flexibility. In *Conditioning for strength and human performance*, edited by T.J. Chandler and L.E. Brown. Philadelphia: Lippincott Williams & Wilkins, 166-181.

Knudson, D., and C. Morrison. 1996. An integrated qualitative analysis of overarm throwing. *Journal of Physical Education, Recreation and Dance* 67 (6): 31-36.

Knudson, D., and C. Morrison. 2000. Visual ratings of the vertical jump are weakly correlated with perceptual style. *Journal of Human Movement Studies* 39: 33-44.

Knudson, D., and D. Kluka. 1997. The impact of vision and vision training on sport performance. *Journal of Physical Education, Recreation and Dance* 68 (4): 17-24.

Knudson, D., B. Elliott, and T. Ackland. 2012. Citation of evidence for research and application in kinesiology. *Kinesiology Review*, 1: 129-136.

Knudson, D., C. Morrison. 1997. *Qualitative analysis of human movement*. Champaign, IL: Human Kinetics.

Knudson, D., C. Morrison. 2002. *Qualitative Analysis of Human Movement*, 2nd ed. Champaign, IL: Human Kinetics.

Knudson, D., C. Morrison, and J. Reeve. 1991. Effect of undergraduate kinesiology courses on qualitative analysis ability. In *Teaching kinesiology and biomechanics in sports*, edited by J. Wilkerson, E. Kreighbaum, and C. Tant. Ames, IA: NASPE Kinesiology Academy, 17-20.

Knudson, D., D. Luedtke, and J. Faribault. 1994. How to analyze the serve. *Strategies* 7 (8): 19-22.

Kociolek, A.M., and P.J. Keir. 2010. Reliability of distal upper extremity posture matching using slow-motion and frame-by-frame video methods. *Human Factors* 52: 441-455.

Kovar, S.K., H.M. Matthews, K.L. Ermler, and J.H. Hehrhof. 1992. Feedback: How to teach how. *Strategies* 5 (1): 21-25.

Kowler, E. 2011. Eye movements: The past 25 years. *Vision Research* 51: 1457-1483.

Kozak, W. 1989. Skill analysis. In *Proceedings: National coaching certification program advanced II seminar*, edited by J. Almstedt et al. Gloucester, ON: Canadian Amateur Hockey Association, 51-77.

Kraft, R.E., and J.A. Smith. 1993. Throwing and catching: How to do it right. *Strategies* 6 (5): 24-27, 29.

Krebs, D.E., J.E. Edelstein, and S. Fishman. 1985. Reliability of observational kinematic gait analysis. *Physical Therapy* 65: 1027-1033.

Kreighbaum, E., and K.M. Barthels. 1985. *Biomechanics: A qualitative approach for studying human movement*. 2nd ed. Minneapolis: Burgess.

Kretchmar, R.T., H. Sherman, and R. Mooney. 1949. A survey of research in the teaching of sports. *Research Quarterly* 20: 238-249.

Krosshaug, T., A. Nakamae, B. Boden, L. Engebretsen, G. Smith, J. Slauterbeck, T.E. Hewitt, and R. Bahr. 2007. Estimating 3D joint kinematics from video sequences of running and cutting maneuvers—assessing the accuracy of simple visual inspection. *Gait & Posture* 26: 378-385.

Kwak, E.C. 1994. The initial effects of various task presentation conditions on students' performance of the lacrosse throw. Ph.D. diss., University of South Carolina, 1993. Abstract in *Dissertation Abstracts International* 54: 2507A.

Lachowetz, T., J. Evon, and J. Pastiglione. 1998. The effect of an upper body strength program on intercollegiate baseball throwing velocity. *Journal of Strength and Conditioning Research* 12: 116-119.

Lafortune, M.A., and P.R. Cavanagh. 1983. Effectiveness and efficiency in bicycle riding. In *Biomechanics VIII-B*, edited by H. Matsui and K. Kobayshi. Champaign, IL: Human Kinetics, 928-936.

Landers, D. 1969. Effect of the numbers of categories systematically observed on individual and group performance ratings. *Perceptual and Motor Skills* 29: 731-735.

Landers, D.M. 1970. A review of research on gymnastic judging. *Journal of Health, Physical Education, Recreation* 41 (7): 85-88.

Landin, D. 1994. The role of verbal cues in skill learning. *Quest* 46: 299-313.

Landin, D.K., E. Hebert, and D.L. Cutton. 1989. Analyzing the augmented feedback patterns of professional tennis instructors. *Journal of Applied Research in Coaching and Athletics* 4: 255-271.

Langley, D. 1993. Teaching new motor patterns: Overcoming student resistance toward change. *Journal of Physical Education, Recreation and Dance* 63 (1): 27-31.

Lappin, J.S., and M.A. Fuqua. 1983. Accurate visual measurement of three-dimensional moving patterns. *Science* 221: 480-482.

Larsson, L.E., M. Miller, R. Norlin, and H. Thaczuk. 1986. Changes in gait patterns after operations in children with spastic cerebral palsy. *International Orthopedics* 10: 155-162.

Latham, A., and H. Cassady. 1997. Analysis of performance: How to nurture pupils' analytical skills. *British Journal of Physical Education* (winter): 28-32.

Lee, A.M., N.C. Keh, and R.A. Magill. 1993. Instructional effects of teacher feedback in physical education. *Journal of Teaching in Physical Education* 12: 228-243.

Lee, S., and L.J. Stoner. 1985. Skill analysis through computer graphic feedback. In *Biomechanics in sports II: Proceedings of ISBS 1985*, edited by J. Terauds and J.N. Barham. Del Mar, CA: Research Center for Sports, 346-353.

Lees, A. 2007. Technique analysis in sports: A critical review. *Journal of Sports Sciences* 20: 813-828.

Lees, A. 2008. Qualitative biomechanical analysis of technique. In *The essentials of performance analysis*, edited by M. Hughes and I.M. Franks. London: Routledge, 162-179.

Lehmann, J.F. 1982. Gait analysis: Diagnosis and management. In *Krusen's handbook of physical medicine and rehabilitation*, edited by F.J. Kottke, G.K. Stillwell, and J.F. Lehmann. Philadelphia: Saunders, 86-101.

Leis, H.H. 1994. The effects of two instructional conditions on sport skill specific analytic proficiency of physical education majors. Ph.D. diss., University of Southern Mississippi, 1993. Abstract in *Dissertation Abstracts International* 54 (8): 2946A.

Lenoir, M., L. Crevits, M. Goethals, J. Wildenbeest, and E. Musch. 2000. Are better eye movements an advantage in ball games? A study of prosaccadic and antisaccadic eye movements. *Perceptual and Motor Skills* 91: 546-552.

Levanon, J., and J. Dapena. 1998. Comparison of the kinematics of the full-instep and pass kicks in soccer. *Medicine and Science in Sports and Exercise* 30: 917-927.

Liang, G. 2001. Teaching children qualitative analysis of fundamental motor skill. Ph.D. diss., West Virginia University, Morgantown.

Liebermann, D.G., L. Katz, M. Hughes, R. Bartlett, J. McClements, and I. Hughes. 2002. Advances in the application of information technology to sport performance. *Journal of Sports Sciences* 20: 755-769.

Lindeman, B., T. Libkuman, D. King, and B. Krause. 2000. Development of an instrument to assess jump-shooting form. *Journal of Sport Behavior* 23: 335-348.

Liu, S., and A.W. Burton. 1999. Changes in basketball shooting patterns as a function of distance. *Perceptual and Motor Skills* 89: 831-845.

Locke, L. 1972. Implications for physical education. *Research Quarterly* 43: 374-386.

Locke, L.F. 1984. Research on teaching teachers: Where are we now? *Journal of Teaching Physical Education* 9 (summer): 63-85.

Lockhart, A. 1966. Communicating with the learner. *Quest* 6: 57-67.

Logan, G.A., and W.C. McKinney. 1970. *Kinesiology*. Dubuque, IA: William C. Brown.

Long, G.M. 1994. Exercises for training vision and dynamic visual acuity among college students. *Perceptual and Motor Skills* 78: 1049-1050.

Long, G.M., and D.A. Rourke. 1989. Training effects on the resolution of moving targets: Dynamic visual acuity. *Human Factors* 31: 443-451.

Lorson, K.M., and J.D. Goodway. 2007. Influence of critical cues and task constraints on overarm throwing performance in elementary age children. *Perceptual and Motor Skills* 105: 753-767.

Lounsbery, M., and C. Coker. 2008. Developing skill-analysis competency in physical education teachers. *Quest* 60: 255-267.

Lowe, B.D. 2004. Accuracy and validity of observational estimates of wrist and forearm posture. *Ergonomics* 47: 527-555.

Luttgens, K., and K.F. Wells. 1982. *Kinesiology: Scientific basis of human movement*. 7th ed. Philadelphia: Saunders.

Lyons, K. 2003. Technique analysis. *Sports Coach* 25 (4): 22-23.

MacLeod, B. 1991. Effects of eyerobics visual skills training on selected performance measures of female varsity soccer players. *Perceptual and Motor Skills* 72: 863-866.

Magill, R.A. 1993. Augmented feedback in skill acquisition. In *Handbook of research on sport psychology*, edited by R.N. Singer, M. Murphey, and L.K. Tennant. New York: Macmillan, 193-212.

Magill, R.A. 1994. The influence of augmented feedback on skill learning depends on characteristics of the skill and learner. *Quest* 46: 314-327.

Magill, R.A., and P.F. Parks. 1983. The psychophysics of kinesthesis for positioning response: The physical stimulus-psychological response relationship. *Research Quarterly for Exercise and Sport* 54: 346-351.

Malina, R.M., and C. Bouchard. 1991. *Growth, maturation, and physical activity.* Champaign, IL: Human Kinetics.

Malkia, E., J. Huhtinen, and P. Luthanen. 1991. A qualitative analysis of walking in children with CP. In *The 11th international congress of the World Confederation for Physical Therapy: Proceedings.* London: World Confederation for Physical Therapy, 1160.

Marett, J.R., J.A. Pavlacka, W.L. Siler, and R. Shapiro. 1984. Kinesiology status update: A national survey. In *Proceedings: Second national symposium on teaching kinesiology and biomechanics in sports,* edited by R. Shapiro and J.R. Marett. Colorado Springs, CO: NASPE, 7-15.

Marey, E.J. 1972. *Movement.* New York: Arno Press.

Marino, G.W. 1982. Qualitative biomechanical analysis of sports skills. *Coaching Science Update* 9: 20-22.

Marks, L.E. 1987. On cross-modal similarity: Auditory-visual interactions in speeded discrimination. *Journal of Experimental Psychology: Human Perception and Performance* 13: 384-394.

Marques-Bruna, P., A. Lees, and M. Scott. 2005. An integrated analytical model for the qualitative assessment of kicking effectiveness in football. In *Science and football V,* edited by M.T. Reilly, J. Cabri, and D. Araugo. London: Routledge, 60-65.

Marques-Bruna, P., A. Lees, and P. Grimshaw. 2008. Structural principle components analysis of the kinematics of the soccer kick using different types of rating scales. *International Journal of Sports Science & Coaching* 3: 73-85.

Martens, R., L. Burwitz, and J. Zuckerman. 1976. Modeling effects on motor performance. *Research Quarterly* 47: 277-291.

Martino, G., and L.E. Marks. 1999. Perceptual and linguistic interactions in speedec clarification: Tests of the semantic coding hypothesis. *Perception* 28: 903-923.

Martino, G., and L.E. Marks. 2000. Cross-modal interaction between vision and touch: The role of synesthetic correspondence. *Perception* 29: 746-754.

Maschette, W. 1985. Correcting technique problems of a successful junior athlete. *Sports Coach* 9 (1): 14-17.

Masser, L. 1985. The effect of refinement on student achievement in a fundamental motor skill K-6. *Journal of Teaching in Physical Education* 6: 174-182.

Masser, L.S. 1993. Critical cues help first-grade students' achievement in handstands and forward rolls. *Journal of Teaching in Physical Education* 12: 301-312.

Matanin, M.J. 1993. Effects of performance principle training on correct analysis and diagnosis of motor skills. Ph.D. diss., Ohio State University. Abstract in *Dissertation Abstracts International* 54: 1724A.

Mather, G. 2008. Perceptual uncertainty and line-call challenges in professional tennis. *Proceedings of the Royal Society: Part B* 275: 1645-1651.

Mathers, S. 1990. Training your eyes: A method of learning ski movement analysis. *Professional Skier* (winter): 30-31.

Mathias, K.E. 1991. A comparison of the effectiveness of interactive video in teaching the ability to analyze two motor skills in swimming. Ed.D. diss., University of Northern Colorado. Abstract in *Dissertation Abstracts International* 51: 3676A.

Matlin, M. 1983. *Cognition.* New York: Holt, Rinehart & Winston.

McCallister, S.G., and G. Napper-Owen. 1999. Observation and analysis of skills of student teachers in physical education. *Physical Educator* 56: 19-32.

McClenaghan, B.A., and D.L. Gallahue. 1978. *Fundamental movement: A developmental and remedial approach.* Philadelphia: Saunders.

McCormick, E.J., and M.S. Sanders. 1982. *Human factors in engineering and design.* 5th ed. New York: McGraw-Hill.

McCraw, P. 1995. The qualitative analysis of motor skills using multi-media software engineering techniques. Master's thesis, Deakin University, Melbourne, Australia.

McCullagh, P. 1986. Model status as a determinant of observational learning and performance. *Journal of Sport Psychology* 8: 319-331.

McCullagh, P. 1987. Model similarity effects on motor performance. *Journal of Sport Psychology* 9: 249-260.

McCullagh, P., J. Stiehl, and M.R. Weiss. 1990. Developmental modeling effects on the quantitative and qualitative aspects of motor performance. *Research Quarterly for Exercise and Sport* 61: 344-350.

McCullagh, P.M., and J.K. Caird. 1990. Correct and learning models and the use of model knowledge of results in the acquisition and retention of a motor skill. *Journal of Human Movement Studies* 18: 107-116.

McCullagh, P.M., and W.S. Little. 1990. Demonstrations and knowledge of results in motor skill acquisition. *Perceptual and Motor Skills* 71: 735-742.

McCullagh, P., M.R. Weiss, and D. Ross. 1989. Modeling considerations in motor skill acquisition and performance: An integrated approach. *Exercise and Sport Sciences Reviews* 17: 475-513.

McCullick, B., P. Schempp, S. Hsu, J. Jung, B. Vickers, and G. Schuknecht. 2006. An analysis of the working memories of expert sport instructors. *Journal of Teaching in Physical Education* 25: 149-165.

McGibbon, C.A., and D.E. Krebs. 1998. The influence of segment endpoint kinematics on segmental power calculations. *Gait & Posture* 7: 237-242.

McGinley, J.L., P.A. Goldie, K.M. Greenwood, and S.J. Olney. 2003. Accuracy and reliability of observational gait analysis data: Judgments of push-off in gait after stroke. *Physical Therapy* 83: 146-160.

McGinley, J.L., M.E. Morris, K.M. Greenwood, P.A. Goldie, and S.J. Olney. 2006. Accuracy of clinical observations of push-off during gait after stroke. *Archives of Physical Medicine and Rehabilitation* 87: 779-785.

McGinnis, P.M. 2005. *Biomechanics of sport and exercise.* 2nd ed. Champaign, IL: Human Kinetics.

McIntyre, D.R., & Pfautsch, E.W. 1982. A kinematic analysis of the baseball batting swings involved in opposite-field and same-field hitting. Research Quarterly for Exercise and Sport, 53, 206-213.

McKethan, R.N., and E.T. Turner. 1999. Using multimedia programming to teach sport skills. Journal of Physical Education, Recreation and Dance 70 (3): 22-25.

McKethan, R.N., M.W. Kernodle, D. Brantz, and J. Fischer. 2003. Qualitative analysis of the overhand throw by undergraduates in education using a distance learning computer program. Perceptual and Motor Skills 97: 979-989.

McLean, S.P., and M.S. Reeder. 2000. Upper extremity kinematics of dominant and non-dominant side batting. Journal of Human Movement Studies 38: 201-212.

McLeod, P., J. Driver, Z. Dienes, and J. Crisp. 1991. Filtering by movement and visual search. Journal of Experimental Psychology: Human Perception and Performance 17 (1): 55-64.

McNaughton, L. 1986. Some drills to improve visual perception abilities in team sport players. Sports Coach 9 (2): 47-49.

McPherson, M.N. 1987. The development, implementation, and evaluation of a program designed to promote competence in skill analysis. Ph.D. diss., University of Alberta.

McPherson, M.N. 1988b. Who: The physical education teacher as diagnostician. Paper presented at the national convention of the American Alliance for Health, Physical Education, Recreation and Dance, April, Kansas City, MO.

McPherson, M.N. 1990. A systematic approach to skill analysis. Sports Science Periodical on Research and Technology in Sport 11 (1): 1-10.

McPherson, M.N. 1996. Qualitative and quantitative analysis in sports. American Journal of Sports Medicine 24: 585-588.

McPherson, M.N., and E.W. Bedingfield. 1985. Development of instructional videotape for qualitative analysis. In Biomechanics in sports II: Proceedings of ISBS 1985, edited by J. Terauds and J.N. Barham. Del Mar, CA: Research Center for Sports, 385-389.

McPherson, M.N., and J. Walsh. 1990. Application of the skill analysis approach to the nordic two skate. Sports Coach 13 (3): 3-7.

Mead, T.P., and J.N. Drowatzky. 1997. Interdependence of vision and audition among inexperienced and experienced tennis players. Perceptual and Motor Skills 85: 163-166.

Meehan, J.W., and R.H. Day. 1995. Visual accommodation as a cue for size. Ergonomics 38: 1239-1249.

Melville, D.S. 1993. Videotaping: An assist for large classes. Strategies 6 (4): 26-28.

Messick, J.A. 1991. Prelongitudinal screening of hypothesized developmental sequences for the overhead tennis serve in experienced tennis players. Research Quarterly for Exercise and Sport 62: 249-256.

Messier, S.P., and K.J. Cirillo. 1989. Effects of a verbal and visual feedback system on running technique, perceived exertion and running economy in female novice runners. Journal of Sports Sciences 7: 113-126.

Metzler, M. 1989. A review of research on time in sport pedagogy. Journal of Teaching in Physical Education 6: 271-285.

Meyer, C.H., A.G. Lasker, and D.A. Robinson. 1985. The upper limit of human smooth pursuit velocity. Vision Research 25: 561-563.

Meyer, N.H. 2002. The use of digital video recording for observational analysis of gait. Physical Medicine and Rehabilitation 16: 179-214.

Michigan Education Assessment Program. 1984. Physical education assessment administration manual, 1984-1985. Lansing, MI: Michigan Department of Education.

Mielke, D., and C. Morrison. 1985. Motor development and skill analysis: Connections to elementary physical education. Journal of Physical Education, Recreation and Dance 56 (9): 48-51.

Mielke, D., and D. Chapman. 1987. Effectiveness of education majors in assessing children on the test of gross motor development. Perceptual and Motor Skills 64: 1249-1250.

Millar, C. 1996. In-line skating basics. New York: Sterling.

Miller, D.I. 1980. Body segment contributions to sport skill performance: Two contrasting approaches. Research Quarterly for Exercise and Sport 51: 219-233.

Miller, G., and C. Gabbard. 1988. Effects of visual aids on acquisition of selected tennis skills. Perceptual and Motor Skills 67: 603-606.

Miller, L. 1998. Get rolling. 2nd ed. New York: McGraw Hill.

Miller, S., and R. Bartlett. 1993. The effects of increased shooting distance in the basketball jump shot. Journal of Sports Sciences 11: 285-293.

Miller, S., and R. Bartlett. 1996. The relationship between basketball shooting kinematics, distance and playing position. Journal of Sports Sciences 14: 243-253.

Minas, S. 1977. Memory coding for movement. Perceptual and Motor Skills 45: 787-790.

Minick, K.I., K.B. Kiesel, L. Burton, A. Taylor, P. Plisky, and R.J. Butler. 2010. Intrrater reliability of the functional movement screen. Journal of Strength and Conditioning Research 24: 479-486.

Miyashita, M., S. Fukashiro, and Y. Hirano. 1986. Feedback of biomechanics data. In Biomechanics: The 1984 Olympic scientific congress proceedings, edited by M. Adrian and H. Deutsch. Eugene, OR: Microform Publications, 47-54.

Miyashita, M., T. Tsunoda, S. Sakurai, H. Nishizono, and T. Mizuno. 1979. The tennis serve as compared with overarm throwing. In Proceedings of a national symposium on the racket sports, edited by J. Groppel. Champaign, IL: University of Illinois, 125-140.

Miyazaki, S., and T. Kubota. 1984. Quantification of gait abnormalities on the basis of continuous foot-force measurement: Correlation between quantification indices and visual rating. *Medical and Biological Engineering and Computing* 22: 70-76.

Mohnsen, B., and C. Thompson. 1997. Using video technology in physical education, part II. *Strategies* 10 (6): 8-11.

Montagne, G., M. Laurent, and H. Ripoll. 1993. Visual information pick-up in ball-catching. *Human Movement Science* 12: 273-297.

Montgomery, J., and D. Knudson. 2002. A method to determine the stride length for baseball pitching. *Applied Research in Coaching and Athletics Annual* 17: 75-84.

Moody, D.L. 1967. Imagery differences among women of varying levels of experience, interests, and abilities in motor skills. *Research Quarterly* 43: 55-61.

Moore, J.S. 1993. The effects of an instructional strategy in naked eye analysis on elementary students' performance of the overarm throw. Ph.D. diss., University of New Mexico.

Moreno, F.J., J.M. Saavedra, R. Sabido, V. Luis, and R. Reina. 2006. Visual search strategies of experiences and nonexperienced swimming coaches. *Perceptual and Motor Skills* 103: 861-872.

Moreno, F.J., R. Reina, L. Vicente, and R. Sabido. 2002. Visual search strategies in experienced and inexperienced gymnastic coaches. *Perceptual and Motor Skills* 95: 901-902.

Morris, G. 1977. Dynamic visual acuity: Implications for the physical educator and coach. *Motor Skills: Theory into Practice* 2: 15-20.

Morrison, C.S. 1976. The effect of vision, motion and laterality in wrist shooting accuracy in ice hockey. Master's thesis, Springfield College, Springfield, MA.

Morrison, C.S. 1994. Comparison of nationality, gender and type of instruction on the acquisition and retention of qualitative analysis of movement ability. In *Proceedings for the 10th commonwealth and international scientific congress: Access to active living*, edited by F.I. Bell and G.H. Van Gyn. Victoria, BC: University of Victoria, 169-174.

Morrison, C.S. 2000. Why don't you analyze the way I analyze? *Journal of Physical Education, Recreation and Dance* 71 (1): 22-25.

Morrison, C. S. 2004. Refinement of terminology in qualitative analysis of human movement. *Perceptual and Motor Skills* 99: 105-106.

Morrison, C.S., and C.M. Frederick. 1998. Relationship of initial and final scores on a qualitative analysis of movement test. *Perceptual and Motor Skills* 87: 651-655.

Morrison, C.S., and E.J. Reeve. 1986. Effect of instruction units on the analysis of related and unrelated skills. *Perceptual and Motor Skills* 62: 563-566.

Morrison, C.S., and E.J. Reeve. 1988b. Effect of undergraduate major and instruction on qualitative skill analysis. *Journal of Human Movement Studies* 15: 291-297.

Morrison, C.S., and J.M. Harrison. 1985. Movement analysis and the classroom teacher. *CAHPER Journal* 51 (5): 16-19.

Morrison, C.S., and J.M. Harrison. 1997. Integrating qualitative analysis of movement in the university physical education curriculum. *Physical Educator* 54: 64-71.

Morrison, C.S., and J. Reeve. 1988a. Effect of different instructional videotape units on undergraduate physical education majors' skill analysis ability. Paper presented at the Texas Association for Health, Physical Education, Recreation and Dance convention, November, San Antonio.

Morrison, C.S., and J. Reeve. 1989. Effect of different videotape instructional units on undergraduate physical education majors' qualitative analysis of skill. *Perceptual and Motor Skills* 69: 111-114.

Morrison, C.S., and J. Reeve. 1992. Perceptual style and instruction in the acquisition of qualitative analysis of movement by majors in elementary education. *Perceptual and Motor Skills* 74: 579-583.

Morrison, C.S., and J. Reeve. 1993. A framework for writing and evaluating critical performance cues in instructional materials for physical education. *Physical Educator* 50 (3): 132-135.

Morrison, C.S., D. Knudson, C. Clayburn, and P. Haywood. 2005. Accuracy of visual estimates of joint angle and speed of movement using reference movements. *Perceptual and Motor Skills* 100: 599-606.

Morrison, C.S., E.J. Reeve, and J. Harrison. 1992. The effect of instruction on the ability to qualitatively analyze and perform movement skills. *CAHPER Journal* 58 (2): 18-20.

Morrison, C.S., E.J. Reeve, and J.M. Harrison. 1984. The effect of two methods of teaching skill analysis and skill performance. Paper presented to the southern district American Alliance for Health, Physical Education, Recreation and Dance, February, Biloxi, MS.

Morrison, C.S., S.K. Gangstead, and J. Reeve. 1990. Two approaches to qualitative analysis: Implications for future directions. Paper presented at the American Alliance for Health, Physical Education, Recreation and Dance national convention, March, New Orleans.

Morrison, J.P., and B.D. Wilson. 1996. Development of a video qualitative analysis system. Paper presented at the first Australasian biomechanics conference, January, Sydney, Australia.

Morton, P. 1990. Effects of training in skill analysis on generalization across age levels. Ph.D. diss., Ohio State University, 1989. Abstract in *Dissertation Abstracts International* 50: 2424A.

Mosher, R.E., and R.W. Schutz. 1983. The development of a test of overarm throwing: An application of generalizability theory. *Canadian Journal of Applied Sport Science* 8 (1): 1-8.

Mosston, M., and S. Ashworth. 1986. *Teaching physical education*. 3rd ed. Columbus, OH: Merrill.

Murphy, S., P. Buckle, and D. Stubbs. 2002. The use of portable ergonomic observation method (PEO) to monitor the sitting posture of schoolchildren in the classroom. *Applied Ergonomics* 33: 365-370.

Naatanen, R. 1990. The role of attention in auditory information processing as revealed by event-related potentials and other brain measures of cognitive function. *Behavioral and Brain Sciences* 13: 201-288.

Naito, K., and T. Maruyama. 2008. Contributions of the muscular torques and motion-dependent torques to generate rapid elbow extension during overhand baseball pitching. *Sports Engineering* 11: 47-56.

Nakatsuka, M., T. Ueda, Y. Nawa, E. Yukawa, T. Hara, and Y. Hara. 2006. Effect of static visual acuity on dynamic visual acuity: A pilot study. *Perceptual and Motor Skills* 103: 160-164.

National Association for Sport and Physical Education. 1992. *NASPE/NCATE physical education guidelines: An instructional manual.* 3rd ed. Reston, VA: AAHPERD.

National Association for Sport and Physical Education. 2008. *National standards and guidelines for physical education teacher education.* 3rd ed. Reston, VA: AAHPERD.

Nelson, M.A. 1991. Developmental skills and children's sports. *Physician and Sportsmedicine* 19 (2): 67-79.

Neptune, R.R., F.E. Zajac, and S.A. Kautz. 2004. Muscle force redistributes power for body progression during walking. *Gait & Posture* 19: 194-205.

Netelenbos, J.B. 2005. Teachers' ratings of gross motor skills suffer from low concurrent validity. *Human Movement Science* 24: 116-137.

Neumaier, A. 1982. Unterschung zur funktion des blickverhaltens bei visuellen wahrnehmungsprozessen im sport. *Sportweissenschaft* 12 (1): 78-91.

Newell, K.M. 1976. Knowledge of results and motor learning. *Exercise and Sport Sciences Reviews* 4: 195-228.

Newell, K.M. 1990. Kinesiology: The label for the study of physical activity in higher education. *Quest* 42: 269-278.

Newell, K.M., J.T. Quinn, W.A. Sparrow, and C.B. Walter. 1983. Kinematic information feedback for learning a rapid arm movement. *Human Movement Science* 2: 235-269.

Newell, K.M., L.R. Morris, and D.M. Scully. 1985. Augmented information and the acquisition of skill in physical activity. *Exercise and Sport Sciences Reviews* 13: 235-261.

Newell, K.M., W.A. Sparrow, and J.T. Quinn. 1985. Kinetic information feedback for learning isometric tasks. *Journal of Human Movement Studies* 11: 113-123.

Newtson, D. 1976. The process of behavior observation. *Journal of Human Movement Studies* 2: 114-122.

Nicholls, R.L., G.S. Fleisig, B.C. Elliott, S.L. Lyman, and E.D. Osinski. 1999. Biomechanical validation of a qualitative analysis of baseball pitching. In *Scientific proceedings of the XVII international symposium on biomechanics in sports,* edited by R.H. Sanders and B.J. Gibson. Perth, Western Australia: Edith Cowan University, 297-300.

Nielsen, A.B., and L. Beauchamp. 1992. The effect of training in conceptual kinesiology on feedback provision patterns. *Journal of Teaching in Physical Education* 11: 126-138.

Nissen, C.W., M. Westwell, S. Ounpuu, M. Patel., J.P. Tate, K. Pierz, J.P. Burns, and J. Bicos. 2007. Adolescent baseball pitching technique: A detailed three-dimensional biomechanical analysis. *Medicine and Science in Sports and Exercise* 39:1347-1357.

Noe, A., L. Pesoa, and E. Thompson. 2000. Beyond the grand illusion: What change blindness tells us about vision. *Visual Cognition* 7: 93-106.

Norman, R.W. 1975. Biomechanics for the community coach. *Journal of Physical Education, Recreation and Dance* 46 (3) (March): 49-52.

Norman, R.W. 1977. An approach to teaching the mechanics of human motion at the undergraduate level. In *Proceedings: Kinesiology: A national conference on teaching,* edited by C.J. Dillman and R.G. Sears. Champaign, IL: University of Illinois, 113-123.

Nunome, H., Y. Ikegami, R. Kozakai, T. Apriantono, and S. Sano. 2006. Segmental dynamics of soccer instep kicking with the preferred and non-preferred leg. *Journal of Sports Sciences* 24: 529-541.

Oltman, P.K., E. Raskin, H.A. Witkin, and S.A. Karp. 1971. *The group embedded figures test.* Palo Alto, CA: Consulting Psychologists Press.

Onate, J.A., K.M. Guskiewicz, S.W. Marshall, C. Giulianni, B. Yu, and W.E. Garrett. 2005. Instruction of jumplanding technique using videotape feedback. *American Journal of Sports Medicine* 33: 831-842.

Ormond, T.C. 1992. The prompt/feedback package in physical education. *Journal of Physical Education, Recreation and Dance* 63 (1): 64-67.

Osborne, M.M., and M.E. Gordon. 1972. An investigation into the accuracy of rating of a gross motor skill. *Research Quarterly* 43: 55-61.

Oslin, J.L., S. Stroot, and D. Siedentop. 1997. Use of component-specific instruction to promote development of the overarm throw. *Journal of Teaching in Physical Education* 16: 340-356.

O'Sullivan, M. 1988. How: The Ohio State University model. Paper presented at the national convention of the American Alliance for Health, Physical Education, Recreation and Dance, April, Kansas City, MO.

Overdorf, V.G. 1990. Timing—in life and in sports—is everything. *Journal of Physical Education, Recreation and Dance* 61 (7): 66-69.

Painter, M.A. 1990. A generalizability analysis of observational abilities in the assessment of hopping using two developmental approaches to motor skill sequencing. Ph.D. diss., Michigan State University, 1989. Abstract in *Dissertation Abstracts International* 50: 3888A.

Painter, M.A. 1994. Developmental sequences for hopping as assessment instruments: A generalizability analysis. *Research Quarterly for Exercise and Sport* 65: 1-10.

Palmer, S., and I. Rock. 1994. Rethinking perceptual organization: The role of uniform connectedness. *Psychonomic Bulletin and Review* 1 (1): 29-55.

Palmer, S.E. 1992. Common region: A new principle of perceptual grouping. *Cognitive Psychology* 24: 436-447.

Paquet, V.L., L. Punnett, and B. Buchholz. 2001. Validity of fixed-interval observations for postural assessment in construction work. *Applied Ergonomics* 32: 215-224.

Parry, R. 2005. A video analysis of how physiotherapists communicate with patients about errors of performance: Insights for practice and policy. *Physiotherapy* 91: 204-214.

Parson, M.L. 1998. Focus student attention with verbal cues. *Strategies* 11 (3): 30-33.

Partridge, D., and I.M. Franks. 1986. Analyzing and modifying coaching behaviors by means of computer aided observation. *Physical Educator* (winter): 8-23.

Patla, A.E., and S.D. Clouse. 1988. Visual assessment of human gait: Reliability and validity. *Rehabilitation Research* (October): 87-96.

Patrick, J., and B.J. Lowdon. 1987. Computer controlled video replay of player activity in sport. *Sports Coach* 10 (3): 20-22.

Pellecchia, G.L., and G.E. Garrett. 2002. Qualitative analysis of lumbar stabilization using point light and normal video displays. *Perceptual and Motor Skills* 94: 1219-1229.

Pellett, T.L., H.A. Henschel-Pellett, and J.M. Harrison. 1994. Feedback effects: Field-based findings. *Journal of Physical Education, Recreation and Dance* 65 (9): 75-78.

Perry, J. 1992. *Gait analysis: Normal and pathological function.* Thorofare, NJ: Slack.

Petrakis, E. 1986. Visual observation patterns of tennis teachers. *Research Quarterly for Exercise and Sport* 57: 254-259.

Petrakis, E. 1987. Analysis of visual search patterns of dance teachers. *Journal of Teaching in Physical Education* 6: 149-156.

Petrakis, E. 1993. Analysis of visual search patterns of tennis teachers. In *Perception and cognition: Advances in eye movement research,* edited by G. Ydewalle and J. Van Rensbergen. Amsterdam: North-Holland/Elsevier, 159-168.

Petrakis, E., and M.K. Romjue. 1990. Cognitive processing of tennis teachers/coaches during skill observation. Paper presented to the central district American Alliance for Health, Physical Education, Recreation and Dance, Denver.

Philipp, J.A., and J.W. Wilkerson. 1990. *Teaching team sports: A coeducational approach.* Champaign, IL: Human Kinetics.

Phillips, S.J., and J.E. Clark. 1984. An integrative approach to teaching kinesiology: A lifespan approach. In *Proceedings: Second national symposium on teaching kinesiology and biomechanics in sports,* edited by R. Shapiro and J.R. Marett. Colorado Springs, CO: NASPE, 19-23.

Phillips, S.J., E.M. Roberts, and T.C. Huang. 1983. Quantification of intersegmental reactions during rapid swing motion. *Journal of Biomechanics* 16: 411-417.

Pinheiro, V. 1994. Diagnosing motor skills: A practical approach. *Journal of Physical Education, Recreation and Dance* 65 (2): 49-54.

Pinheiro, V., and H.A. Simon. 1992. An operational model of motor skill diagnosis. *Journal of Teaching in Physical Education* 11: 288-302.

Pinheiro, V., and S. Cai. 1999. Preservice teachers diagnosing live motor performance. *Research Quarterly for Exercise and Sport* 70 (suppl.): A98-99. (abstract)

Pinheiro, V.E.D. 1990. Motor skill diagnosis: Diagnostic processes of expert and novice coaches. Ph.D. diss., University of Pittsburgh, 1989. Abstract in *Dissertation Abstracts International* 50 (11): 3516A.

Pinheiro, V.E.D. 2000. Qualitative analysis for the elementary grades. *Journal of Physical Education, Recreation and Dance* 71 (1): 18-20, 25.

Piscopo, J., and J.A. Bailey. 1981. *Kinesiology: The science of movement.* New York: Wiley.

Plagenhoef, S. 1971. *Patterns of human motion: A cinematographic analysis.* Englewood Cliffs, NJ: Prentice-Hall.

Platt, B.B., and D.H. Warren. 1972. Auditory localization: The importance of eye movements and a textured visual environment. *Perception and Psychophysics* 12: 245-248.

Pomeroy, V. 1990. Development of an ADL oriented assessment-of-mobility scale suitable for use with elderly people with dementia. *Physiotherapy* 76: 446-448.

Portman, P.A. 1989. Parent intervention program. *Strategies* 3 (2): 13-19.

Powell, M., and J. Svensson. 1998. *In-line skating.* 2nd ed. Champaign, IL: Human Kinetics.

Prinzmetal, W., and L. Gettleman. 1993. Vertical-horizontal illusion: One eye is better than two. *Perception and Psychophysics* 53: 81-88.

Proctor, R.W., and A. Dutta. 1995. *Skill acquisition and human performance.* Thousand Oaks, CA: Sage.

Putnam, C.A. 1991. A segment interaction analysis of proximal-to-distal sequential segment motion patterns. *Medicine and Science in Sports and Exercise* 23: 130-144.

Radford, K.W. 1988. Observation: A neglected teaching skill. *CAHPER Journal* 54 (6): 45-47.

Radford, K.W. 1989. Movement observation in physical education: A definitional effort. *Journal of Teaching in Physical Education* 9: 1-24.

Radford, K.W. 1991. For increased teacher effectiveness: Link observation, feedback and assessment. *CAHPER Journal* 57 (2): 4-9.

Raudensky, J. 1999. Effects of a critical element training package using self-instruction on elementary in service teachers' ability to analyze, diagnose, and provide feedback for the striking skill of batting. Ph.D. diss., Ohio State University, 1998. Abstract in *Dissertation Abstracts International* 59: 3773A.

Raudsepp, J. 2002. Horizontal-vertical illusion: Continuous decrement or the deviant first guess? *Perceptual and Motor Skills* 94: 599-604.

Reagan, J.L. 2002. Videotape instruction versus illustration for influencing quality of performance, motivation, and confidence to perform simple and complex exercises in healthy subjects. *Physiotherapy Theory and Practice* 18: 65-73.

Reeve, J., and C. Morrison. 1986. Teaching for learning: The application of systematic evaluation. *Journal of Physical Education, Recreation and Dance* 57 (6): 37-39.

Regan, D. 1997. Visual factors in hitting and catching. *Journal of Sports Sciences* 15: 533-558.

Reid, M., D. Whiteside, and B. Elliott. 2010. Effect of skill decomposition on racket and ball kinematics of the elite junior tennis serve. *Sports Biomechanics* 9: 296-303.

Reiken, G.B. 1982. Description of women's gymnastic coaches' observations of movement. Ph.D. diss., Teachers College, Columbia University. Abstract in *Dissertation Abstracts International* 43: 397A.

Revlen, L., and M. Gabor. 1981. *Sports vision*. New York: Workman.

Reynolds, A. 1992. What is competent beginning teaching? A review of the literature. *Review of Educational Research* 2 (1): 1-35.

Riggs, L.A. 1971. Vision. In *Woodworth and Schlosberg's experimental psychology*, edited by J.W. Kling and L.A. Riggs. 3rd ed. New York: Holt, Rinehart & Winston.

Rink, J.E., and T.J. Hall. 2008. Research on effective teaching in elementary school physical education. *Elementary School Journal* 108: 207-218.

Ripoll, H., and P. Fleurance. 1988. What does keeping one's eye on the ball mean? *Ergonomics* 31: 1647-1654.

Roberton, M.A. 1978. Longitudinal evidence of developmental stages in the forceful overarm throw. *Journal of Human Movement Studies* 4: 153-167.

Roberton, M.A. 1983. Changing motor patterns during childhood. In *Motor development during childhood and adolescence*, edited by J.R. Thomas. Minneapolis: Burgess, 48-90.

Roberton, M.A. 1989. Future directions in motor development research: Applied aspects. In *Future directions in exercise and sport science research*, edited by J. Skinner et al. Champaign, IL: Human Kinetics, 369-391.

Roberton, M.A., and L.E. Halverson. 1984. *Developing children: Their changing movement*. Philadelphia: Lea & Febiger.

Roberts, E.M. 1971. Cinematography in biomechanical investigation. In *Selected topics on biomechanics: Proceedings of CIC symposium on biomechanics*, edited by J.M. Cooper. Chicago: Athletic Institute, 41-50.

Roberts, G., and D. Treasure, eds. 2012. *Advances in motivation in sport and exercise*. 3rd ed. Champaign, IL: Human Kinetics.

Robinson, D.A. 1981. Control of eye movements. In *Handbook of physiology. Section 1: The nervous system*, vol. 2, part 2. Bethesda, MD: American Physiological Society, 1275-1320.

Robinson, S.M. 1974. Visual assessment of children's gross motor patterns by adults with backgrounds in teacher education. Ph.D. diss., University of Wisconsin, Madison.

Robson, T. 2003. *The hitting edge*. Champaign, IL: Human Kinetics.

Roemmich, J.N., and A.D. Rogol. 1995. Physiology of growth and development: Its relationship to performance in the young athlete. *Clinics in Sports Medicine* 14: 483-502.

Rojas, F.J., M. Cepero, A. Ona, and M. Gutierrez. 2000. Kinematic adjustments in the basketball jump shot against an opponent. *Ergonomics* 43: 1651-1660.

Romance, T.J. 1985. Observing for confidence. *Journal of Physical Education, Recreation and Dance* 56 (6): 47-49.

Rose, D.J., and E.M. Heath. 1990. The contribution of a fundamental motor skill to the performance and learning of a complex sport skill. *Journal of Human Movement Studies* 19: 75-84.

Rose, D.J., E.M. Heath, and D. Megale. 1990. Development of a diagnostic instrument for evaluating tennis serving performance. *Perceptual and Motor Skills* 71: 355-363.

Rose, G.K. 1983. Clinical gait assessment: A personal view. *Journal of Medical Engineering and Technology* 7: 273-279.

Ross, D., A.M. Bird, S.G. Doody, and M. Zoeller. 1985. Effects of modeling and videotape feedback with knowledge of results on motor performance. *Human Movement Science* 4: 149-157.

Rothstein, A.L. 1980. Effective use of videotape replay in learning motor skills. *Journal of Physical Education, Recreation and Dance* 51 (2): 59-60.

Rothstein, A.L., and R.K. Arnold. 1976. Bridging the gap: Application of research on videotape feedback and bowling. *Motor Skills: Theory into Practice* 1: 35-62.

Runeson, S., and G. Frykholm. 1981. Visual perception of lifted weight. *Journal of Experimental Psychology: Human Perception and Performance* 7: 733-740.

Rush, D.A. 1991. Improving skill analysis for diving. Ph.D. diss., Ohio State University, 1990. Abstract in *Dissertation Abstracts International* 51: 2313A.

Russell, P.J., and S.J. Phillips. 1989. A preliminary comparison on front and back squat exercises. *Research Quarterly for Exercise and Sport* 60: 201-208.

Sage, G.H. 1984. *Motor learning and control: A neurophysiological approach*. Dubuque, IA: William C. Brown.

Saleh, M., and G. Murdoch. 1985. In defense of gait analysis. *Journal of Bone and Joint Surgery (Br)* 67: 237-241.

Sanders, R., and B. Wilson. 1989. Some biomechanical tips for better teaching and coaching: Part 1. *New Zealand Journal of Health, Physical Education and Recreation* 23 (4): 14-15.

Sanders, R., and B. Wilson. 1990a. Some biomechanical tips for better teaching and coaching: Part 2. *New Zealand Journal of Health, Physical Education and Recreation* 24 (1): 16-17.

Sanders, R., and B. Wilson. 1990b. Some biomechanical tips for better teaching and coaching: Part 3. *New Zealand Journal of Health, Physical Education and Recreation* 24 (2): 19-21.

Sanders, R.H. 1995. Can skilled performers readily change technique? An example, conventional to wave action breaststroke. *Human Movement Science* 14: 665-679.

Sanderson, D.J., and P.R. Cavanagh. 1990. Use of augmented feedback for the modification of the pedaling mechanics of cyclists. *Canadian Journal of Sport Sciences* 15: 38-42.

Sanderson, F.H., and H.T.A. Whiting. 1974. Dynamic visual acuity and performance in a catching task. *Journal of Motor Behavior* 6: 87-94.

Sanny, J. 1999. Measuring the diameter of your blind spot. *Physics Teacher* 37: 348-349.

Satern, M.N. 1986. Apparent and actual use of observational frameworks by experienced teachers. Paper presented at the national convention of the American Alliance for Health, Physical Education, Recreation and Dance, April, Cincinnati. ERIC Document Reproduction Service, ED 273-588.

Satern, M.N. 1999. Teaching undergraduate biomechanics/-kinesiology: A national survey. Paper presented to the Biomechanics Academy at the American Alliance for Health, Physical Education, Recreation and Dance national convention, April, Boston.

Satern, M.N. 2011. Defining the "correct form": Using biomechanics to develop reliable and valid assessments. *Strategies* 25 (2): 32-34.

Satern, M.N., M.M. Coleman, and M.H. Matsakis. 1991. The effect of observational training on the frequency of skill-related feedback given by pre-service teachers during two peer teaching experiences. *KAHPERD Journal* 60 (2): 12-16.

Saunders, J., V. Inman, and H. Eberhart. 1953. The major determinants in normal and pathological gait. *Journal of Bone and Joint Surgery* 35A: 543-558.

Schempp, P.G., and S. Woorons-Johnson. 2006. Learning to see: Developing the perception of an expert teacher. *Journal of Physical Education, Recreation and Dance* 77 (6): 29-33.

Schleihauf, R.E. 1983. An analysis of skill acquisition in swimming. In *Collected papers on sports biomechanics,* edited by G.A. Wood. Perth, Western Australia: University of Western Australia Press, 117-141.

Schmidt, R.A., and C.A. Wrisberg. 2008. *Motor learning and performance.* 4th ed. Champaign, IL: Human Kinetics.

Schneider, W., and R.W. Shiffrin. 1977. Controlled and automatic human information processing: Decision research and attention. *Psychological Review* 84: 1-66.

Scott, M.G. 1942. *Analysis of human motion.* New York: Crofts.

Scully, D.M. 1986. Visual perception of technical execution and aesthetic quality in biological motion. *Human Movement Science* 5: 185-206.

Seat, J.E., and C.A. Wrisberg. 1996. The visual instruction system. *Research Quarterly for Exercise and Sport* 67: 106-108.

Secrist, G.E., and B.O. Hartman. 1993. Situated awareness: The trainability of the near-threshold information acquisition dimension. *Aviation, Space and Environmental Medicine* 64: 885-897.

Seefeldt, V.D., and J.L. Haubenstricker. 1982. Patterns, phases, or stages: An analytical model for study of developmental movement. In *The development of movement control and coordination,* edited by J.A.S. Kelso and J.E. Clark. New York: Wiley, 309-318.

Sharpe, T. 1993. What are some guidelines on giving feedback to students in physical education? *Journal of Physical Education, Recreation and Dance* 64 (9): 13.

Shea, C.H., and C. Northan. 1982. Discrimination of visual linear velocities. *Research Quarterly for Exercise and Sport* 53: 222-225.

Shea, C.H., W.L. Shebilske, and S. Worchel. 1993. *Motor learning and control.* Englewood Cliffs, NJ: Prentice-Hall.

Sheets, A.L., G.D. Abrams, S. Corazza, M.R. Safran, and T.P. Andriacchi. 2011. Kinematics differences between the flat, kick, and slice serves measured using a markerless motion capture method. *Annals of Biomedical Engineering* 39: 3011-3020.

Sherman, A. 1980. Overview of research information regarding vision and sports. *Journal of the American Optometric Association* 51: 661-666.

Sherman, C.A., and B. Crassini. 1999. The golf swing scale: A study of the quality and outcomes of golf shots for elite and novice players. *Applied Research in Coaching and Athletics Annual* 14: 1-16.

Sherman, C.A., and B.S. Rushall. 1993. Improving swimming stroke technique: A case study. *Applied Research in Coaching and Athletics Annual* 8: 123-143.

Sherman, C.A., W.A. Sparrow, D. Jolley, and J. Eldering. 2001. Coaches' perceptions of golf swing kinematics. *International Journal of Sport Psychology* 31: 257-270.

Shields, B.C. 1995. Successful "Q"munication. *IDEA Today* (September): 62-63.

Shiffrar, M. 1994. When what meets where. *Current Directions in Psychological Science* 3: 96-100.

Shiffrar, M., and J.J. Freyd. 1990. Apparent motion of the human body. *Psychological Science* 1: 257-264.

Shiffrar, M., and J.J. Freyd. 1993. Timing and apparent motion path choice with human body photographs. *Psychological Science* 4: 379-384.

Shigehisa, P.M.J., T. Shigehisa, and J.R. Symons. 1973. Effects of intensity of auditory stimulation on photopic visual sensitivity in relation to personality. *Japanese Psychological Research* 15: 164-172.

Shigehisa, T., and J.R. Symons. 1973. Effect of intensity of visual stimulation on auditory sensitivity in relation to personality. *British Journal of Psychology* 64: 205-213.

Shim, J., and L.G. Carlton. 1997. Perception of kinematic characteristics in the motion of lifted weight. *Journal of Motor Behavior* 29: 131-146.

Siedentop, D. 1991. *Developing teaching skills in physical education.* 3rd ed. Mountain View, CA: Mayfield.

Siedentop, D., and L. Locke. 1997. Making a difference for physical education: What professors and practitioners must build together. *Journal of Physical Education, Recreation and Dance* 68 (4): 25-33.

Silverman, S. 1994. Communication and motor skill learning: What we learn from research in the gymnasium. *Quest* 46: 345-355.

Simmons, R.W., and H.A. King. 1994. Expertise in the observation and subjective analysis of motor performance: A review of empirical research. *Journal of Human Movement Studies* 27: 49-74.

Simon, H.A. 1979. *Models of thought*. New Haven, CT: Yale University Press.

Simons, D.J., and D.T. Levin. 1997. Change blindness. *Trends in Cognitive Science* 1 (17): 261-267.

Simons, D.J., and D.T. Levin. 1998. Failure to detect changes to people during real-world interaction. *Psychonomic Bulletin and Review* 4 (5): 644-649.

Sinclair, G.D. 1988. Pedagogical considerations. *CAHPER Journal* 54 (3): 32-36.

Skrinar, G.S., and S.J. Hoffman. 1979. Effect of outcome information on analytic ability of golf teachers. *Perceptual and Motor Skills* 48: 703-708.

Slettum, B., C. Fox, M.A. Looney, and D.M. Jay. 2001. Validity and reliability of a folk-dance performance checklist for children. *Measurement in Physical Education and Exercise Science* 5: 35-55.

Smolensky, P. 1986. Formal modeling of sub-symbolic processes: An introduction to harmony theory. In *Advances in cognitive science 1*, edited by N.E. Sharkey. Chichester, England: Ellis Horwood, 204-235.

Solso, R.L. 1979. *Cognitive psychology*. New York: Harcourt Brace Jovanovich.

Southard, D. 2002. Change in throwing pattern: Critical values for control parameter of velocity. *Research Quarterly for Exercise and Sport* 73: 396-407.

Southard, D. 2009. Throwing pattern: Changes in timing of joint lag according to age between and within skill level. *Research Quarterly for Exercise and Sport* 80: 213-222.

Spaeth, R.K. 1972. Maximizing goal attainment. *Research Quarterly* 43: 337-361.

Sparrow, W.A., and C. Sherman. 2001. Visual expertise in the perception of action. *Exercise and Sport Sciences Reviews* 29: 124-128.

Sparrow, W.A., J. Shemmell, and A.J. Shinkfield. 2001. Visual perception of action categories and the "bowl-throw" decision in cricket. *Journal of Science and Medicine in Sport* 4: 233-244.

Starek, J., and P. McCullagh. 1999. Effect of self-modeling on the performance of beginning swimmers. *Sport Psychologist* 13: 269-287.

Steinberg, G.M., S.G. Frehlich, and L.K. Tennant. 1995. Dextrality and eye position in putting performance. *Perceptual and Motor Skills* 80: 635-640.

Steindler, A. 1955. *Kinesiology of the human body under normal and pathological conditions*. Springfield, IL: Charles C Thomas.

Ste-Marie, D.M. 2000. Expertise in women's gymnastic judging: An observational approach. *Perceptual and Motor Skills* 90: 543-546.

Ste-Marie, D.M., and S.M. Valiquette. 1996. Enduring memory-influenced biases in gymnastic judging. *Journal of Experimental Psychology: Learning, Memory, and Cognition* 22: 1498-1502.

Ste-Marie, D.M., and T.D. Lee. 1991. Prior processing effects on gymnastic judging. *Journal of Experimental Psychology: Learning, Memory, and Cognition* 17: 126-136.

Stensrud, S., G. Myklebust, E. Kristianslund, R. Bahr, and T. Krosshaug. 2011. Correlation between two-dimensional video analysis and subjective assessment in evaluating knee control among elite female team handball players. *British Journal of Sports Medicine* 45: 589-595.

Stephenson, D.A., and A.S. Jackson. 1977. The effects of training on judges' ratings of a gymnastic event. *Research Quarterly* 48: 177-180.

Stodden, D.F., D.L. Fuhrhop, and S.J. Langendorfer. 2004. Comparison of biomechanical and component throwing analyses. *Research Quarterly for Exercise and Sport* 75: A52.

Stodden, D.F., J.D. Goodway, S.J. Langendorfer, M.A. Roberton, M.E. Rudisill, C. Garcia, and L.E. Garcia. 2008. A developmental perspective on the role of motor skill competence in physical activity: An emergent relationship. *Quest* 60: 290-306.

Stoner, L.J. 1984. Is this performer skilled or unskilled? In *Proceedings: Second national symposium on teaching kinesiology and biomechanics in sports*, edited by R. Shapiro and J.R. Marett. Colorado Springs, CO: NASPE, 233-234.

Strand, B. 1988. The development of checkpoints for skill observation. *New Jersey Journal of Physical Education, Recreation and Dance* 62 (1): 19-21.

Strohmeyer, H.S., K. Williams, and D. Schaub-George. 1991. Developmental sequences for catching a small ball: A prelongitudinal screening. *Research Quarterly for Exercise and Sport* 62: 257-266.

Stroot, S.A., and J.L. Oslin. 1993. Use of instructional statements by preservice teachers for overhand throwing performance of children. *Journal of Teaching in Physical Education* 13: 24-45.

Stuberg, W., L. Straw, and L. Deuine. 1990. Validity of visually recorded temporal-distance measures at selected walking velocities for gait analysis. *Perceptual and Motor Skills* 70: 323-333.

Suomi, R., and J. Suomi. 1997. Effectiveness of a training program with physical education students and experienced physical education teachers in scoring the test of gross motor development. *Perceptual and Motor Skills* 84: 771-778.

Swinnen, S. 1984a. Field dependence/independence as a factor in learning complex motor skills and underlying sex differences. *International Journal of Sports Psychology* 15: 236-249.

Swinnen, S. 1984b. Some evidence to the hemispheric asymmetry model of lateral eye movements. *Perceptual and Motor Skills* 57: 319-325.

Swinnen, S. 1984c. Some evidence to the hemispheric symmetry model of lateral eye movements. *Perceptual and Motor Skills* 58: 79-88.

Tant, C. 1990. A kick is a kick—or is it? *Strategies* 4 (2): 19-22.

Taylor, J.K. 1995. Developing observational abilities in preservice physical education teachers. Ph.D. diss., University of South Carolina, 1994. Abstract in *Dissertation Abstracts International* 55: 1872A.

Taylor, J.K., K.G. Hussey, P.H. Werner, J.E. Rink, and K.E. French. 1993. The effects of strategy, skill and strategy and skill instruction on skill and knowledge in ninth grade badminton. *Research Quarterly for Exercise and Sport* 64 (suppl.): 96A. (abstract)

Terwee, C.B., A.F. deWinter, R.J. Scholten, M.P. Jans, W. Deville, D. van Schaardenburg, and L.M. Bouter. 2005. Interobserver reproducibility of the visual estimation of range of motion of the shoulder. *Archives of Physical Medicine and Rehabilitation* 86: 1356-1361.

Theios, J., and P.C. Amarhein. 1989. Theoretical analysis of the cognitive processing of lexical and pictorial stimuli: Reading, naming and visual conceptual comparisons. *Psychological Review* 96 (1): 5-24.

Thorndike, E.L. 1927. The law of effect. *American Journal of Psychology* 39: 212-222.

Thornton, I.M., J. Pinto, and M. Shiffrar. 1998. The visual perception of human locomotion. *Cognitive Neuropsychology* 15: 535-552.

Tieg, D. 1983. Eyes on the PGA tour. *Golf Digest* (July): 85-89.

Tobey, C. 1992. The best kind of feedback. *Strategies* 6 (2): 19-20.

Todorov, E., R. Shadmehr, and E. Bizzi. 1997. Augmented feedback presented in a virtual environment accelerates learning of a difficult motor task. *Journal of Motor Behavior* 29: 147-158.

Toro, B., C. Nester, and P. Farren. 2003. A review of observational gait assessment in clinical practice. *Physiotherapy Theory and Practice* 19: 137-150.

Toro, B., C.J. Nester, and P.C. Farren. 2007a. The development and validity of the Salford gait tool: An observation-based clinical gait assessment tool. *Archives of Physical Medicine and Rehabilitation* 88: 321-327.

Toro, B., C.J. Nester, and P.C. Farren. 2007b. Inter- and intraobserver repeatability of the Salford gait tool: An observation-based clinical gait assessment tool. *Archives of Physical Medicine and Rehabilitation* 88: 328-332.

Torrey, L. 1985. *Stretching the limits: Breakthroughs in sports science that create superathletes.* New York: Dodd, Mead.

Triesman, A. 1986. Features and objects in visual processing. *Scientific American* 255 (5): 114-125.

Treisman, A.M., and G.L. Gelade. 1980. A feature integration theory of attention. *Cognitive Psychology* 12: 97-136.

Trinity, J., and J.J. Annesi. 1996. Coaching with video. *Strategies* 9 (8): 23-25.

Trower, P., and B. Kiely. 1983. Video feedback: Help or hindrance? A review and analysis. In *Using video: Psychological and social applications,* edited by P. Dowrick and S. Briggs. Chichester, England: Wiley, 181-197.

Tzetzis, G., E. Kioumourtzoglou, A.Y. Laiso, and N. Stergiou. 1999. The effect of different feedback models on acquisition and retention of technique in basketball. *Journal of Human Movement Studies* 37: 163-181.

Ulrich, B.G. 1977. A module of instruction for golf swing error detection. In *Research and practice in physical education,* edited by R.E. Stadulis. Champaign, IL: Human Kinetics, 19-27.

Ulrich, D.A. 1984. The reliability of classification decisions made with the objectives-based motor skill assessment instrument. *Adapted Physical Activity Quarterly* 1: 52-60.

Ulrich, D.A. 2000. *Test of gross motor development.* 2nd ed. Austin, TX: PRO-ED Inc.

Ulrich, D.A., B.D. Ulrich, and C.R. Branta. 1988. Developmental gross motor skill ratings: A generalizability analysis. *Research Quarterly for Exercise and Sport* 59: 203-209.

Valenti, S.S., and A. Costall. 1997. Visual perception of lifted weight from kinematic and static (photographic) displays. *Journal of Experimental Psychology: Human Perception and Performance* 24 (1): 181-198.

Vandenberg, G.S. 1971. *Mental rotations test.* Boulder, CO: Institute of Behavioral Genetics, University of Colorado.

van den Tillaar, R. 2004. Effect of different training programs on the velocity of overarm throwing: A brief review. *Journal of Strength and Conditioning Research* 18: 388-396.

Vanderbeck, E. 1979. "It isn't right but I don't know what's wrong with it": An approach to error identification. *Journal of Physical Education and Recreation* 50 (5): 54-56.

Van Dijk, H., M. Jannink, and H. Hermens. 2005. Effect of augmented feedback on motor function of the affected upper extremity in rehabilitation patients: A systematic review of randomized controlled trials. *Journal of Rehabilitation Medicine* 37: 202-211.

van Wieringen, P.C.W., H.H. Emmen, R.J. Bootsma, M. Hoogesteger, and H.T.A. Whiting. 1989. The effect of video-feedback on the learning of the tennis service by intermediate players. *Journal of Sports Sciences* 7: 153-162.

Vickers, J.N. 1989. *Instructional design for teaching physical activities: A knowledge structures approach.* Champaign, IL: Human Kinetics.

Vickers, J.N. 2007. *Perception, cognition, decision training: The quiet eye in action.* Champaign, IL: Human Kinetics.

Vickers, J.N. 2011. Skill acquisition: Designing optimal environments. In *Performance psychology: A practitioner's guide,* edited by D. Collins, A. Button, and H. Richards. Edinburgh: Elsevier, 191-206.

Village, J., C. Trask, N. Luong, Y. Chow, P. Johnson, M. Koehoorn, and K. Teschke. 2009. Development and evaluation of an observational back-exposure sampling tool (Back-EST) for work-related back injury risk factors. *Applied Ergonomics* 40: 538-544.

Vincent, R.H. 1984. In or out? See if you can make this line call. *Tennis* (March): 35-37.

Walkley, J.W., and C.E. Kelley. 1989. The effectiveness of an interactive videodisk qualitative assessment training program. *Research Quarterly for Exercise and Sport* 60: 280-285.

Wang, J., and M. Griffin. 1998. Early correction of movement errors can help student performance. *Journal of Physical Education, Recreation and Dance* 69 (4): 50-52.

Warren, D.H. 1970. Inter-modality interactions in spatial localization. *Cognitive Psychology* 1: 114-133.

Watkins, M.A., D.L. Riddle, R.L. Lamb, and W.J. Personius. 1991. Reliability of goniometric measurements and visual estimates of knee range of motion obtained in a clinical setting. *Physical Therapy* 71: 90-96.

Watts, R.G., and A.T. Bahill. 1990. *Keep your eye on the ball: The science and folklore of baseball.* New York: Freeman.

Weiss, M.R. 1982. Developmental modeling enhancing children's motor skill acquisition. *Journal of Physical Education, Recreation and Dance* 53 (9): 49-50, 67.

Welch, C.M., S.A. Banks, F.F. Cook, and P. Dravovitch. 1995. Hitting a baseball: A biomechanical description. *Journal of Orthopaedic and Sports Physical Therapy* 22: 193-201.

Welch, R.B., and D.H. Warren. 1980. Immediate perceptual response to intersensory discrepancy. *Psychological Bulletin* 88: 638-667.

Werder, J.K., and L.H. Kalakian. 1985. *Assessment in adapted physical education.* Minneapolis: Burgess.

Werner, P., and J.E. Rink. 1987. Case studies of teacher effectiveness in second grade physical education. *Journal of Teaching in Physical Education* 8: 280-297.

Whittle, M.W. 2007. *Gait analysis: An introduction.* 4th ed. Oxford: Butterworth-Heinemann.

Wiart, L., and J. Darrah. 2001. Review of four tests of gross motor development. *Developmental Medicine and Child Neurology* 43: 279-285.

Wickens, C.D. 1981. Processing resources in attention, dual task performance, and workload assessment. Engineering-Psychology Research Laboratory, University of Illinois at Urbana. Technical Report EPL-81-3/ONR-81-3.

Wickens, C.D. 1984a. *Engineering psychology.* Columbus, OH: Merrill.

Wickens, C.D. 1984b. Processing resources in attention. In *Varieties of attention,* edited by R. Parasuraman and R. Davies. New York: Academic Press, 63-102.

Wickstrom, R.L. 1983. *Fundamental motor patterns.* 3rd ed. Philadelphia: Lea & Febiger.

Wiese-Bjornstal, D.M. 1993. Giving and evaluating demonstrations. *Strategies* 6 (7): 13-15.

Wild, M. 1938. The behavior pattern of throwing and some observations concerning its course of development in children. *Research Quarterly* 9: 20-24.

Wilkerson, J.D. 1985. Application of concepts: A second look. Paper presented at the American Alliance for Health, Physical Education, Recreation and Dance national convention, April, Atlanta.

Wilkerson, J.D., E. Kreighbaum, and C.L. Tant, eds. 1991. *Teaching kinesiology and biomechanics.* Ames, IA: Iowa State University.

Wilkerson, J., K. Ludwig, and M. Butcher, eds. 1997. *Proceedings of the fourth national symposium on teaching biomechanics.* Denton, TX: Texas Woman's University.

Wilkinson, S. 1986. The effects of a visual discrimination training program on the acquisition and maintenance of physical education students' volleyball skill analytic ability. Ph.D. diss., Ohio State University. Abstract in *Dissertation Abstracts International* 47: 1650A.

Wilkinson, S. 1990. *Skill analysis: Past, present and future perspectives.* Paper presented at the American Alliance for Health, Physical Education, Recreation and Dance national convention, March, New Orleans.

Wilkinson, S. 1991. The effect of an instructional videotape on the ability of physical education majors to diagnose errors in the overarm throwing pattern. In *Abstracts of research papers 1991,* edited by W. Liemohn. Reston, VA: AAHPERD, 74.

Wilkinson, S. 1992a. Effects of training in visual discrimination after one year: Visual analysis of volleyball skills. *Perceptual and Motor Skills* 75: 19-24.

Wilkinson, S. 1992b. A training program for improving undergraduates' analytic skill in volleyball. *Journal of Teaching in Physical Education* 11: 177-194.

Wilkinson, S. 1996. Visual analysis of the overarm throw and related sport skills: Training and transfer effects. *Journal of Teaching in Physical Education* 16: 66-78.

Williams, E. 1996. Effects of a multimedia performance principle training program on correct analysis and diagnosis of throwlike movements. Ph.D. diss., Ohio State University, 1995. Abstract in *Dissertation Abstracts International* 56: 3504A.

Williams, E.U., and D. Tannehill. 1999. Effects of a multimedia performance principle training program on correct analysis and diagnosis of throwlike movements. *Physical Educator* 56: 143-154.

Williams, J.G. 1987. Visual demonstration and movement sequencing: Effects of instructional control of the eyes. *Perceptual and Motor Skills* 65: 366.

Williams, J.G. 1989a. Motor skill instruction, visual demonstration and eye movements. *Physical Education Review* 12 (1): 49-55.

Williams, J.G. 1989b. Throwing action from full-cue and motion-only video-models of an arm movement sequence. *Perceptual and Motor Skills* 68: 259-266.

Williams, J.G. 1989c. Visual demonstration and movement production: Effects of timing variations in a models action. *Perceptual and Motor Skills* 68: 891-896.

Williams, J.G. 1992. Catching action: Visuomotor adaptations in children. *Perceptual and Motor Skills* 75: 211-219.

Williams, J.G., and M. Callaghan. 1990. Comparison of visual estimation and goniometry in determination of a shoulder joint angle. *Physiotherapy* 76: 655-657.

Williams, K. 1980. Developmental characteristics of a forward roll. *Research Quarterly for Exercise and Sport* 51: 703-713.

Williams, K., K. Haywood, and A. Van Sant. 1996. Force and accuracy throws by older adults: II. *Journal of Aging and Physical Activity* 4: 194-202.

Wilson, B.D. 2008. Development in video technology for coaching. *Sports Technology* 1: 34-40.

Wilson, S.J., P. Glue, D. Ball, and D. Nutt. 1993. Saccadic eye movement parameters in normal subjects. *Electroencephalography and Clinical Neurophysiology* 86: 69-74.

Wilson, V.E. 1976. Objectivity, validity, and reliability of gymnastic judging. *Research Quarterly* 47: 169-173.

Winter, D.A. 1984. Kinematic and kinetic patterns in human gait: Variability and compensating effects. *Human Movement Science* 3: 51-76.

Winter, D.A. 1987. *Biomechanics and motor control of human gait.* Waterloo, ON: University of Waterloo Press.

Winter, D.A. 1989. Biomechanics of normal and pathological gait: Implications for understanding human locomotor control. *Journal of Motor Behavior* 21: 337-355.

Witkin, H.A. 1954. *Personality through perception: An experimental and clinical study.* Westport, CT: Greenwood Press.

Witkin, H.A., P.K. Oltman, E. Raskin, and S.A. Karp. 1971. *A manual for the group embedded figures test.* Palo Alto, CA: Consulting Psychologists Press.

Wood, C.A., J.D. Gallagher, P.V. Martino, and M. Ross. 1992. Alternate forms of knowledge of results: Interaction of augmented feedback modality on learning. *Journal of Human Movement Studies* 22: 213-230.

Woollacott, M.H., and A. Shumway-Cook, eds. 1989. *Development of posture and gait across the life span.* Columbia, SC: University of South Carolina Press.

Wretenberg, P., Y. Feng, and U.P. Arborelius. 1996. High- and low-bar squatting techniques during weight training. *Medicine and Science in Sports and Exercise* 28: 218-224.

Wulf, G., and C. Shea. 2004. Understanding the role of augmented feedback: The good, bad, and the ugly. In *Skill acquisition in sport: Research, theory, and practice,* edited by A.M. Williams and N.J. Hodges. London: Routledge, 121-145.

Wulf, G., C. Shea, and R. Lewthwaite. 2010. Motor skill learning and performance: A review of influential factors. *Medical Education* 44: 75-84.

Wulf, G., N. McConnel, M. Gartner, and A. Schwarz. 2002. Enhancing learning of sport skills through external-focus feedback. *Journal of Motor Behavior* 34: 171-182.

Yantis, S. 1992. Multielement visual tracking: Attention and perceptual organization. *Cognitive Psychology* 24: 295-340.

Youndas, J.W., C.L. Bogard, and V.J. Suman. 1993. Reliability of goniometric measurements and visual estimates of ankle joint active range of motion obtained in a clinical setting. *Archives of Physical Medicine and Rehabilitation* 74: 1113-1118.

Youndas, J.W., J.R. Carey, and T.R. Garrett. 1991. Reliability of measurements of cervical spine range of motion: Comparison of three methods. *Physical Therapy* 71: 90-96.

Zajac, F.E. 2002. Understanding muscle coordination of the human leg with dynamic simulations. *Journal of Biomechanics* 35: 1011-1018.

Zajac, F.E., and M.E. Gordon. 1989. Determining muscle's force and action in multi-articular movement. *Exercise and Sport Sciences Reviews* 17: 187-230.

Zebas, C., and H.M. Johnson. 1989. Transfer of learning from the overhand throw to the tennis serve. *Strategies* 2 (6): 17-18, 27.

Ziegler, S.G. 1987. Effects of stimulus cueing on the acquisition of ground strokes of beginning tennis players. *Journal of Applied Behavior Analysis* 20: 405-411.

Zollman, D., and R.G. Fuller. 1984. Interactive videodisks: New technology for the analysis of human motion. In *Proceedings: Second national symposium on teaching kinesiology and biomechanics in sports,* edited by R. Shapiro and J.R. Marett. Colorado Springs, CO: NASPE, 53-56.

Index

Note: The italicized *f* and *t* following page numbers refer to figures and tables, respectively.

Reuschlein, P. 190
Reynolds, A. 93
Roberton, M.A. 25, 39, 192
Roberts, E.M. 126
rod and frame test 65
Romjue, M.K. 30
Rose, D.J. 25
Rothstein, A.L. 146, 169

S

saccadic eye movements 51, 52-53
Sage, G.H. 4
Saleh, M. 36
sampling rates of human vision 163
Sanders, R. 19
scanning strategy
  about 102
  observation based on importance 105
  observation by phases of movement 102-103, 104f
  observation from general to specific 105-106
  observation of balance 104-105
Schleihauf, B. 84, 147, 176
Schmidt, R.A. 27, 55
Schutz, R.W. 40
scientific research as a source of knowledge 78-80
Scout 172
Secrist, G.E. 46
Sembiante, J.L. 64
senses and perception in QMD
  audition 57-58
  development of spatial perceptual ability studies 65-66
  discussion questions 71
  expert versus novice studies 66
  gestalt representation of movement 68-70
  good and bad examples of performance 66
  imagery research 64
  integration of information and 47
  integration of senses 59-60
  kinesthetic proprioception 58
  perceptual process 58-59
  perceptual tasks 62-63
  psychological research on perception of biological motion 66-68
  research on perception in QMD 64-68
  sensory detection of an event 60-61
  spatial ability studies 64-66
  spatial stimuli 61
  summary 70-71
  temporal information 61
  theoretical background for 46
  touch 58
  vision. *See* vision perception system; visual acuity
sequential method of evaluation 21, 117-118
Shea, C. 135
Sherman, H. 94
Shiffrar, M. 66, 67
Siedentop, D. 29

Siliconcoach 172, 174
SIMI software 172-173
Simon, H.A. 31, 111, 124
SimulCam 174
skill analysis 9
SkillCapture 176
skills 80
SkillSpector 176
slow-motion video 110
smooth pursuit eye movements 51, 52
soccer instep kick analysis
  critical features 184, 185t
  observational models for kicking 187
  rolling ball kick 190-191
  running kick 188-189
  SOS 185-186
  stationary ball kick 188-189
software availability for QMD 172-174, 176
SOS. *See* systematic observational strategy
SPORT 79
Sport and Leisure Index 79
Sports Coach 79
Sports Illustrated 53
Sportsvision Magazine 57
Sports Vision Section of American Optometric Association 57
squat exercise analysis
  critical features 202-203
  single-leg squat technique 206-207
  SOS 203
  squat with Olympic bar technique 206-207
  warm-up squat technique 204-205
Stamm, C.L. 109
static visual acuity (SVA) 48, 49-50
Stodden, D.F. 36
Strand, B. 90, 91
Strategies (journal) 33, 77
Strength and Conditioning Journal 79
style 80
subjective analysis 9
summary feedback 141
SVA (static visual acuity) 48, 49-50
swimming strokes instruction 233
systematic observation 9
systematic observational strategy (SOS)
  about 94, 98
  analysis tutorials 181, 185-186, 193, 199, 203
  discussion questions 112-113
  extended observation 110-111
  focus of observation and 102-105
  integrated use of all senses 111-112
  number of observations 109-110
  observation defined 98-99
  observation of movement steps by James and Dufek 100-101
  perception and 99
  practical applications 101
  purpose and goal of 98
  situation for observation 106-107
  skill development system by Barret 99-100

summary 112
theoretical framework by Radford 100
vantage points for observation 107-109
written observational plans suggestions 106

T

task sheets 91, 92f
technique analysis 9
techniques 80
Temporal and Spatial Model 187
tennis serve analysis 210, 223
theory-into-practice situations
  athlete's confidence level guidance 231, 236
  basketball analyses 221, 228
  batting analysis 222
  bench pressing technique instruction 224
  discussion questions 239
  free weights versus machine weights discussion 227
  further practice 239
  hockey checking technique instruction 234
  impatient client interactions 238
  injury recognition 232, 237
  inline skating technique analysis 230
  judging a diving demonstration preparation 226
  observation position selection 235
  overzealous parent intervention 225
  players' responses to external stimuli 229
  post-injury play readiness analysis 218-220
  summary 239
  swimming strokes instruction 233
  tennis serve analysis 223
Thorndike's law of effect 135
time and motion studies in QMD 32
TimeWarp 174
Tobey, C. 138
Toro, B. 37
touch 58
Track and Field Coaches Review 79
tutorials
  discussion questions 215
  movement sequence presentation 180
  nonbiomechanical factors affecting performance 180t
  QMD Explorations 208-215
  QMD of catching 181-184
  QMD of frisbee throw 198-203
  QMD of overarm throwing 190-199
  QMD of the soccer instep kick 184-191
  QMD of the squat exercise 202-207
  summary 215
  web resources 179

U

Ulrich, B.D. 39
Ulrich, D.A. 39, 121
Urquia, F. 109

About the Author

Duane V. Knudson, PhD, is a professor and chair of the department of health and human performance at Texas State University in San Marcos. He earned his PhD in biomechanics from the University of Wisconsin at Madison and has held tenured faculty positions at Baylor University in Waco, Texas, and California State University at Chico. Previously he served as associate dean and interim chair of the department of kinesiology at Chico State.

Knudson's research in qualitative movement diagnosis (QMD) has garnered him an international reputation as an expert on the topic. He coauthored the first text on QMD, *Qualitative Analysis of Human Movement*, which has been translated into five languages. In addition to research on QMD, Knudson researches the biomechanics of tennis and stretching and the teaching and learning of biomechanical concepts.

Knudson has authored 3 books, 11 book chapters, and 24 refereed articles in scientific proceedings. He has received numerous grants for his work and has published over 70 peer-reviewed articles in journals, including the *Journal of Applied Biomechanics*, *International Journal of Sport Biomechanics*, *Sports Biomechanics*, *International Journal of Sports Medicine*, *Sports Medicine*, *British Journal of Sports Medicine*, *Journal of Sports Sciences*, *Journal of Science and Medicine in Sport*, *European Journal of Applied Physiology*, *Sports Engineering*, *Gait & Posture*, *Research Quarterly for Exercise in Sport*, and *Journal of Strength and Conditioning Research*.

An elected fellow of three scholarly societies, Knudson served on the editorial board of five journals and was the 2011 recipient of the Ruth B. Glassow Honor Award from the National Association for Sport and Physical Education (NASPE). He has also served as vice president of publications for the International Society of Biomechanics in Sports (ISBS). He is also a member of the American Society of Biomechanics, American College of Sports Medicine (ACSM), and the American Alliance for Health, Physical Education, Recreation and Dance (AAHPERD).

He and his wife, Lois, reside in San Marcos. In his free time, Knudson enjoys reading and playing tennis. Photo courtesy of Duane Knudson.